Secret Science

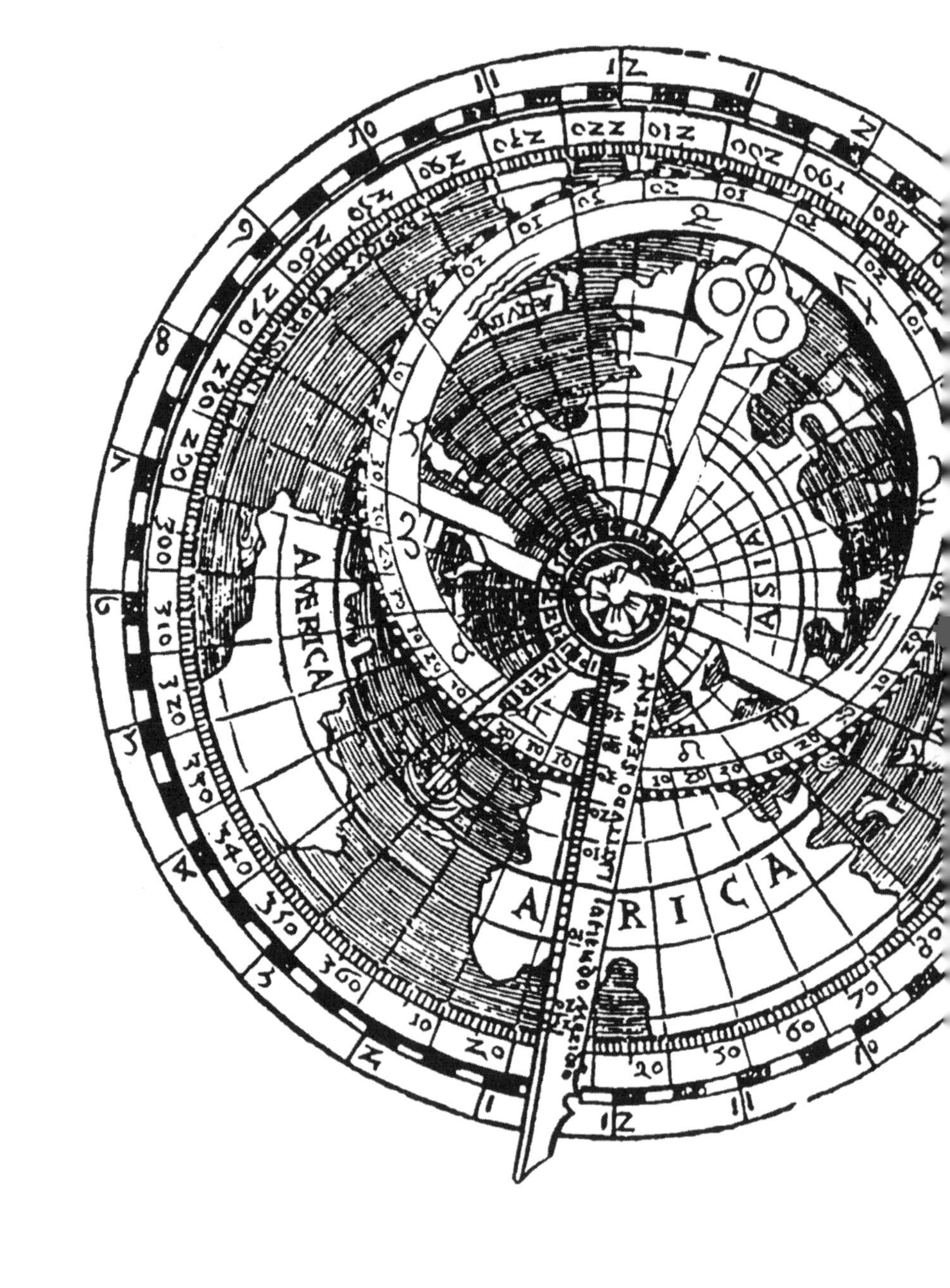
AMERICA
ASIA

Secret Science

Spanish Cosmography and the New World

MARÍA M. PORTUONDO

The University of Chicago Press *Chicago and London*

María M. Portuondo is an assistant professor at The Johns Hopkins University. She has a bachelor of science in electrical engineering and a PhD in the history of science and technology.

The University of Chicago Press, Chicago 60637
The University of Chicago Press, Ltd., London

Printed in the United States of America

18 17 16 15 14 13 12 11 10 09   1 2 3 4 5

ISBN-13: 978-0-226-67534-3 (cloth)
ISBN-10: 0-226-67534-3 (cloth)

Library of Congress Cataloging-in-Publication Data

Portuondo, María M.
Secret science : Spanish cosmography and the new world / María M. Portuondo.
p. cm.
Includes bibliographical references and index.
ISBN-13: 978-0-226-67534-3 (cloth : alk. paper)
ISBN-10: 0-226-67534-3 (cloth : alk. paper) 1. Cosmography—History. 2. Science—Historiography. 3. America—Discovery and exploration—Spanish. 4. Spain—History. I. Title.
QB29.P67 2009
912.09–dc22
2008034282

∞ The paper used in this publication meets the minimum requirements of the American National Standard for Information Sciences—Permanence of Paper for Printed Library Materials, ANSI Z39.48-1992.

Pintaremos nuevo cielo nunca visto de nuestros pasados, nueva tierra nunca imaginada, con la extrañeza que tiene, donde no hallaremos cosa que parezca a las nuestras; nuevos Arboles, yerbas, fieras, aves y pescados; nuevos hombres, costumbres y religión; grandes acaecimientos en la conquista y la posesión de lo conquistado.

We will paint a new sky never seen by our ancestors, a new land never imagined, in all its strangeness, where we will not find anything resembling ours; new trees, plants, wild animals, birds and fish; new men, customs and religion; the great events of the conquest and the possession of the conquered.

JUAN PÁEZ DE CASTRO, *De las cosas necesarias para escribir historia* (1555), Library of the Royal Monastery of San Lorenzo de El Escorial, manuscript &.III.10.

Contents

Illustrations

FIGURES

COLOR PLATES

These images can be found on the unnumbered pages following page no. 234.

Tables

Acknowledgments

This book originated at the Department of the History of Science and Technology at The Johns Hopkins University, and it is my professors and my fellow graduate students there whom I must first thank for making it possible. They provided me with a fascinating learning environment and a forum from which to explore the many ways of practicing history of science. Professors Sharon Kingsland and Bill Leslie, both as chairs and as teachers, sponsored my changing interests with financial support and free time. Larry Principe's contagious enthusiasm for the historian's craft was a constant source of motivation. The seed for the work that developed into this book was planted by Richard Kagan, who encouraged me to work on a topic dear to his heart and who for two years indulged my history of science digressions during his Early Modern Spanish History seminar.

The research for this book would not have been possible without financial support from several endowed grants. Portions of my research were underwritten by two grants from The Johns Hopkins University: the Leonard and Helen R. Stulman Fellowship in the Humanities and the J. Brien Key Graduate Student Assistance Award. The Atlantic History Seminar at Harvard University provided me with a short-term research grant that allowed me an extended stay at various archives in Spain. I would like to express my gratitude to the underwriters and administrators of these programs for their generous support. The History Department and the College of Liberal Arts and Sciences at the University of Florida granted me travel research funds and a semester of leave that allowed me to revise the manuscript. They also provided a generous subvention so that the manuscript maps in this book could appear in color. My colleagues Fred Gregory and Betty Smocovitis read an early version of the manuscript and provided invaluable suggestions and insight. Kip D. Kuntz patiently assisted me with some aspects of the astronomy, but I am wholly responsible for any errors that might appear in the computations. At the University of Chicago Press I've had the pleasure of working with Karen Darling, who assembled a learned team of anonymous readers. I thank them for their sugestions and criticism.

I am also indebted to the staff of the numerous archives and libraries I visited in Spain. Two individuals stand out among them. At the Library of the Royal Monastery of El Escorial, José Luis del Valle Merino helped chase down sometimes-elusive volumes, and Mercedes Noviembre at the Zabálburu Archive generously gave of her time during building renovations to accommodate my travel schedule. I must also thank Francisco Fernández de Navarrete, Marquis of Legarda, who graciously searched through his family's archive and provided me with a copy of an important source.

This book would not have been possible without the unwavering support of Janet Patacca, who encouraged me to undertake the intellectual adventure that led to these pages.

Segments from the second and fifth chapters appear in "Spanish Cosmography and the New World Crisis," in *Beyond the Black Legend: Spain and the Scientific Revolution / Mas allá de la Leyenda Negra: España y la Revolución Científica*, edited by William Eamon and Victor Navarro Brotóns (Valencia: Soler, 2007). Portions of the second chapter appear as "Cosmography at the *Casa, Consejo y Corte* during the Century of Discovery" in *Science in the Spanish and Portuguese Empires (1500–1800)* (Stanford, Calif.: Stanford University Press, 2008).

Abbreviations

AGI Archive of Indies, Seville
AGS General Archives of Simancas
AHN National Historical Archive, Madrid
APR Archive of Protocols, Madrid
BAH Library of the Academy of History, Madrid
BCC Biblioteca Capitular y Colombina, Catedral de Sevilla
BL British Library
BME Library of the Royal Monastery of San Lorenzo of El Escorial
BMN Library of the Naval Museum, Madrid
BN National Library, Madrid
BUS Library of the University of Salamanca
CSIC Consejo Superior de Investigaciones Científicas
DIA Joaquín Francisco Pacheco, Francisco de Cárdenas y Espejo, and Luis Torres de Mendoza, eds., *Colección de documentos inéditos, relativos al descubrimiento conquista y organización de las antiguas posesiones españolas de América y Oceanía, sacados de los archivos del reino, y muy especialmente del de Indias*, 42 vols. (Vaduz, Liechtenstein: Kraus Reprint, 1964)
DIU Academia de la Historia Real, ed., *Colección de documentos inéditos relativos al descubrimiento, conquista y organización de las antiguas posesiones españolas de ultramar*, 25 vols. (Nendeln, Liechtenstein: Kraus Reprint, 1967)
DIE Martín Fernández de Navarrete, ed., *Colección de documentos inéditos para la historia de España* (Vaduz, Liechtenstein: Kraus Reprint, 1964–75)
IVDJ Instituto de Valencia de Don Juan
JCB John Carter Brown Library
UTX Benson Latin American Collection, University of Texas at Austin
ZAB Zabálburu Archive, Madrid

A Note on Translations

I have chosen to modernize the spelling and, in some instances, the phrasing of all Spanish quotations except in instances where the original was particularly enlightening. When a primary source was listed in a catalog under a descriptive title in Spanish, I have retained it, but otherwise I have created a descriptive but fictitious title. All other translations, except where otherwise indicated, are mine. For repeated references to a person with a compound Castilian surname, I use only the second surname. For example, "Juan López de Velasco" appears as "Velasco" and "Andrés García de Céspedes" as "Céspedes." It has been my intention to point out the place of publication of all previously published primary sources and the name of the historian responsible for a manuscript's publication. I offer my sincerest apologies should I have overlooked another historian's archival work.

INTRODUCTION

A New World and Spanish Cosmography

During the century following the discovery of America, Europeans tried to come to terms with the reality of the vast, newly discovered lands of what they so aptly named the New World. The earliest descriptions of the land and its people only opened the floodgates to new questions about the nature of the New World. What did this new world contain? Where did these lands lie in relation to Europe? Who lived there? Were they like us or different? These questions vexed the minds of the few men in Europe who claimed to have expert knowledge of the natural world and its peoples. They recognized in the questions posed by the discovery one of the greatest challenges of their time.

Many of those who took up this challenge bore the title *cosmographer* and practiced a new discipline with roots deep in Renaissance humanism. The discipline of Renaissance cosmographers drew from what we might recognize from our modern perspective as geography, cartography, ethnography, natural history, history, and certain elements of astronomy. For them, this new sky and new land had to be reconciled with an image of the world imprinted on European minds by biblical and classical narratives. Thus cosmographers first sought structure and guidance from the classical texts that for centuries had defined the contours of the known world. But the edifice constructed by the classics and nurtured by humanists soon proved to be resting on a shaky foundation. It did not take long for sixteenth-century cosmographers to privilege eyewitness reports over classical accounts and to embark on their own empirical investigations.

Nowhere was the determination to create a new framework to explain the reality of the New World more steadfast than in sixteenth-century Spain. Spanish cosmographers brought to their discipline alternative epistemologies and new methodologies that eventually changed how

Europeans understood the natural world. They resorted to voyages of scientific exploration, descriptive geographies, new cartographic methods, new navigation technologies, and an unrelenting questioning of those living in the new lands and those arriving from beyond the "Ocean Sea" to formulate an accurate and useful description of the world. By the reign of Philip II (r. 1556–98), these efforts had been institutionalized and functioned under the direction of the king's cosmographers.

By the 1570s, the onus on these cosmographers to describe their nation's expanding empire resulted in a series of ambitious large-scale scientific projects. Some of these projects have been studied attentively by historians, such as the work of naturalist Francisco Hernández in Mexico and the questionnaires associated with the *relaciones de Indias*. Other projects are less known, such as the project to use lunar eclipses to determine the longitude of Spain's overseas domains or the reform of navigational charts used by pilots sailing to the Indies. For the most part, historical work on these projects has approached them as independent enterprises, answering loosely to administrative needs of the Spanish monarchy or the king's curiosity. More recently, Jesús Bustamante has characterized these projects as the culmination of Spanish humanism.[1] In contrast, my book approaches these projects from the perspective of coordinated scientific practice.[2] From this vantage point, the practitioners—royal cosmographers—and the practice of cosmography emerge as one of the principal threads binding together, bureaucratically as well as intellectually, the scientific activities undertaken in Spain to explicate the New World.

I should qualify what I mean by the word *science* in this book. Practices that from our modern perspective might seem scientific and that we associate with post-Newtonian methodologies either did not exist during the early modern era or belonged to wholly different approaches to explaining nature. Therefore, when early modern historians refer to *science* we are using the word anachronistically but also as an expedient way of referring to a group of quite distinct ways of producing knowledge about the natural world. These approaches included natural philosophy, experimentalism, natural history, natural magic, and mixed mathematics. In this book, I also

1. Jesús Bustamante García, "Los círculos intelectuales y las empresas culturales de Felipe II: Tiempos, lugares y ritmos del humanismo en la España del siglo XVI," in *Élites intelectuales y modelos colectivos: Mundo ibérico (siglos XVI–XIX)*, ed. Mónica Quijada and Jesús Bustamante García (Madrid: CSIC, 2003).

2. My historiographical approach coincides with that of some recent scholarship on early modern history of science in Spain; see the works of Antonio Barrera-Osorio and Jorge Cañizares Esguerra cited throughout this book (n. 29).

use *science* as a shorthand to refer to the specific theoretical framework and practices of cosmographers, as in "the science of cosmography."[3]

Renaissance cosmography was still a new discipline during the century of discovery. It had been reconstituted only during the fifteenth century with the rediscovery of Ptolemy's *Geography* and a humanist revalorization of the work of classical geographers such as Strabo and Pomponius Mela. The discipline of cosmography—understood as a set of theories, language, and practices that are taught and which therefore has disciples[4]—found an institutional home at Spanish universities, where it gained its disciplinary status when it adopted a corpus of classical works as a theoretical foundation. It also assimilated many of the methodological and epistemological approaches characteristic of humanism and reduced these to a set of practices that came to identify the cosmographical practitioner. The humanistic fountainhead that produced cosmography during the late fifteenth century also deeded the discipline two principal modes of representing cosmographical knowledge: Ptolemaic cartography and the descriptive cosmographical *opus*.

The work of Spanish royal cosmographers was science with a mission, deployed solely for the benefit of the state, meaning specifically the Habsburg monarchy. At every step royal cosmographers obeyed a

3. The word *ciencia* was in common usage in sixteenth-century Spain, carrying a variety of meanings: referring to a specific discipline, denoting a way of gaining knowledge, and describing how well something was known. For example, royal cosmographer Alonso de Santa Cruz referred specifically to astrology and cosmography as *ciencias*, as in, for example, "Despues de esto yo me di a saber las ciencias de Astrología y Cosmografía" (AGI, P-260, N. 2, R. 6, "Borrador y apuntaciones para el prólogo del libro intitulado 'Islario General'"). While Covarrubias in the supplement to his *Tesoro de la lengua castellana* (1611) defines *ciencia* as the "certain knowledge of something through its causes" ("el conocimiento cierto de alguna cosa por su causa") or "to know with certainty" ("saber con certeza"). Sebastián de Covarrubias Horozco, *Tesoro de la lengua castellana o española: Edición integral e ilustrada de Ignacio Arellano y Rafael Zafra* (Madrid, Iberoamericana [Frankfurt am Main]: Vervuert, 2006), 528.

4. I follow Donald Kelley's definition of a discipline as an endeavor with "a characteristic method, specialized terminology, a community of practitioners, a canon of authorities, an agenda of problems to be addressed, and perhaps more formal signs of professional condition, such as journals, textbooks, courses of study, libraries, rituals, and social gatherings" (Donald R. Kelley, introduction to *History and the Disciplines: The Reclassification of Knowledge in Early Modern Europe*, ed. Kelley [Rochester, N.Y.: University of Rochester Press, 1997], 1). For a recent survey of the role of the university in shaping early modern scientific disciplines, see Mordechai Feingold and Víctor Navarro Brotóns, eds., *Universities and Science in the Early Modern Period* (Dordrecht, Netherlands: Springer, 2006).

utilitarian mandate that demanded a specific product, a timely deliverable and a cost-effective implementation. In this utilitarian context, a royal cosmographer's personal aspirations or intellectual inclinations that did not directly serve institutional ends had to be subsumed to the interests of the state or be abandoned. Royal cosmographers were often associated with the royal court, but most worked at the two principal institutions established to coordinate the colonization and exploitation of the New World: the Casa de la Contratación or House of Trade in Seville and the Council of Indies.

The Casa de la Contratación was established in 1503 to regulate all commerce and navigation to the New World or the Indies. It coordinated all business enterprises involving Spain's possessions in the New World, which included acting as the tax collector, providing judicial oversight to business ventures, and regulating travel to the colonies. To carry out this enterprise, the Casa needed persons skilled in the art of navigation, who could map the coastlines of this New World, prepare the instruments and charts required for sailing safely to these shores, and train generations of pilots on how to guide a ship. The Casa de la Contratación thus needed sea charts, instruments, and astronomical tables that closely reflected reality—a ship could not be sent to sail on an imagined ocean to an imagined land.[5]

Cosmographical production at the Council of Indies was no less pragmatic than at the Casa but addressed very different needs. As the monarch's advisory council responsible for governing the Spanish territories in the New World, its members oversaw the exploitation of its natural and human resources. They also seated the judicial tribunal adjudicating all legal and political matters in the New World, as well as coordinated the appointment of administrators and religious leaders serving abroad. The members of the Council of Indies needed accurate information about the New World to help them govern the empire effectively. Philosophical speculation had no place in this program, not because of fear of deviating from orthodoxy and running afoul of the Inquisition[6] but because such

5. The classic work on the Council of Indies and the Casa de la Contratación was written in the 1930s by Ernst Schäfer. Ernesto Schäfer, *El Consejo Real y Supremo de las Indias: Su historia, organización y labor administrativa hasta la terminación de la Casa de Austria*, 2 vols. (Madrid: Junta de Castilla y León, Marcial Pons, 2003). For a recent assessment of the Casa's nautical activities, see Antonio Acosta Rodríguez, Adolfo Luis González Rodríguez, and Enriqueta Vila Vilar, eds., *La Casa de la Contratación y la navegación entre España y las Indias* (Sevilla: Universidad de Sevilla, CSIC, Fundación El Monte, 2003).

6. José Pardo Tomás concludes that regulations governing the importation of books written by Protestants contributed to the isolation of scientific men in late sixteenth-century Spain but did not necessarily obstruct local scientific practice, since the Holy

musings rarely yielded tangible results that could be applied to the pressing problems of the expanding empire.

The royal cosmographers in this study served under three Habsburg monarchs, beginning with Charles V (r. 1516–56), who deeded his son Philip II (r. 1556–98) the Habsburg possessions in northern Europe, Italy, and Spain, as well as the all the lands in the New World. Under Philip II, the Spanish Habsburg monarchy attained its maximum territorial expansion, controlling the Low Countries, large portions of Italy, most of the known American continent, the Philippines, and, after the 1580 annexation of Portugal, the entire Iberian Peninsula. This assemblage of territories which formed part of Philip's patrimony determined the composite nature of the Spanish monarchy during the era of discovery.[7] Local nobles, including former royal houses, regional assemblies, and colonial viceroyalties, each with distinct governance and traditions, jostled within the complex administrative structure to assert rights and privileges. When not specifically overruled by royal decree, different laws and administrative procedures applied within each constituent group. This assured some acquiescence to monarchical power by preserving local laws but also made the consistent application of certain laws—as in the case of secrecy regulations—almost impossible throughout the empire. The kingdom of Castile, became the center of power under Philip II, with Castilian nobles and *letrados* (men of letters) dominating the religious and governmental spheres.

In the climate of rapid territorial expansion of the sixteenth century, knowing with certainty the location of newly discovered territories had significant geopolitical implications. Even after their unification, Spain and Portugal continued to debate the location of the line of demarcation established by the pope in 1493 and revised under the Treaty of Tordesillas in 1494. This division of the newly discovered lands into two spheres of influence could not but vex other European nations, especially when news of the nature and extent of the potential wealth of the Indies became known

---

Office in Spain rarely censured scientific books. José Pardo Tomás, *Ciencia y censura: La Inquisición española y los libros científicos en los siglos XVI y XVII* (Madrid: CSIC, 1991), 334–39.

7. John H. Elliott, "A Europe of Composite Monarchies," *Past and Present* 137 (1992): 51. For a concise introduction to the Spanish Habsburg monarchy during the sixteenth century and its relationship with the New World, see John H. Elliott, *Spain and Its World, 1500-1700* (New Haven, Conn.: Yale University Press, 1989), 7–26. For a general history of sixteenth- and seventeenth-century Spanish monarchy, see Antonio Domínguez Ortiz, *Desde Carlos V a la paz de los Pirineos, 1517–1660* (Barcelona: Ediciones Grijalbo, 1974), and John H. Elliott, *Imperial Spain, 1469–1716* (London: Penguin Books, 1990).

after the conquest of Mexico and Peru. The French and English, and later the Dutch, and countless pirates and privateers found the poorly defended borderlands of the Spanish monarchy overseas tempting targets.

The consolidation of the largest empire the world had ever known also coincided with the confessional conflicts that plagued Europe during the sixteenth century. The midcentury saw an entrenchment of religious positions around the Catholic orthodoxy of Philip II and the rise of the Inquisition in Iberia and the Indies as a powerful arbiter on matters of faith. In the Indies, the spiritual life of the native population was cause of great concern among the religious orders—Franciscans, Dominicans, Augustinians, and later, Jesuits—responsible for their conversion and welfare. Repeatedly throughout the sixteenth century, the Council of Indies sought to strengthen the tenuous hold the Spanish monarchy had over both the spiritual life and the political choices of an often-rebellious native population.

The Spanish monarchy had at best nominal hegemony over the myriad ethnic groups, nationalities, and confessions that formed part of its European domains. Keeping at least a semblance of unity required a delicate balance of military force, patronage, and local autonomy. In the Indies, the logistical difficulties imposed by distance required a finely tuned apparatus of monarchical power. Control of the Spanish viceroyalties and *audiencias* of the Indies was rarely achieved through the exercise of military power; rather, it generally required complicated and lengthy negotiations with settlers.[8] The Council of Indies carefully balanced the settlers' demands for political and fiscal accommodations with their fiduciary duty to the Spanish monarch to administer what the king considered his personal possessions. By the late sixteenth century, the crown's repeated bankruptcies brought to the fore the real or imagined difficulties of maintaining Spain's vast overseas domains, as did the continued toll on Castile's resources of the ongoing struggle to pacify the Low Countries.

Royal cosmographers in sixteenth-century Spain operated within a bureaucratic structure built to administer this complex empire. By the midcentury, the threats posed by foreign and internal enemies moved the monarchy to consider cosmographical work the equivalent of today's state secret. Within the early modern conception of the state, the concept of a state secret was defined by the potential of certain kinds of knowledge to harm to the monarchy. The Spanish monarchy took the defensive posture of censoring and prohibiting the circulation of maps, geographic descriptions,

8. John H. Elliott, *Empires of the Atlantic World: Britain and Spain in America, 1492–1830* (New Haven, Conn.: Yale University Press, 2006), 130–33.

and historical account about the Indies for strategic military and political reasons. The rationale behind the secrecy policy was very straightforward. If documents that revealed the geodesic coordinates, geographical features, coastal outlines, hydrography, and natural resources of the New World were produced and circulated publicly, these, in the hands of enemies, could be used to reach the New World and inflict harm on the crown's patrimony and the peoples the state had the obligation to protect. Thus this type of knowledge was considered to have strategic, defensive, and monetary value and needed to be safeguarded from foreign and internal enemies alike. Like anything of value, cosmographical information was hidden, stolen, intentionally misrepresented, used to leverage personal merits, employed to further institutional or royal agendas, and, of course, bought and sold.

From our modern perspective, it might seem naive to think that by keeping geographical information secret one nation could effectively hide the New World from eager interlopers. The Spanish monarchy's effort to maintain secrecy regarding the location of new and profitable geographical discoveries, however, was hardly new. The Portuguese monarchy and its *Côrtes* acted in 1481 to prohibit the dissemination of nautical charts and descriptive histories that told of recent Portuguese discoveries.[9] By 1527 the Casa de la Contratación prohibited foreign pilots from owning navigational charts.[10] Richard Hakluyt, in the dedication of his *Divers Voyages* (1582), left testimony of how effective, or at least troublesome, Spain's efforts to contain cosmographical information about the New World had been: "Portingales time to be out of date and that the nakedness of the Spaniards and ther long hidden secretes are now at length espied, whereby they went about to delude the worlde, I conceive gret hope that the time approcheth and nowe is, that we of England may share and part stakes . . . bothe with the Spaniarde and the Portingale in part of America, and other regions as yet undiscovered."[11]

How then did this environment of secrecy influence a Spanish royal cosmographer's practice? For example, the need to safeguard geographical information from potential foreign interlopers meant that many cosmographical works remained unpublished. This was not, as some

9. Jaime Cortesão, "The Pre-Columbian Discovery of America," *Geographical Journal* 89, no. 1 (1937): 30–32.

10. José Pulido Rubio, *El piloto mayor de la Casa de la Contratación de Sevilla: Pilotos mayores, catedráticos de cosmografía y cosmógrafos* (Sevilla: Publicaciones de la Escuela de Estudios Hispano-Americanos, 1950), 141.

11. As cited in J. N. Hillgarth, *The Mirror of Spain, 1500–1700: The Formation of a Myth* (Ann Arbor: University of Michigan Press, 2000), 374.

historiography suggests, an indication that Spain as a colonial power wanted to suppress Amerindian identities or that there was a lack of interest on the monarch's part. Instead, and quite to the contrary, it meant that the material contained in these works was considered valuable and strategically sensitive. Furthermore, secrecy provisions stipulated that cosmographical works circulate only among a select group of government officials, yet this did not mean the practice took place in a vacuum. Instead, there is clear evidence of a lively tradition of critique. Secrecy also meant that within the closed boundaries it imposed on the small circle of practitioners who were privy to this information, one man's talent, disposition, and inclination could have tremendous influence on cosmographical practice.

The transition between the reigns of the two Philips, Philip II and Philip III (r. 1598–1621), presents the opportunity to study cosmography under two monarchs who differed significantly in character and style, who at times had diametrically opposed imperial programs, and who operated different systems of royal patronage. Throughout his forty-year reign, Philip II demanded that his bureaucrat-cosmographers diligently gather information about the New World but also that they treat this information as secret, while under Philip III royal cosmographers were free, and even encouraged, to publish their findings and divulge what had been considered state secrets just a decade before. This inversion in the secrecy policy was associated with a fundamental change in the patronage structure and courtly context in which the cosmographer of the Council of Indies operated. Royal cosmographers were quick to adapt to these changes, particularly when the new patronage equation and repeated foreign incursions into Spanish territories began eroding the established policies concerning secrecy.

The cartographic silences (as J. B. Harley so apply described them) that resulted from this context of secrecy speaks to the complex historical situation surrounding the construction of an early modern map of the New World.[12] Only a few maps drawn by sixteenth-century Spanish cosmographers survive. This is precisely what we should expect to find given the secrecy restrictions in place to control cartographic production at the Council of Indies and at the Casa de la Contratación. Two aspects operated to deny us these maps. The nautical charts prepared or authorized by the Casa de la Contratación were by their very nature ephemeral documents. They were traced upon, pricked repeatedly with dividers, sprayed with saltwater, and routinely discarded when new charts became available.

12. J. B. Harley, *The New Nature of Maps*, ed. Paul Laxton (Baltimore: Johns Hopkins University Press, 2001), 59–60, 105.

At the Council of Indies, the maps were considered secret reference documents that were not originally intended for publication. The apparent silence that the paucity of maps suggests, however, lies in contrast to the vigorous and novel cosmographical practices that took place at these institutions throughout the sixteenth century.

Parallel to the historical narrative that chronicles the changing standards of secrecy, this book also explores fundamental changes to the discipline of cosmography that came about as a result of the discovery of the New World. *Cosmography*, as formulated by the humanist scholars who defined the practices associated with the discipline during the late fifteenth century, also denoted a particular mode of representation associated with the discipline. A cosmography consisted of both maps and descriptive texts. These required distinct skills of the cosmographer, the one mathematical and graphic, the other descriptive and textual. As a direct consequence of the discovery of the New World, Spanish cosmographers found themselves struggling to reconcile the mode of representation of typical Renaissance cosmographies with an imperial context that demanded accuracy, timeliness, and the production of useful works. Likewise, the bookish epistemology associated with Renaissance cosmography gradually shifted in favor of an empirical and mathematical approach.

This book argues that by the later part of the sixteenth century, Renaissance cosmography was failing to provide an adequate conceptual framework from which to approach the New World and as a result cosmographical practice dissolved along epistemic and methodological lines. The descriptive part of geography became the domain of the historian and chronicler. Cosmography proper now came under the command of mathematical practitioners who focused on mathematical cartography and other complementary empirical practices. The events in this book anticipate the move toward specialization and the development of new methodologies that historians have noted among English, French, and German cosmographers.[13]

13. For a general discussion of the changing nature of cosmography during the sixteenth century, see Klaus A. Vogel, "Cosmography," in *The Cambridge History of Science*, ed. Lorraine Daston and Katherine Park (Cambridge: Cambridge University Press, 2006), 470–71. For a study of French cosmography, see Frank Lestringant, *Mapping the Renaissance World*, trans. David Fausset (Cambridge: Polity, 1994), 129. The case of England is discussed by Lesley B. Cormack, *Charting an Empire* (Chicago: University of Chicago Press, 1997), 37–42, and Robert J. Mayhew, "Geography, Print Culture, and the Renaissance: 'The Road Less Travelled By,'" *History of European Ideas* 27 (2001): 349–69.

I interpret historical evidence that shows a period of changing practices as an indication of pressures operating within a group of practitioners that could result in the reconceptualization of the methodologies and epistemology associated with a scientific discipline.[14] The discovery and colonization of the New World is an example of such a pressure, and the activities of a particular group of practitioners—royal cosmographers—constitute historical evidence of how scientific practitioners responded to this transcendental event. This book shows how the very process of answering the challenging questions posed by the discovery of the New World obligated Spanish cosmographers to develop new knowledge-making practices that eventually redrew the intellectual boundaries of the cosmographical discipline.

For interpretive background, this book relies on the historiography of science that studies the role of mathematical practitioners and cosmographers in early modern Europe.[15] By exploring the development of empirical practices, mathematical ways of rationalizing space, and new ways of collecting knowledge about nature, I believe it is possible to situate the practice of

14. I borrow this historical approach from the literature on laboratory practice and, more specifically, the use of early modern laboratory notebooks to reconstruct a scientist's intellectual universe. For examples of works sensitive to this kind of historical interpretation, see William R. Newman and Lawrence M. Principe, *Alchemy Tried in the Fire: Starkey, Boyle, and the Fate of Helmontian Chymistry* (Chicago: University of Chicago Press, 2002), and Frederic L. Holmes, Jürgen Renn, and Hans-Jörg Rheinberger, eds., *Reworking the Bench: Research Notebooks in the History of Science* (Dordrecht, Netherlands: Kluwer Academic, 2003).

15. For some works in the history of science that discuss the role of the cosmographer and mathematical practitioner in early modern Europe, see José María López Piñero, *Ciencia y técnica en la sociedad española de los siglos XVI y XVII* (Barcelona: Labor, 1979); Robert S. Westman, "The Astronomer's Role in the Sixteenth Century: A Preliminary Study," *History of Science* 18 (1980); J. A. Bennett, "The 'Mechanics' Philosophy and the Mechanical Philosophy," *History of Science* 24 (1986); and Peter R. Dear, *Discipline and Experience: The Mathematical Way in the Scientific Revolution* (Chicago: University of Chicago Press, 1995). For a biographical approach to the practice of cosmography, in this case French royal cosmographer André Thevet, see Lestringant, *Mapping the Renaissance World*. For an English practitioner, see William H. Sherman, *John Dee: The Politics of Reading and Writing in the English Renaissance* (Amherst: University of Massachusetts Press, 1995). For an educational perspective on early modern cosmographical practice in England and France, see Cormack, *Charting an Empire*, and François de Dainville, *La géographie des humanistes: Les Jésuites et l'éducation de la société française* (Paris: Beauchesne et Ses Fils, 1940).

cosmography, and more specifically that of Spanish cosmographers, within the broader literature of the early modern history of science. Cosmography has been underrepresented in this field, but recent studies are carving a well-deserved niche for the geographical disciplines in the historiography of the Scientific Revolution.[16] The work of the Spanish royal cosmographers is likewise understood to be the product of cultural and social interactions within a court milieu and to respond to the particular demands of the institutional patrons who commissioned the works.[17] Thus, cosmographical practice and production are contextualized within humanist circles, court culture, and bureaucratic structures of late sixteenth-century Spain, with the goal of fashioning a historically sensitive interpretation.[18]

The organizing scheme of the book is a chronological exploration of what proved to be cosmography's malleable epistemological foundation during the century of discovery—the methods employed and the criteria used to ascertain cosmographical facts. I begin with a survey of the intellectual history of Renaissance cosmography in order to define the methods associated with the discipline after the discovery of the New World. To delve further into the complex epistemological, methodological, and interpretive elements of sixteenth-century Spanish cosmography, I seek to reconstruct the scientific practices that guided cosmographers in the collection, translation, and codification of knowledge about the New World. I understand "scientific practices" to be specialized activities that ultimately contribute to the production of knowledge about the natural world.[19] More specifically for

16. See Charles Whithers's contribution, in addition to the essays by Peter Dear, David Livingstone, and John Henry, in *Geography and Revolution*, ed. David N. Livingstone and Charles W. J. Whithers (Chicago: University of Chicago Press, 2005), 99.

17. For two fascinating but distinct approaches to the role of patronage in shaping scientific practice, see Richard S. Westfall, "Science and Patronage: Galileo and the Telescope," *Isis* 76, no. 1 (1985), and Mario Biagioli, *Galileo, Courtier: The Practice of Science in the Culture of Absolutism* (Chicago: Chicago University Press, 1993). For a survey of patronage in a variety of early modern settings, see Bruce T. Moran, ed., *Patronage and Institutions: Science, Technology, and Medicine at the European Court, 1500–1750* (Rochester, N.Y.: Boydell, 1991).

18. The literature on science within early modern bureaucracies is understandably scarce, as few European countries had sophisticated governmental structures similar to the ones Spain developed in response to the discovery and colonization of the New World. For a general survey of Spanish, American, and French bureaucracies, see Mark A. Burkholder, ed., *Administrators of Empire* (Aldershot, UK: Ashgate/Variorum, 1998).

19. For examples of a sociological approach to the study of scientific practices which considers cultural aspects and social interaction central to scientific activity, see Andrew Pickering, ed., *Science as Practice and Culture* (Chicago: University of Chicago

the subject of this book, they are what scientists "did" and that led to the production of knowledge of the nature, geography, and the people of the New World. This broad definition is both necessary and helpful in light of the permeable boundaries that existed among sixteenth-century disciplines we now, admittedly somewhat anachronistically, label *scientific*.

The scientific practices that concern this study can be procedural, conceptual, material, literary, or social in nature, but they have as a common mission inquiry into the natural world and the ordering of the knowledge produced. Disciplines may share certain practices; thus a cosmographer calculating a latitude coordinate engaged in a series of activities not unlike those used by an astronomer to determine the position of a star. Likewise, the literary practices associated with composing a sixteenth-century history have a lot in common with those of an early modern naturalist writing a description of an animal's behavior and habitat.[20] Moreover, the set of practices associated with a particular scientific discipline, such as cosmography, were far from static. For whereas they originated from the experience and knowledge base associated with a community of practitioners, they changed in response to a number of factors, among which could be a challenge to the theoretical underpinnings of the discipline or the perceived shortcomings of a given set of practices to yield the desired body of knowledge.

Rather than addressing historiographical questions about specific cosmographical works in this introduction, I have chosen to take these up in each pertinent chapter. Here, however, I would like to address some issues concerning the historiography of science that to this day taint historical perceptions about science in Spain. Historians of Spanish science have long wrestled with a historical narrative that maintained that during the seventeenth century, science in Spain began a gradual decline that left Spanish natural philosophers at the margins of the theories that led to the

---

Press, 1992), and Jed Z. Buchwald, ed., *Scientific Practice: Theories and Stories of Doing Physics* (Chicago: University of Chicago Press, 1995). For a general introduction to the many and often contentious strands of practice theory, see Jan Golinski, "The Theory of Practice and the Practice of Theory: Sociological Approaches in the History of Science," *Isis* 81 (1990), and Theodore R. Schatzki, Karin Knorr Cetina, and Eike von Savigny, eds., *The Practice Turn in Contemporary Theory* (London: Routledge, 2001), 1–16.

20. Laurent Pinon, "Conrad Gessner and the Historical Depth of Renaissance Natural History," in *Historia: Empiricism and Erudition in Early Modern Europe*, ed. Gianna Pomata and Nancy G. Siriasi (Cambridge, Mass.: MIT Press, 2005), 241–67.

Scientific Revolution. Since the late seventeenth century, social commentators and historians—both Spanish and Continental—have tried to explain the discontinuity between the perceived vitality of the sixteenth century and the "decline" of the following century. The resulting debate posited a number of conflicting answers to the question of Spain's seventeenth-century scientific decline and is known as "the polemic of Spanish science."[21]

Various theories have been posited to answer the principal question behind the polemic—why did Spain remain at the margins of the Scientific Revolution during the seventeenth century? The way the question was formulated, with its tacit acceptance of the notion of a Scientific Revolution and of a Spanish decline, contributed to what became a persistent and at times acrimonious debate.[22] Attempts to explain the decline relied on an acknowledged contrast between the intellectual climate of the early sixteenth century and that of the seventeenth century. Whereas the first half of the sixteenth century marked the high point of Spain's openness toward Europe and the intellectual currents of the Renaissance, the end of the century marked the high point of theological repression and intolerance. Historians of Spanish science hotly debate to what extent theological repression during the latter half of the sixteenth century influenced the practice of science.

The polemic's historiography presents several different and heavily politicized answers to this question. Historian José María López Piñero traces the polemic's origin to the *Novatores* or reformers of the later part of the seventeenth century. They tried zealously to introduce into Spain the ideas of the Scientific Revolution and the Enlightenment. Faced with a nation in frank decay, and disillusioned by the scholasticism entrenched at the universities, they broke with the past and rejected the work of previous Spanish thinkers. One camp of historians, then, whose primary motive was to attack the Inquisition and the institutions that supported it, portrayed Spain as a

21. Ernesto García Camarero, ed., *La polémica de la ciencia española* (Madrid: Editorial Alianza, 1970).

22. The validity of the term *scientific revolution* to characterize the changes that took place between 1500 and 1700 in various scientific fields has been challenged over the last two decades. Whereas the traditional historiography of science described the abandonment of Aristotelian natural philosophy during this time as a "major discontinuity," others argue that the change was far more gradual and diffused. For a concise survey of this debate, see the following edited volumes: Margaret J. Osler, ed., *Rethinking the Scientific Revolution* (Cambridge: Cambridge University Press, 2000), 4–5, and David C. Lindberg and Robert S. Westman, eds., *Reappraisals of the Scientific Revolution* (Cambridge: Cambridge University Press, 1990).

scientific wasteland and carelessly extended their bias to previous centuries. In reaction, the other faction, seeking to downplay the power of the church, produced panegyric-style biographies of Spanish men of science. Both groups, however, did little detailed investigation into Spain's scientific past and left future generations with the impression that little of scientific value had been accomplished. Other apologists chose to frame Spain's lack of interest in scientific pursuits in terms of "national character." They argued that Spaniards during the seventeenth century—*la Edad de Oro*—simply valued art, literature, and the humanities over science. Under this cloak of exceptionalism they located the discussion of scientific practice in Spain outside of any synthetic narratives of the history of Western science.[23]

Further complicating the polemic of Spanish science has been the persistent negative attitudes about Spain inculcated by centuries of retelling of the Black Legend.[24] The legend is not one story but rather a conglomeration of stories that paint a none-too-flattering picture of sixteenth-century Spain and Spaniards. As the dialectic of the Reformation and Counterreformation migrated to the popular pamphlet literature of the late sixteenth century,[25] foreign critics portrayed Spain as the Leviathan against which all Protestant nations should battle. One of its most important vehicles of dissemination was William of Orange's *Apologie* (1580). It was a thinly veiled exposé of all that was wrong with Spain: the hegemony of the Inquisition over the every facet of Spanish life, the unbridled commercial greed of Castilians, militarism and cruelty, and, quoting from Bartolomé de Las Casas, the slaughter of Indians in the New World. Finally, it accused Philip II of incest and of the murder of his son, Prince Carlos.

By the late seventeenth century and into the eighteenth, the Black Legend unfolded into a negative assessment of Spanish character, which

23. López Piñero, *Ciencia y técnica*, 16–17, 21–24; and Mauricio Jalón, "Empresas científicas: Sobre las políticas de la ciencia en el siglo XVII," in *Madrid, ciencia y corte*, ed. Antonio Lafuente and Javier Moscoso (Madrid: CSIC, 1999), 159–63. On intellectual life during the sixteenth century, see Elliott, *Imperial Spain*, 225–27.

24. For a dispassionate historiographical survey, see Ricardo García Cárcel, *La leyenda negra: Historia y opinión* (Madrid: Alianza Editorial, 1998). For an example of a Spanish response to the Black Legend, this one penned after the 1898 loss of Spain's last American and Pacific colonies, see Julián Juderías, *La leyenda negra: Estudios acerca del concepto de España en el extranjero* (Castilla y León: Junta de Castilla y León, Consejería de Educación y Cultura, Caja Salamanaca y Soria, 1997). For a study of European attitudes about Spain, see William S. Maltby, *The Black Legend in England: The Development of Anti-Spanish Sentiment, 1558–1660* (Durham, N.C.: Duke University Press, 1971), and Hillgarth, *Mirror of Spain*.

25. García Cárcel, *Leyenda negra*, 28.

included accusations of ignorance and backwardness. The watershed point in this debate as it pertained to science was the article on Spain written by Nicholas Masson de Marvilliers in the modern geography section of the *Encyclopédie methodique* in 1782. He described Spaniards as having some aptitude for science but nonetheless as populating the most ignorant nation in Europe.[26] The Latin American wars of independence and the 1898 Spanish-American War reenergized the legend, but in these instances the focus was on Spanish cruelty toward the native population of the New World. This negative characterization was aided by a tradition of self-critique among Spanish intellectuals that foreign observers, predisposed by the Black Legend, subsumed into their views.[27]

Returning to the polemic of Spanish science, in my opinion, the debate itself and the lack of unbiased historical analysis that resulted from it have stubbornly undermined efforts to relate the practice of science in Spain, whether during its zenith or its decline, to that of the rest of Europe. For historians of science writing positivistic accounts in the twentieth century—even those able to overcome centuries-old prejudices inculcated by the Black Legend—the apparent discontinuity in scientific practice in Spain proved so insurmountable that they simply omitted Spain from their accounts.[28] No doubt the invisibility of early modern Spanish science has been exacerbated further by the fact that the scientific projects undertaken during the reign of Phillip II were considered state secrets and the material remained scattered and unstudied until recently.

Despite the marked lack of interest in Spanish science beyond the Pyrenees, historians in Spain have—using the words of one of its greatest exponents, López Piñero—"tortured archives" for centuries in search of evidence of scientific activity. The work of these eighteenth- and nineteenth-century historians is invaluable for the modern researcher, as my frequent references to the works of Martín Fernández de Navarrete, Marcos Jiménez de la Espada, and Felipe Picatoste y Rodríguez will attest. Not only does their work provide a useful compilation of primary sources,

26. García Camarero, ed., *Polémica de la ciencia española*, 9.

27. "The Decline of Spain," in Elliott, *Spain and Its World*, 219.

28. For a survey of the historiographical issues involved, see Víctor Navarro Brotóns and William Eamon, "Spain and the Scientific Revolution: Historiographical Questions and Conjectures," in *Más allá de la leyenda negra. España y la Revolución Científica*, ed. Víctor Navarro Brotóns and William Eamon (Valencia: Soler, 2007), 27–38. For a different interpretation of the invisibility of Spanish science, see Jorge Cañizares Esguerra, "Renaissance Iberian Science: Ignored How Much Longer?" *Perspectives on Science* 12, no. 1 (2004): 86–125.

many of which no longer exist, but their narratives piece together the important "who-what-where-and-when" of the lives of the principal Spanish practitioners of science.

Largely because of López Piñero's pioneering work since the 1970s, a nucleus of historians of science working for the Consejo Superior de Investigaciones Científicas (CSIC) and at Spanish universities continue to tackle archives with the same zeal as their forerunners but with a historical approach that avoids the positivistic narratives of yesteryear. They also skirt the old history's nationalistic and apologetic interpretations. This study draws heavily on their work. Throughout the book I have tried to acknowledge as many of their writings as possible, but I must especially highlight Mariano Esteban Piñeiro, Víctor Navarro Brotóns, and María Isabel Vicente, whose indefatigable work on early modern Spanish astronomy and cosmography constitute impressive contributions to the field.

The last three decades have also seen a small number of historians outside Spain producing works on early modern Spanish cosmography, such as David Goodman, Richard Kagan, Ursula Lamb, and Alison Sandman. This book joins a small but growing literature on Spanish science written in English that eschews exceptionalist interpretation of scientific practice in early modern Spain and maintains that to understand the origins of the Scientific Revolution, the historian must turn precisely to those nations—Spain and Portugal—most intimately involved with the discovery of the New World. Recently, Jorge Cañizares Esguerra pleaded for the inclusion of Iberian science in the narrative of the Scientific Revolution, and Antonio Barrera-Osorio has argued from this perspective in his study of empirical practices in sixteenth-century Spain.[29] It is my hope that the story between the covers of this book provides a new perspective to the historical narrative of the Scientific Revolution on how practitioners of Western science responded to the New World.

The point of departure for this book is the question, what is cosmography? Therefore the book begins with a study of how several intellectual traditions coalesced into the discipline of Renaissance cosmography between the years 1450 and 1530. This involves a survey of the discipline's principal texts, particularly those that dealt with the New World, and includes an analysis that

29. Jorge Cañizares Esguerra, *Nature, Empire, and Nation: Explorations of the History of Science in the Iberian World* (Stanford, Calif.: Stanford University Press, 2006), 26–45, and Antonio Barrera-Osorio, *Experiencing Nature: The Spanish American Empire and the Early Scientific Revolution* (Austin: University of Texas Press, 2006).

highlights methodological, epistemological, and stylistic aspects of these important texts. To illustrate how the discipline was practiced in Spain, I discuss two sites of knowledge production: the University of Salamanca and the House of Trade (Casa de la Contratación) in Seville. I draw upon the cosmographical curriculum at the University of Salamanca and surviving classroom material from astronomer Jerónimo Muñoz to identify methodological and epistemic aspects of what contemporaries took to be ideal cosmographical practice. I juxtapose these to the utilitarian practices at the House of Trade in Seville. Chapter 1 points to an interesting appropriation on the part of Spanish cosmographers, both academic and "practical," of classical and Continental models of practice and representation in light of the discovery and initial colonization of the New World. By the 1550s, cosmographical practitioners in Spain had developed an autochthonous practice and even a new genre: the navigation manual. However, the cosmography practiced in Spain in the 1550s was fundamentally different from that practiced in 1600. The rest of the book explores the reasons behind this transformation.

The second chapter finds cosmography at a turning point. Fifty years of trying to incorporate knowledge of the New World into the cosmographical corpus had highlighted some deficiencies within cosmographical practice. Two aspects receive particular attention: the inadequacy of the traditional modes of textual representation associated with Renaissance cosmography and the question of personal agency in cosmographical work. The chapter focuses on how royal cosmographer Alonso de Santa Cruz (c. 1505–67) and Philip II's right-hand man on scientific matters, Juan de Herrera (1530–97), addressed what they perceived as a tension between traditional modes of cosmographical production and the knowledge needed for the imperial program. In response they developed what I describe as distinctive styles of cosmographical practice, molded in large part by each institution's administrative duties and, increasingly, by a preference for either mathematical or didactic cosmographical tools to explicate the New World.

Beginning with the third chapter, the book's focus shifts to cosmographical practice at the Council of Indies and explores the influence of legal culture on scientific practice. Between 1570 and 1575, the president of the Council of Indies, Juan de Ovando (d. 1575), instituted several legal measures that incorporated into the bureaucracy of the empire mechanisms for collecting and organizing cosmographical information about the Indies and—crucially—keeping this knowledge secure within the council's confines. His actions were a consequence of increasing problems with colonial administration attributed to lack of information about the New World.

Ovando's actions incorporated legalistic methodologies and juridical ways of establishing matters of fact—not unlike the approach Francis Bacon would later adopt—into cosmographical practice that subverted key aspects of traditional Renaissance cosmography. I explore parallels between these measures and similar epistemic approaches that historians have identified in other European nations, principally in the work of Francis Bacon.

Ovando's reforms created the post of cosmographer-chronicler major of the Council of Indies, and the fourth chapter is thus a study of the life and work of the first man to serve in this post, Juan López de Velasco (c. 1530–98). Over the twenty years he held the post, his humanistic interests and self-taught cosmographical knowledge dictated cosmographical practice in the Council of Indies. His personal correspondence offers new insights into Velasco's role as the locus of information exchange between the Old and New worlds and how these functioned in relation to the patronage dynamics of the court of Philip II. The fifth chapter examines Velasco's principal cosmographical work, the *Geografía y descripción de las Indias* (1574), and a subsequent shorter work, the *Sumario* (c. 1580), two works that because of their strategic value remained unpublished during the reign of Philip II. I use critiques from Italian cosmographer Juan Bautista Gesio (d. 1581) to illustrate ways in which the works are emblematic of the epistemological and representational deficiencies that certain practitioners perceived as having overtaken the cosmographical genre.

In response to his critics, López de Velasco set in motion the most ambitious and successful program ever attempted during the early modern era for collecting geographical, ethnographical, and natural historical information about the New World. Chapter 6 considers the genesis, implementation, and results of two major projects: the questionnaires of the Indies and a project to systematically observe lunar eclipses to determine longitude. I present Velasco's projects as the result of a revalorization of the epistemological criteria used to ascertain cosmographical facts and as the reconceptualization of the modes of representation necessary to encompass the fluid nature of knowledge about the New World. The book concludes with a chapter on the transition between the monarchies of Philip II and his son. It shows how Velasco's successors adapted to the new patronage situation in the court of Philip III and the concomitant relaxation of secrecy restrictions and new valorization of geographical knowledge. The cosmographical work of Andrés García de Céspedes, royal cosmographer of the Council of Indies from 1596 to 1611, is shown to be emblematic of the definitive divorce of the descriptive and mathematical practices associated with Renaissance cosmography.

ONE

Renaissance Cosmography in the Era of Discovery

It did not take long for the news that Christopher Columbus had discovered a westward route to Cathay to travel the length and breadth of Spain. As the astounding news spread, many began to question the location and nature of the lands Columbus claimed for the Spanish crown. The first chronicler of the discoveries, Peter Martyr d'Anghiera, writing soon after Columbus's first voyage, thought that Columbus had discovered the Antipodes rather than a westward passage to China. Columbus died without having fully realized—or perhaps acknowledging publicly—that he had discovered heretofore-unknown lands.[1] These questions about the location and novelty of the New World were but the first of countless others that arose throughout the century of discovery. The task of answering these questions, particularly in Spain, fell to a group of scientific practitioners who had expertise in what was then a relatively new discipline, cosmography.

This chapter traces the intellectual history of cosmography, from its classical origins as exemplified by Ptolemy's *Geography* and until, when reinvented during the fifteenth century as a humanistic discipline, it became part of the arts faculty of early modern universities. A close examination of the curriculum at Spanish universities illustrates the epistemic and methodological approaches that were in use at the moment of the discovery of the New World and that were later employed as cosmographers began to assimilate the new geographical knowledge into this humanistic discipline.

1. Felipe Fernández-Armesto, *Columbus* (New York: Oxford University Press, 1991), 95–97. For more on Peter Martyr's interpretation of Columbus's cosmographical thought, see Juan Gil, "Pedro Mártir de Anglería, intérprete de la cosmografía colombina," *Anuario de estudios Americanos* 39 (1982).

I use university statutes, curriculum, and class notes from the Universities of Salamanca and Valencia to delineate the boundaries of what constituted ideal cosmographical practice. This will serve as the departure point from which to explore cosmographical practice at the institutions set up by the Spanish monarchy to administer its overseas possessions.

After mapping the intellectual boundaries of Renaissance cosmography and how it responded initially to the discovery of the New World, the chapter turns to an aspect that defined cosmographical practice in Spain: navigation. Under the conceptual umbrella of cosmography, the mathematical rationalism inherent in the Ptolemaic conception of space found its expression in the navigation manual, a new genre that promised readers a happy marriage of theory and practice. Thus we find that by the mid-sixteenth century, cosmographical practitioners in Spain straddled the difficult divide between ideal practice as taught at the universities and the demands of applying scientific theory to what had traditionally been the craft of piloting a ship. This and other tensions strained Renaissance cosmography, so that by the mid-sixteenth century the discipline was poised to undergo a significant reorientation.

Humanists Adopt Ptolemy: European Practitioners Create a New Discipline

Renaissance cosmography rested on a foundation of Aristotelian natural philosophy, but in a context where the terrestrial and celestial spheres were conceived as spaces that could be described mathematically and charted accordingly.[2] A complete description of the universe—the cosmographer's ultimate goal—also required integrating human and natural elements into the mathematized geographical landscape. Thus cosmographical production during the Renaissance integrated three classical intellectual

2. The historical study of Renaissance cosmography beyond Spain has traditionally focused on the study of maps and descriptive geographies from different perspectives, from meticulous study of map authorship and the development of the underlying mathematical principles of cartographic projections (Karrow) to the study of maps as emblematic representations of the state (Harley and Buisseret). Other studies find that artistic trends in the Renaissance, such as the perspective grid, also found expression in cartographic design, as did the "esoteric culture and cosmology" associated with geometrizing the landscape (Edgerton and Cosgrove). Descriptive geographies, except for the ones about the New World, have been somewhat neglected, as the texts have generally been discussed in the context of a comprehensive survey (Bowen and Grafton). For biographical studies of cosmographical practitioners, see Lestringant's work on André Thevet and Blair's on Jean Bodin.

traditions: Aristotelian natural philosophy, which contributed the framework for the fundamental understanding of the natural world; Euclidean geometry and Ptolemaic geography, which provided the tools for mathematical representation of the universe; and Pomponius Mela's and Pliny's works, which served as models to incorporate the human, animal, and plant kingdoms into this universe. Cosmography was, in short, the science that explained the earthly sphere by locating it within a mathematical grid bounding space and time. Natural phenomena and human actions that defied mathematization were described; words were the tools that took the phenomena from the realm of the unknown into the known.

In Spain, the discovery of the New World presented an unprecedented challenge and opportunity. The men answering this challenge came from a variety of places. Some were university educated and practiced as doctors or teachers, but many were self-taught and found work at the new institutions created to serve the needs of the expanding territorial claims of the Spanish monarchy. From the halls of the universities of Salamanca, Valencia, and Alcala to the busy ports of Cadiz and Seville, Spanish monarchs sought men with a particular blend of academic and practical knowledge about the natural world to help them understand the New World. In addition, the cosmographer, a mathematical practitioner at a time when few could claim that expertise, was often called upon to tackle engineering problems in fields as varied as navigation, artillery, and architecture.

Judged solely by their reliance on empirical methods and use of mathematics as a tool for achieving utilitarian results, Spanish cosmographers fall squarely within the tradition of Italian and English mathematical practitioners considered by some to be the first exponents of a nascent Scientific Revolution.[3] Perhaps their closest counterparts in continental Europe were mathematical practitioners, along the lines of Nícolo Tartaglia (1500–57), Simon Stevin (1548–1620), John Dee (1527–1608), and of course Galileo. These were generally practical men who earned their living serving noble patrons and practiced cosmography outside the university environment. In late Renaissance Spain, cosmographers were the men best prepared to answer the pressing questions "what are the discovered lands?"

3. J. A. Bennett, "The Challenge of Practical Mathematics," in *Science, Culture, and Popular Belief in Renaissance Europe*, ed. Maurice Slawinski, Paolo L. Rossi, and Stephen Pumfrey (Manchester: Manchester University Press, 1991), 176–90; Eric H. Ash, *Power, Knowledge, and Expertise in Elizabethan England* (Baltimore: Johns Hopkins University Press, 2004), chaps. 3 and 4; and John Henry, *The Scientific Revolution and the Origins of Modern Science* (New York: St. Martin's, 1997), 14–16.

and "where are they located?" To address these questions, cosmographers turned to the Ptolemaic corpus for guidance.

Cosmography as a discipline began to coalesce during the fifteenth century as part of the humanistic revival of classical learning that characterized that century. Its foundational text, Ptolemy's *Geography*, was written in the second century AD, but the discipline remained dormant for most of the Middle Ages. Only a few medieval scholars, primarily Byzantine, seem to have been familiar with segments of the Greek text of the *Geography*, and its popularity as a reference text paled in comparison to that of Ptolemy's influential astronomical treatise, the *Almagest*, which included some of the theoretical material contained in the *Geography*.[4] The 1406 Latin translation of the *Geography* by Jacopo d'Angelo renewed interest in the text and earned cosmography recognition as a humanistic discipline.[5] After the first printed edition appeared in 1475, the *Geography* became the unchallenged model for most subsequent cosmographical works during the Renaissance. What appears to have been "lost" during the Middle Ages is Ptolemy's explanation of the epistemological approach and methodological aspects of cosmographical practice. These included instructions on how to compile geographical information from seemingly conflicting accounts, how to translate geographical coordinates into degrees of latitude and longitude, and how to draw maps using various cartographic projections. The book defined a set of practices that became fundamental to the discipline as it was practiced during the Renaissance. For cosmographers reading the *Geography* twelve hundred years after it was written, Ptolemy's method became canonical.

At first glance the *Geography* is a large and unwieldy text. Most of it consists of tables listing the latitude and longitude of cities in antiquity, with topographical descriptions of important cities relegated to brief passages. Perhaps this contributed to the perception that the *Geography* was a reference text rather than a theoretical treatise on mapmaking. Only two short chapters explain the mechanics of cartography. Humanistic scholars, however, recognized the pedagogical nature of the book and set out to understand and mimic the Great Astronomer's methods. The reason for the

4. Ptolemy, *Ptolemy's Geography: An Annotated Translation of the Theoretical Chapters*, trans. J. Lennart Berggren and Alexander Jones (Princeton, N.J.: Princeton University Press, 2000), 50. I use this recent translation of the *Geography*'s theoretical chapters throughout my discussion; I cite Berggren's and Jones's introduction using page numbers, while I cite the translated text by book and chapter.

5. The "rediscovery" of the Greek text is attributed to Maximos Planudes circa 1300 (ibid., 43, 49).

text's organization becomes apparent when the *Geography* is read as a mapmaking guide.[6] The tables listing geographic coordinates were intended to serve as the raw data for the set of instructions on how to draw maps using the various cartographic projections discussed in the book's second section.

Ptolemy's brief introductory theoretical chapter, however, is the work's greatest contribution to the cosmographical art. In it he discussed the theory of using degrees of latitude and longitude as coordinate points to determine geographic location on a sphere. Celestially referenced latitude and longitude established a definitive correspondence between earthly and universal coordinates. Projected onto the earth's surface, the celestial equator became the equinoctial line, the celestial poles our Arctic and Antarctic poles, and the same for the tropics. The Ptolemaic system located the earth unambiguously within the universe, and to complete the earth's inclusion in the cosmos, Ptolemy explained how to locate the accidents of the earth within this universal grid.

The *Geography* guided the reader through the steps necessary for drawing a map of the *oikoumenē* or "known part of the world."[7] Ptolemy first made a distinction between geography and chorography or, as modern translators have defined these terms, "world cartography" and "regional cartography." The chorographer's task was "registering practically everything down to the least thing therein," while the author of a "world cartography" or *geographia* sought "to show the known world as a single and continuous entity ... [taking account] only of the things that are associated with it in its broader, general outlines."[8] A *geographia* required text to supplement the maps. This could be in the form of brief descriptions of significant geographical features and major cities, or it could take the form of a longer exposition. The accompanying text also included accounts of "notable peoples" that inhabited the described regions and their history.[9] The descriptive text was meant to work hand in hand with the map. The

6. A reinterpretation of the *Geography*'s theoretical chapters has led scholars to characterize the book as a how-to cartographic manual and to suggest that Renaissance humanists read it as such (ibid., 3–4).

7. Ibid., 20–22.

8. *Geography*, 1.1. Berggren and Jones also note that Strabo used the word "geography" to mean a textual work or a descriptive geography, rather than a map (ibid., 57, note 1).

9. Ptolemy worked within a Greek tradition of geographical works that considered man an integral part of the natural world. Elizabeth Rawson, *Intellectual Life in the Late Roman Republic* (Baltimore: Johns Hopkins University Press, 1985), 250.

function of the map went beyond graphic representation of the *oikoumenē*; it also served as a visual aid to the accompanying text—a map for a verbal discourse.

Ptolemy criticized earlier Greek geographical works for their way of spatially locating places by describing their relative direction and distance from an arbitrary reference point. Ptolemy found this method, akin to dead reckoning, unsatisfactory. To a reader in the Renaissance familiar with the shortcomings of locating a ship's position at sea using dead reckoning, Ptolemy offered a promising solution. The best way to improve location data, he said, was through mathematics and developing a good understanding of geographical coordinates.[10] If the geographer's goal was to maintain the proportions of a map of the *oikoumenē* as close as possible to those of the real world, the only suitable frame of reference for Ptolemy was the immutable heaven. By using astronomical observations to fix geographical locations on earth, Ptolemy placed geography on a mathematical and—as he saw it—more accurate and objective foundation. Using mathematical rationalism as the point of departure, Ptolemy then laid out his research program for the would-be cartographer. The first step, he advised, was "systematic research, assembling the maximum of knowledge from the reports of people with scientific training who have toured the individual countries; and . . . the inquiry and reporting is partly a matter of surveying, and partly of astronomical observation."[11]

Coming from the author of the *Almagest*, Ptolemy's underscoring the advantages of geographical coordinates based on astronomical observations over those determined by surveying is not surprising. The resulting measurements could then be drawn upon a grid delimited by celestial points and circles projected against the terrestrial sphere, with the intervening sections divided into equally spaced latitudes and longitudes. Ptolemy therefore devoted large sections of the *Geography* to resolving inconsistencies in longitudinal distances from voyages taken on land and sea given in his varied sources, a set of problems that would plague geographers until the early nineteenth century. For whereas the latitudinal grid is located on earth relative to "fixed" points in the sky—the celestial poles—and its intervals are determined by the position of the sun at noon, the longitudinal grid is solely a function of the earth's diurnal motion and therefore of time.[12] Without an ability to calculate the passage of time accurately, earth-bound methods of calculating longitude are futile. Ptolemy, although never addressing this subject directly, clearly understood this problem and

10. *Geography*, 1.1. 11. Ibid., 1.2. 12. Ibid., 1.2–3.

thus strongly advocated the one celestial observation known to him that could yield terrestrial longitudinal distances—simultaneous viewing of lunar eclipses. He complained, however, that the practice was not widespread enough to provide him with significant data.

Lunar eclipses offer observers the chance to determine the difference in local time between two points on earth where the eclipse is visible. Since the eclipse occurs at the same sidereal time for both locations, observers need only note the difference in local time from one location to another to calculate longitudinal distance. The difference in time corresponds to the difference in longitude, since one hour equals 15 degrees. Other than expressing frustration that over the years only a few eclipses had been carefully observed and recorded in different places of the *oikoumenē*, in the *Geography* Ptolemy did not address the many problems associated with the eclipse method, nor did he propose any other methods for calculating longitudinal distances. For Spanish cosmographers of the age of discovery, the Ptolemy of the *Geography* appeared to have supreme confidence in the eclipse method, and as we will see later in this work, they went to extraordinary lengths to follow his advice to record simultaneous eclipse observations over their extensive empire.

Ptolemy advised would-be cartographers to reconcile—to fit—into their maps geographical features described from credible sources, even if the coordinates they provided were not astronomically determined. Thus the map itself served as a tool to determine the mathematical coherency of a narrative account. A well-constructed map was the best way to organize vast amounts of geographical information—advice Renaissance cosmographers took to heart.[13] It is clear throughout the *Geography* that the author was well aware that the inhabited parts of the world extended beyond known borders, and he cautioned that what was presumed known had likely been described incorrectly. Therefore, the geographer needed to maintain an equally critical attitude toward old and new accounts.[14] When astronomical data were not available, he warned, the cartographer must examine descriptive accounts for inconsistencies, inaccuracies, and even exaggerations.

He purposely limited his project to preparing a map of the "known world," because his epistemic approach did not allow him to depict areas of the globe for which he lacked verifiable information (figure 1.1). A Ptolemaic map was an honest map, one where the design faithfully represented the geographical coordinates derived from "truthful" and coherent descriptions.

13. Ibid., 1.17. 14. Ibid., 1.5.

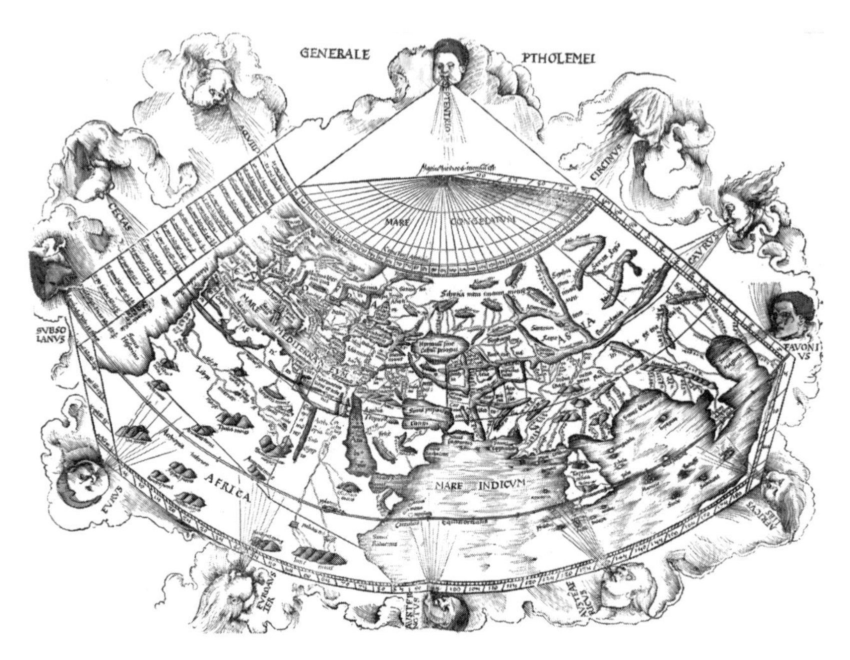

Figure 1.1. Ptolemaic world map in a simple conical projection. Claudius Ptolemy, *Cosmographia* (Strasburg: J. Scotus, 1520). Library of Congress.

A cosmographer during the Renaissance reading Ptolemy with the intention of imitating his work learned from the master that the geographic canon could not be taken for granted. As we study the work of sixteenth-century Spanish cosmographers, we will see that despite their respect for Ptolemy's text, they rarely hesitated to correct the master in matters of fact.

Yet for all of the *Geography's* importance in formulating Renaissance cosmography, it was through Sacrobosco's *Sphere* and Regiomontanus's *De triangulis* (1464) that the discipline found its most useful interpretive tools.[15] Regiomontanus's book taught generations of cosmographers the fundamentals of Euclidean geometry with a slew of practical examples and applications, while Sacrobosco's *Sphere* was, according to historian Lynn Thorndike, "the clearest, most elementary, and most used textbook in astronomy and cosmography from the thirteenth to the seventeenth century."[16]

15. Sacrobosco, or John of Holywood or Halifax (thirteenth century). Regiomontanus was the Latinate name of Johann Müller (1436–76), a German astronomer and mathematician who was chiefly responsible for the revival of plane geometry and trigonometry in Europe.

16. Lynn Thorndike, *The Sphere of Sacrobosco and Its Commentators* (Chicago: University of Chicago Press, 1949), 1.

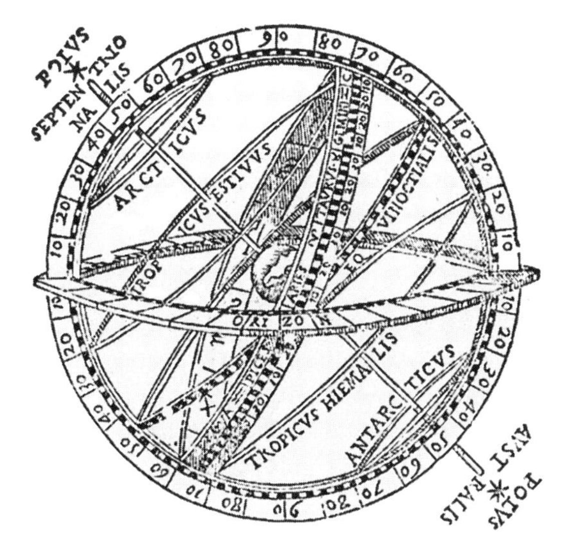

Figure 1.2. Armillary sphere typical of Sacrobosco's *Sphere*. Peter Apian and Gemma Frisius, *Cosmographia Petri Apiani* (Antwerp: Gregorio Bontio, 1545). Image courtesy History of Science Collections, University of Oklahoma Libraries; copyright the Board of Regents of the University of Oklahoma.

Sacrobosco's description of the relationship between the earth and the cosmos in the *Sphere* derives from the medieval Ptolemaic tradition of the *Almagest* and not from the *Geography* and therefore lacks the epistemic and methodological aspects discussed in the latter (figure 1.2). The *Sphere* was probably written in the early thirteenth century for use as a textbook at the University of Paris. It was immensely popular, likely due to its straightforward narrative and simple explanation of Ptolemaic astronomical fundamentals.[17]

Sacrobosco excerpted from a number of sources—Alfargani, Aristotle, and Macrobius—what he deemed most important and arranged the material in a clear and simple exposition suited for pedagogy. The short book explained just enough natural philosophy and astronomy for the reader to understand the nature and arrangement of the celestial spheres. Throughout his exposition Sacrobosco also enjoyed explaining the physical

17. Owen Gingerich, "Sacrobosco as a Textbook," *Journal for the History of Astronomy* 19 (1988): 273, and Olaf Pedersen, "In Quest of Sacrobosco," *Journal for the History of Astronomy* 16 (1985).

phenomena behind astronomical references in famous passages from Ovid, Virgil, and Lucan. Should the textbook not inspire a student to pursue astronomical studies further, at least he would be able to comprehend Virgil! By setting aside, however, the more esoteric aspects of peripatetic thought, the *Sphere* provided generations of readers with a synthesized and unquestioning description of the cosmos.

The *Sphere* taught generations of future cosmographers the principles of projecting celestial circles onto the earthly sphere and defined the parts of the earth considered inhabitable—with a certainty absent from Ptolemy. Sacrobosco explained that only the zones directly beneath the area delimited by the summer tropic and the Arctic Circle and the winter tropic and the Antarctic Circle are inhabitable. The other regions are either too cold because of their proximity to the poles or too hot because the sun beats down on them from directly overhead, as is the case with the equatorial Torrid Zone.[18] During the age of discovery, the *Sphere*'s concise text became the favorite introductory chapters in cosmographical works that discussed new geographic discoveries or the art of navigation. Its language became the language of cosmography, even when sixteenth-century cosmographers relished the opportunity to point out Sacrobosco's and Ptolemy's mistakes concerning life in the Torrid Zone.

While Ptolemy and Sacrobosco explained concepts and provided tools to mathematize the dimensions of space and time in order to describe the natural world, cosmographers during the Renaissance continued to rely on three other classical books: Pliny the Elder's *Naturalis historia*, Strabo's *Geographia*, and Pomponius Mela's *De situ orbis*. Throughout the Middle Ages, the monumental *Naturalis historia* remained the most authoritative reference book on the subject of natural history. Its encyclopedic arrangement, simple yet descriptive prose, and range of topics were widely imitated. During the sixteenth century, however, humanists' interest in Pliny's work shifted from an emphasis on the text as an example of classical Latin rhetoric to a critical assessment that challenged Pliny's facts.[19] Nonetheless, Pliny's work continued to be the model for natural histories well into the seventeenth century.

Strabo's *Geographia* (first century AD), like Ptolemy's later work, contained a theoretical treatise on the principles of geography, but the vast

18. Thorndike, *Sphere*, 17–21.

19. Charles G. Nauert Jr., "Humanists, Scientists, and Pliny: Changing Approaches to a Classical Author," *American Historical Review* 84, no. 1 (1979): 80.

work is best known for its lengthy description of the peoples and regions of the *oikoumenē*. Strabo was fascinated not so much by the geographical landscape or by how to mathematize this landscape to create a map as by the stories and character of its inhabitants. This Hellenic resident of Rome considered that a complete description of a place and it peoples had to include a recounting of their myths and fables. A fervent devotee of Homer, whom he frequently cited as a source of geographical information, Strabo saw the line between fact and fiction as one that a geographer should blur. He explained in the first chapter, "And we might pardon the poet [Homer] even if he has inserted things of a mythical nature in his historical and didactic narrative. That deserves no censure; for Eratosthenes is wrong in his contention that the aim of every poet is to entertain, not to instruct; indeed the wisest of the writers on poetry say, on the contrary, that poetry is a kind of elementary philosophy."[20]

As a model for how to write a descriptive geography, however, few authors—save perhaps Strabo—surpassed Pomponius Mela's popularity in Spain. Written also during the first century AD, *De situ orbis* is a relatively concise prose description of the known world. Long before the revival of Ptolemy's *Geography*, "geographical thinking" in Europe implied a verbal description rather than a cartographic representation of place.[21] Mela's narrative is organized as a traveler would keep a diary, sequentially describing different regions as a traveler would chance upon them during a journey. As historian and translator F. E. Romer observes, "Mela cast his work as a map."[22] The book seems to never have had a map in antiquity, although editions made during the Renaissance tended to include one. Like the *Geography*, Mela's *De situ orbis* was not widely read during the Middle Ages. It was not until it was "rediscovered" by Petrarch and Boccaccio in the fourteenth century that it gained in popularity. The first Spanish translation by Joan Faras appeared in the 1490s.[23] Just as the *Sphere* became the favorite pedagogical text for astronomy, Melas's *De situ orbis* became the

20. Strabo Geographia 1.1.10, in *The Geography of Strabo*, trans. Horace Leonard Jones and John Robert Sitlington Sterrett (London: Heinemann; New York: Putman's, 1917–33), 1:22–23.

21. R. W. Karrow Jr., "Intellectual Foundations of the Cartographic Revolution," PhD diss., Loyola University Chicago, 1999, 53.

22. Pomponius Mela, *Pomponius Mela's Description of the World*, trans. F. E. Romer (Ann Arbor: University of Michigan Press, 1998), 21. I will use page numbers when referring to Romer's introduction and book and chapter when referring to the *De chorographia* text.

23. Ibid., 28–29.

keystone of geographical studies in Spanish universities. Perhaps Mela's roots in the Iberian Peninsula (Mela proudly hailed from Hispania Baetica) and the fact that his description of Iberia surpasses that of Italy's in length made the text popular in Spain.[24] As was the case with Sacrobosco's *Sphere*, the longevity in the classroom of Mela's texts was largely due to the ease with which it lent itself to commentary.

Mela's *De chorographia* (the actual title of the manuscript) begins with the words "de situ orbis" or "a description of the known world."[25] He understood the complexities of organizing the vast amount of geographical material included in his book and repeatedly beckoned the reader to trust him as a guide for this imaginary voyage around the known world. To help the reader form a cohesive picture of the whole, Mela first introduced the major continents and their relationship to one another. He then moved on to describe significant features along each of their coastlines and adjoining territories farther inland. To avoid the potential confusion generated by such a verbal map, he explained, "To start with, in fact, let me untangle what the shape of the whole is, what its greatest parts are, what the condition of its parts taken one at a time is, and how they are inhabited; then, back to the borders and coasts of all lands as they exist to the interior and on the seacoast, to the extent that the sea enters them and washes up around them, and with those additions that, in the nature of the regions and their inhabitants, need to be recorded."[26]

The explanatory scheme Mela selects—from the general to the particular—is not much different from Ptolemy's instructions for preparing a map of the world. The general contours of the land are discussed first, as an exercise to "untangle the shape of the whole," an exercise akin to constructing Ptolemy's universal map. Once this is accomplished, the geographer returned to this landscape to describe it in detail, that is, to write a chorography. The order in which new lands are introduced is significant; coastlines are described first, followed by parts inland. For Mela, as for Ptolemy, the coastline served as an unambiguous boundary that delimited continents, separated or joined peoples, and posed a formidable barrier. The coast had to be described first; inland spaces were secondary to it. This use of coastal itineraries to organize the narrative and help locate and

24. Among Spanish commentators of Mela's *De situ orbis* were Pedro Juan Olivar (1536) with twenty-three editions, Fernando Núñez de Guzmán (1543), and Francisco Sánchez (1574). López Piñero, *Ciencia y técnica*, 214, and Víctor Navarro Brotóns, *Bibliographía physico-mathemática hispánica (1475–1900)* (Valencia: CSIC, 1999), 232–37.

25. Mela, *Pomponius Mela*, 8.

26. *De chorographia* 1.2.

guide the reader through the text remained characteristic of geographical works well into the modern era.

Mela did not limit his description to accidents of the land or significant man-made features but also included descriptions of the peoples who inhabited these lands and brief accounts of their history. He enlivened the somewhat repetitive narrative with tales of the more scandalous activities of the native peoples. Although Mela shows some skepticism regarding the more fantastic legends, he nonetheless repeated them and relishes the stories. When depicting the easternmost fringes of the known world, for example, just beyond the Ganges River, where the islands of Chryse and Argyre lay, he commented, "The first has golden soil—so the old writers have handed down—the other has silver soil. Moreover, as seems to be the case really, either the name comes from the fact, or the legend comes from the designation."[27] Be they truth or legend, cosmography's descriptive texts considered these accounts of human actions an essential component, intrinsic to the nature of the land and thus had to be told.

It was through the tradition of Mela and Strabo that cosmographical writing during the Renaissance preserved the classical tradition of including a historical narrative, as well as ethnographical accounts and natural history, as part of a geographical reckoning of a territory. From our modern perspective, the cosmographer's role as historian is perhaps the most difficult aspect of practice to integrate into the Renaissance cosmographer's intellectual landscape as outlined above. Our difficulty is due in large part to a modern bias that disassociates history from disciplines that we now consider scientific: natural history, ethnography, geography, cartography, and so on. The cosmographers' work as historians needs to be understood in light of their mission to write a complete description of the universe and as the way by which the actions of human inhabitants were recorded in the geographical space being described. Just as a map represented pictorially a geographical space and natural histories described the plant and animal kingdoms populating it, so chronicles served to record human actions in the natural world.

Often, this aspect of the work done by Spanish cosmographers was more in line with the writing of annals, understood as a systematic recording of past events.[28] However, the task of writing chronicles should not be

27. Ibid. 3.70.

28. In the classical era, "writing history" generally meant writing about current events, usually from an eyewitness perspective, while "writing annals" meant recording past events. These definitions changed during the Renaissance, as "writing history" increasingly implied writing about past events using written records as sources. In early modern Spain, "writing *crónicas*" generally referred to producing a chronologically

confused with writing *history*, a discipline that implies interpretation and obeys a different rhetorical style. In classical literature, the historical genre suggests a text that lacked a temporal element but implied a questioning of past events (and therefore interpretation). In its humanist embodiment, "the early modern *historia* straddled the distinction between human and natural subject, embracing accounts of objects in the natural world as well as the record of human action."[29] The task of the historian included singling out events and actions that were worthy of fame and remembrance, and beyond that, as Las Casas wrote in the prologue to *Historia de Indias*, a historian was one who sought the causes and reasons for past events.[30]

By the mid-sixteenth century, a new literary genre developed in Spain that fully incorporated the cosmographical and the historical traditions. Known as *historia natural e historia moral*, or natural and moral history, this genre drew from the Aristotelian hierarchical view of nature, where the natural world sustains human beings' moral actions, that is, actions stemming from God-given free will.[31] Within a history that claimed to be both *natural y moral*, the text was usually divided into two distinct parts. One part, usually at the beginning of the book, described those aspects that were deemed to be perpetual or unchanging, such as geography, the nature of the plants and animals inhabiting the space and other features of the land. The *historia natural* was written in an intensively descriptive, mostly expository style. This first part thus served as an introduction to the principal object of any historical work: to chronicle humankind's actions upon this landscape. Man's moral soul and free will placed him in a category apart from other creatures inhabiting the natural world, at the apex of a natural hierarchy

organized exposition of past events. Walter D. Mignolo, *The Darker Side of the Renaissance: Literacy, Territoriality, and Colonization* (Ann Arbor: University of Michigan Press, 1995), 140.

29. Gianna Pomata and Nancy G. Siriasi, eds., *Historia: Empiricism and Erudition in Early Modern Europe* (Cambridge, Mass.: MIT Press, 2005), 1–2.

30. Roberto González Echevarría, "The Second Discovery of America," *Yale Review* 86, no. 1 (1998): 145–46. Las Casas as cited in Walter D. Mignolo, "Cartas, crónicas y relaciones del descubrimiento y la conquista," in *Historia de la literatura hispanoamericana*, ed. Luis Iñigo Madrigal (Madrid: Cátedra, 1982), 77.

31. See O'Gorman's prologue to Acosta's *Historia natural y moral de las Indias* for an explanation of the genre: José de Acosta, *Historia natural y moral de las Indias, en que se tratan de las cosas notables del cielo, y elementos, metales, plantas y animales dellas: y los ritos, y ceremonias, leyes y gobierno, y guerras de los Indios*, ed. Edmundo O'Gorman, 2nd ed. (Mexico: Fondo de Cultura Económica, 1962), cxl–cxlvi, and Pilar Ponce Leiva, ed., *Relaciones histórico-geográficas de la Audiencia de Quito, s. XVI–XIX*, 2 vols. (Madrid: CSIC, 1991), 1:31.

and yet distinct from it. This was the model used by Gonzalo Fernández de Oviedo in his *Historia general y natural de las Indias Occidentales* (1535).

Fernández de Oviedo's book was also influential in establishing the empirical epistemological criteria used in subsequent New World natural histories. Oviedo's narrative relies on constant references to having personally "seen" and "known" what he describes and on explanations of trials he conducted as means of ascertaining matters of fact. He intended the text for readers hungry to learn about the economic potential of the New World, the temperament of its inhabitants, and the suitability of its environment for Europeans.[32] As such, it is one of the earliest examples of the utilitarian culture that permeated the Spanish imperial project of the New World and that manifested itself in natural histories and cosmographies—two disciplines with methodological tools that could be used effectively to explicate the New World.

The revival of Ptolemy's *Geography* during the Renaissance has been interpreted as a turning point in cartographic history. Maps changed.[33] The medieval *mappamundi*, deeply embedded in the religious consciousness of the era, had been meant to provide a temporal and spatial reckoning of significant events rather than to depict geography. When the geography of a place was described, it took the form of an itinerary, a mode of discourse dating from late antiquity, and was only rarely accompanied by a schematic map. The medieval conception of space was bounded by faith and what was known well enough to depict in a pictogram. This conception of space changed into a geometric space conceived along the mathematical lines of Ptolemaic cartography.

Some historians, however, have taken this characterization further. They argue that the Ptolemaic revival precipitated a transformation of

32. On Oviedo's epistemology, use of witnessing, and firsthand accounts, see Antonello Gerbi, *Nature in the New World: From Christopher Columbus to Gonzalo Fernández de Oviedo*, trans. Jeremy Moyle (Pittsburgh: University of Pittsburgh Press, 1985), 226–31. For a detailed study of Oviedo's utilitarian interest and its associated experimental methodology, see José Pardo Tomás and María Luz López Terrada, *Las primeras noticias sobre plantas americanas en las relaciones de viajes y crónicas de Indias, 1493–1553* (Valencia: Instituto de Estudios Documentales e Históricos sobre la Ciencia, Universitat de València, CSIC, 1993), 86–97.

33. David Buisseret, ed., *Monarchs, Ministers, and Maps: The Emergence of Cartography as a Tool of Government in Early Modern Europe* (Chicago: University of Chicago Press, 1992), 1–4, and J. B. Harley and D. Woodward, eds., *The History of Cartography*, 4 vols. (Chicago: University of Chicago Press, 1987), 1:504–6.

spatial conception that conditioned Europe for the rationality that would later characterize modernity and allowed for new and seemingly empty spaces to be imagined as real. This new spatial visualization of the world, they argue, kindled aspirations for imperialistic expansion.[34] The transformation in spatial thinking is sometimes portrayed as an all-or-nothing equation rather than a century-long process of accommodation between competing intellectual traditions that took place simultaneously during and *after* colonial expansion. I join others, such as Ricardo Padrón, who warn against this kind of sweeping generalization and the deterministic history it implies. Padrón has argued that the changes in spatial thinking brought about by the Ptolemaic grid were far from pervasive and that in the case of Spain, its expansion overseas took place with the aid of maps firmly rooted in the sense of space of the medieval itinerary in the form of the portolan chart.[35]

Any study that privileges "the grid" and spatial thinking as a precondition of modernity needs also to take into account the prevalence and popularity of descriptive geographies as another way of describing space. To suppose that Ptolemaic cartography trumped traditional descriptive geographies in favor of a purely geometrical conception of space, or that descriptive accounts were written to somehow fill a map's "blank spaces," unfortunately ignores the influence the models of Strabo and Mela had on the humanist authors. It also overlooks the contributions of various intellectual traditions that converged in Renaissance cosmography. A humanism-driven adherence to classical models, I believe, did more to influence the content of descriptive geographies than any new "geometric concept of space" in search of meaning. The Ptolemaic revival with its emphasis on graphical maps fits neatly within the tradition of descriptive geographies, only now the reader need not imagine the map as was required during the Middle Ages but could instead consult it at a glance.

34. For some examples of these approaches, see David Harvey, "Between Space and Time: Reflections on the Geographical Imagination," *Annals of the Association of American Geographers* 80, no. 3 (1990): 424; D. Woodward, "Maps and the Rationalization of Geographic Space," in *Circa 1492: Art in the Age of Exploration*, ed. Jay A. Levenson (New Haven, Conn.: Yale University Press, 1991); Samuel Y. Edgerton, "From Mental Matrix to Mappamundi to Christian Empire: The Heritage of Ptolemaic Cartography in the Renaissance," in *Art and Cartography*, ed. David Woodward (Chicago: University of Chicago Press, 1987); and J. B. Harley, *The New Nature of Maps*, ed. Paul Laxton (Baltimore: Johns Hopkins University Press, 2001).

35. Ricardo Padrón, "Mapping Plus Ultra: Cartography, Space, and Hispanic Modernity," *Representation* 79 (2002): 31.

Throughout Europe, Ptolemy's goal to describe the "whole of the known world" became a mission for cosmographical practitioners, and the resulting *magnum opus* became the crowning achievement of their profession. The cosmographies of Sebastian Münster (1488–1552)[36] and André Thevet (1504–92)[37] answered Ptolemy's call. They adopted the classical works of Mela and Strabo as literary models and, to various degrees, Ptolemy's methodology. Their ambitious cosmographical projects required an exhaustive exercise in compilation and erudition aimed at creating an image of the world at a particular moment in time. It was an image constructed of discrete descriptions of places that, once collected and arranged, portrayed the whole. Natural history, Plinian and Aristotelian, served as the departure point for descriptions of the plants and animals in a region.

In Renaissance cosmography, in order for a place to exist, it required a historical narrative, whether mythical or factual, that located the place within the human context. Part of describing also consisted in locating it within the web of symbols and correspondences that defined the European understanding of the world.[38] As an exercise in selection and compilation of knowledge, it borrowed the comprehensive style of the medieval encyclopedia but used as its internal organizing scheme the geography it described. Maps, rather than serving as definitive and literal depictions of the terrain, were used as visual aids and as graphical outlines of the text. The exercise of composing a cosmographical opus required both a historical and an allegorical reading of sources, skills that as humanists these cosmographers possessed.[39]

The task was monumental. As historian Frank Lestringant explains using André Thevet's work as an example, the cosmographical enterprise

36. Münster published a Latin edition of Ptolemy's *Geography* in 1540 and his monumental *Universal Cosmography* in German in 1544 and Latin in 1550. For some studies of Münster's work, see Lucien Louis Gallois, *Les géographes allemands de la Renaissance* (Paris: E. Leroux, 1890), and Jean Bergevin, *Déterminisme et géographie: Hérodote, Strabon, Albert le Grand et Sebastian Münster*, Travaux du Département de géographie de l'Université Laval 8 (Sainte-Foy, Québec: Presses de l'Université Laval, 1992).

37. Thevet was cosmographer and historian to France's Catherine de Medici and subsequent French monarchs. His *Cosmographie universelle* was published in Paris in 1571. For more on Thevet, see Lestringant's *Mapping the Renaissance World*.

38. John Elliott has noted the difficulty of integrating the New World into this worldview and characterized the enterprise as an effort to find similitude, often reduced to constructing a "European dream which had little to do with American reality." John H. Elliott, *The Old World and the New, 1492–1650* (Cambridge: Cambridge University Press, 1970), 27.

39. Anthony Grafton, *Defenders of the Text: The Traditions of Scholarship in an Age of Science, 1450–1800* (Cambridge, Mass.: Harvard University Press, 1991), 41.

was tainted with hubris.[40] For Thevet, the scope of the mission, rather than proving overwhelming, freed him to compose a cosmography that reveled in recounting and describing from the minute to the universal, from the true to the fantastic, from the godly to the blasphemous. While for Münster the scope of the mission served as a platform to illustrate the power of an omnipresent God,[41] Thevet articulated an epistemic criterion that privileged eyewitness sources and relentlessly questioned authorities, including the Bible. In contrast, Münster's encyclopedic approach has been labeled naive.[42] The Renaissance cosmographical opus was not a static or rigid genre. It proved malleable in the hands of practitioners not intimidated by the task of writing a description of the entire world.

Some cosmographers preferred to focus their works on technical and cartographic topics and thus wrote for an extra-academic audience in a European marketplace hungry for information about the new discoveries. The most popular authors of this version of the cosmographical genre were Peter Apian (1495–1552) and his immensely popular *Cosmographicus liber* (1524), and Gemma Frisius (1508–1555) with his many editions of Apian, as well his own influential *De principiis astronomiae et cosmographiae* (1530). Frisius's *De principiis* follows what had by midcentury become the standard format for this type of cosmographical texts: an introductory theoretical section based on Sacrobosco's *Sphere* precedes a practical section that explains the use of various instruments, followed by a final section containing a descriptive geography of the known world. The texts are in essence self-contained courses in the theoretical underpinnings and technical aspects of cosmography.

The didactic nature of the books appealed to an autodidactic audience interested in new geographical discoveries and the cartographic tools that made the new lands commensurable. Apian and later Frisius designed their cosmographies with what we would characterize today as "interactive features" intended for the reader to mimic cosmographical practice. The books are printed with various *volvelles*—computational aids constructed by cutting out and assembling printed pages of the book (see figure 1.3). Along with instructions on how to construct similar wooden or brass instruments, these devices made the cosmographical arts approachable and bridged the gap between theoretical sciences and technological

40. Lestringant, *Mapping the Renaissance World*, 6.

41. Bergevin, *Déterminisme et géographie*, 156–62.

42. Lestringant, *Mapping the Renaissance World*, 11, and Bergevin, *Déterminisme et géographie*, 127.

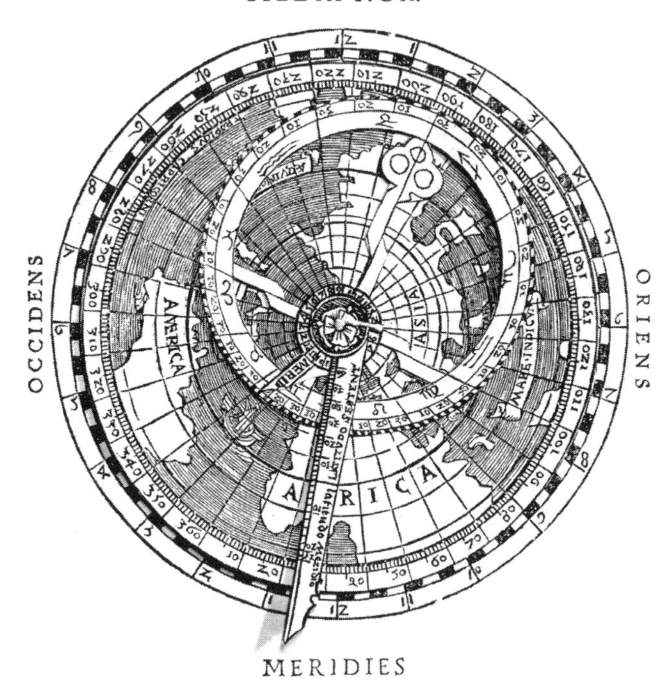

Figure 1.3. *Volvelle* for determining the position of the zodiac. Peter Apian and Gemma Frisius, *Cosmographia Petri Apiani* (Antwerp: Gregorio Bontio, 1545). Image courtesy History of Science Collections, University of Oklahoma Libraries; copyright the Board of Regents of the University of Oklahoma.

tradition.[43] These texts, however, are far from "how-to manuals." The descriptions of cartographic projections and instruments in these books allows readers to understand the theoretical fundamentals behind a map or an apparatus but are seldom explicit enough to permit a reader to replicate them. Despite this, the Apian/Frisius books were immensely popular, as their extended publishing history attests.[44] Their role as important means

43. Steven Vanden Broecke, "The Use of Visual Media in Renaissance Cosmography: The *Cosmography* of Peter Apian and Gemma Frisius," *Paedagogica Historica* 36, no. 1 (2000): 133.

44. There were more than forty editions of the *De principiis astronomiae et cosmographiae* alone. Fernand Gratien van Ortroy, *Bio-bibliographie de Gemma Frisius, fondateur de l'école belge de géographie, de son fils Corneille et de ses neveux les Arsenius* (Brussels: M. Lamertin, 1920). There were at least two Spanish editions of the Apian/Frisius cosmography in

for propagating information about new instruments, geographical information, and cartographic methods is uncontested. For example, Frisius's explanation for the use of mechanical clocks as a means of calculating longitude is recognized as the predecessor of modern means of determining longitude at sea, and his method of surveying using triangulation is used to this day.[45] For all its popularity, however, the Apian/Frisius corpus devotes little space to the geography of the New World; the description of the New World grew to only twenty-four pages in later editions by including translations of the works of Francisco López de Gómara and Gerónimo Girava.[46]

Cosmographers working after the discovery of the New World, especially after the 1522 return of the Magellan/Elcano circumnavigation, took the Ptolemaic toolkit and the advice to approach old and new geographical information with skepticism as a mandate to create the new image of the world. They realized that for the first time in history the sphere of earth and water could be portrayed and described in its entirety. The surface of the globe was in the process of becoming "known," and Ptolemy had showed them how to use this information to "know" it. Would Renaissance cosmography, its interpretive tools, classically styled narrative, humanist erudition, and mathematical methods prove capable of assimilating the new geography into the cosmographical canon?

Ideal Practice: Cosmography at the University

During the one hundred years following the discovery of the New World, scholars at the University of Salamanca were among the many conduits of information about the New World. Salamanca's university was the oldest and most prestigious in Spain; its scholars featured prominently in the Columbian project and later in resolving issues that arose concerning the Treaty of Tordesillas and the line of demarcation. The plan of studies at the university shows that news from the New World was integrated into the curriculum in what essentially constituted an ongoing dialogue with

1548 and 1575 printed in Antwerp. Navarro Brotóns, *Bibliographía physico-mathemática hispánica*, 80–81.

45. A. Pogo, "Gemma Frisius, His Method of Determining Differences of Longitude by Transporting Timepieces (1530), and His Treatise on Triangulation (1533)," *Isis* 22 (1935): 471.

46. Gemma Frisius et al., *Cosmographia, siue Descriptio universi orbis* (Antwerp: Ioan. Bellerum, 1584).

the classics, remaining faithful to the medieval tradition of commentary and turning to these for thematic and stylistic guidance. But the university curriculum—specially toward the midcentury—also reveals a changing conception of the practices associated with cosmography, exemplified by an emphasis in teaching the mathematical aspects of cosmographical practices, principally cartography. This is illustrated in the work of its principal professors and in a series of curriculum reforms that serve to articulate the parameters of what entailed ideal cosmographical practice.

The careers of two of Spain's leading humanist scholars give some insight into how this new discipline responded to the constantly changing geographical landscape during the first decades after the discovery of the New World. Pedro Sánchez Ciruelo (ca. 1470–1548) and Elio Antonio de Nebrija (ca. 1444–1522), career academics at two of Spain's leading universities, are considered paradigms of an intellectual movement labeled scientific humanism (*humanismo científico*) centered at the University of Salamanca.[47] The movement was anchored by a group of university professors interested in the revival along humanistic lines of the classics of natural philosophy and astronomy. Their approach emphasized the linguistic as well as the scientific criticism of classical works. They sought not only to recover ancient texts but also to amend and incorporate into this classical corpus the results of new experiences in the New World.[48]

While studying theology at the University of Paris, Pedro Sánchez Ciruelo published a number of mathematical treatises in the vein of Paris nominalists and Oxford calculators.[49] Among his later works is a commentary on Sacrobosco's *Sphere*, the *Opusculum de Sphera mundi* (Paris, 1508, and Alcala, 1526), one of the first published commentaries on the *Sphere* written by a Spanish professor.[50] Meant to serve as a means for learning

47. Cirilo Flórez Miguel, Pablo García Castillo, and Roberto Albares Albares, *El humanismo científico* (Salamanca: Caja de Ahorros y Monte de Piedad, 1988), 111–12.

48. Roberto Albares Albares, "El humanismo científico de Pedro Ciruelo," in *La Universidad Complutense Cisneriana* (Madrid: Editiorial Complutense, 1996), 178–80.

49. Other nominalists working on mathematics and the physical sciences at the University of Salamanca included Juán Martínez Silíceo (1486–1557), Fernán Pérez de Oliva (1492–1531), and Pedro Margalho (1471–1556). For more on the topic of nominalism at Salamanca, see Fernán Pérez de Oliva, *Cosmografía nueva*, bilingual ed. prepared by Cirilo Flórez Miguel, Acta Salamanticensia (Salamanca: Ediciones Universidad de Salamanca, 1985), 12–18.

50. In addition to Ciruelo's commentary, there were at least eleven works exclusively dedicated to Sacrobosco in Spain, eight circulating as printed editions. Antonio

basic astronomy in order to practice astrology, the text integrates various intellectual currents to explain the physical phenomena described by Sacrobosco. In an interesting exposition at the end of the book framed as a dialogue between Ciruelo and a friend, the author explains that it is the philosopher's duty to search for the truth and, through a critical dialogue with the ancients, correct their work.[51] Ciruelo conveys, as he surely conveyed to the many generations of students who attended his lectures, the view that mathematics was one of the few tools of natural philosophy based on unequivocal principles and thus offered resources to correct the ancients.

In other instances, Sacrobosco's *Sphere* became a launching platform for new astronomical theories. Whereas Ciruelo's principal preoccupation in the *Opusculum de Sphera mundi* was mounting a mathematical defense of astrology, the geographic discoveries of the late fifteenth century propelled other scholars with humanistic sensitivities and a reverence for mathematics to attempt to integrate these discoveries into the accepted geographic canon. The University of Salamanca was the locus of this activity in the years between 1498 and 1530. Works written soon after the discovery of the New World show a marked preoccupation with the topics of measurement and the size of the earth.[52]

Elio Antonio de Nebrija was the first Salamantine scholar to address in print geographical concerns about the discovery of the New World.[53] Nebrija held the professorships of grammar and rhetoric at Salamanca

---

Hurtado Torres, "La 'Esphera' de Sacrobosco en la España de los siglos XVI y XVII: Difusión bibliográfica," *Cuadernos Bibliográficos* 44 (1982): 50–51.

51. Albares Albares, "Humanismo científico de Pedro Ciruelo," 200.

52. Francisco Núñez de la Yerba's book titled *Cosmografia pomponii cum figuris* (Salamanca, 1498) commented on Ptolemy's and Pliny's remarks about the size of the earth in an edition of Mela's *De situ orbis*, yet he still divided the earth into three continents. The first time the newly discovered lands were described in a cosmographical book as a new continent was in Pedro Margalho's (ca. 1473–1537) *Physices compendium* (Salamanca, 1520). Cirilo Flórez Miguel, "Cosmógrafos salamantinos de Renacimiento y cambio de paradigma," in *Ciencia, vida y espacio en Iberoamérica*, ed. José Luis Peset (Madrid: CSIC, 1989), 382. For more on the topic of measurement, see Ana María Carabias Torres, "La medida del espacio en el Renacimiento: La aportación de la Universidad de Salamanca," *Cuadernos de Historia de España* 76 (2000): 198–99.

53. Flórez Miguel, García Castillo, and Albares Albares, *Humanismo científico*, 113–14. For a bio-bibliography, see the introduction to Elio Antonio de Nebrija, *Elio Antonio de Nebrija, cosmógrafo: In cosmographiae libros introductorium*, ed. Hermandad de los Santos de Lebrija, trans. Virginia Bonmatí Sánchez (Cádiz: Agrija Ediciones, 2000).

and later at the University of Alcala. Earlier he had studied Ptolemy's *Geography* while studying in Bologna in the 1490s. His *In cosmographiae libros introductorium* (Salamanca, 1498) relies on the models set by Ptolemy's *Geography* and Sacrobosco's *Sphere* to discuss the new discoveries and how they undermined the geographical works of the ancients. His approach was characteristic of the humanist that he was. Throughout the text, Nebrija takes pains to clarify the meaning of cosmographical terms in different languages, extracting from these definitions all possible meanings and even including a glossary of cosmographical terms at the end of the text. For Nebrija, applying the rigors of humanistic grammar to geographical descriptions demanded an unambiguous correspondence between a geographical name and the geographical location it named. It likewise required that the language of cosmography—words such as *meridian, equinoctial* or *equator*, and *pole*—be defined precisely to correspond to the mathematical model set by Ptolemy, in the process achieving the ultimate in linguistic precision: a correspondence between word and number.[54] Nebrija's preoccupation with linguistic precision has been viewed as a vital component of a science in the process of reconceptualization.[55]

Nebrija does more than simply translate Ptolemy following the rigors of linguistics. Throughout the text, he frequently challenged Ptolemy's conclusions and those of other ancient authorities.[56] In doing so, he became one of the first scholars to assume the attitude historian David N. Livingston has labeled "intellectual anti-authoritarianism," typical of the countries that participated in voyages of discovery.[57] For example, Nebrija

54. For more on Nebrija's preoccupation with mathematical precision and how this translated into efforts to quantify the terrestrial degree and other measures, see Mariano Esteban Piñeiro, "Elio Antonio de Nebrija y la búsqueda de patrones universales de medida," in *El Tratado de Tordesillas y su época*, ed. Luis Antonio Ribot García (Madrid: Junta de Castilla y León, 1995), 572–73.

55. Flórez Miguel, "Cosmógrafos salamantinos," 386.

56. Nebrija continued to wrestle with conflicting notions about the earth's composition. He alternated between the scholastic tradition depicting the terrestrial sphere as consisting of two eccentric spheres one of earth and another of water (Aristotle/Scotus) and the Ptolemaic concept of a single sphere in which the sphere of earth and water shared a center and land masses were considered protrusions. Margalho in his *Physices compendium* (1520) was convinced by accounts of the new geographic discoveries to side with the Ptolemaic notion. W. G. L. Randles, "Science et cartographie: L'image de monde physique à fin du XVe siècle," in *El Tratado de Tordesillas y su época*, ed. Luis Antonio Ribot García (Madrid: Junta de Castilla y León, 1995), 940–41.

57. David N. Livingstone, *The Geographical Tradition: Episodes in the History of a Contested Enterprise* (Oxford: Blackwell, 1992), 56, 63.

explained that the Portuguese had sailed south of Africa and had shown that, contrary to what Ptolemy maintained, there was no "terra incognita" enclosing the Indian Ocean.[58] He also dismissed all that the ancients had said about the Antipodes, adding: "Nothing certain about the existence [of the Antipodes] was transmitted to us by our elders, but today thanks to the courage of the men of our time it will soon happen that they will give us the true description of that land, of the islands as well as the continents; our sailors have already told us about a large part of the coasts."[59]

Absent from Nebrija's translation and commentary, however, are the practical details that made Ptolemy's *Geography* a how-to manual. For example, Nebrija does not discuss how to measure latitude or how to compile geographic information. Nor is his translation of Ptolemy's cylindrical projection comprehensive enough to be useful in drawing a map. Also missing are the tables of latitude and longitude and the geographical description of the continents. Thus we are left with a book that, although it serves as an adequate introduction to the concepts, language, and some of the theory behind Ptolemaic geography, lacks practical instruction. Nebrija—the quintessential Spanish humanist—understood that the mathematic precision promised by Ptolemaic geography required an equally precise language. In the *Introductorium*, he was rehearsing and polishing the cosmographical vocabulary used to mathematize the newly discovered spaces, but he avoided exploring the actual mathematics required for the task.

After Nebrija and Ciruelo, a series of adjustments to the mathematics and astronomy curriculum made throughout the sixteenth century at Salamanca suggest willingness on the part of the professorate to incorporate new methods and technologies into the pedagogical program. They also indicate a gradual conceptual shift in the way cosmography was taught, away from exegesis of the classics and linguistics and toward mathematical rationalism. Before continuing on to the higher faculty of theology, law, or medicine, a student at Salamanca had to complete the course of study dictated by the arts faculty, as was traditional in most European universities of the time. While studying "arts," students learned about the natural world from a variety of sources. During their first two years, they studied Aristotle's *De coelo et mundo* and *Meteorologica*. As part of the study of rhetoric,

58. Nebrija, *Elio Antonio de Nebrija, cosmógrafo*, 97.

59. "Nada cierto sobre su existencia nos fue trasmitido por nuestros mayores, pero hoy en día gracias a la audacia del hombre de nuestro tiempo pronto ocurrirá que nos aporten la verdadera descripción de aquella tierra, tanto de las islas como del continente; de gran parte de la costa nos han informado nuestros marinos" (ibid., 99).

they read commentaries on Pomponio Mela's *De situ orbis* and Pliny's *Historia naturalis*. In the third year they read Sacrobosco's *Sphere* and some of Ptolemy's *Geography*, as well as some rudimentary astrology.[60] Students interested in cosmography would then continue their studies with the professor of astrology and mathematics.

Since the mid-fifteenth century, the University of Salamanca had had a chair of mathematics and astrology.[61] The professor occupying the chair taught the fundamentals of astronomy and astrology to third-year students in the arts curriculum. The university's statutes of 1529, and later those of 1538, describe the subject matter taught by the professor of mathematics and astrology in general terms. He was to teach arithmetic, geometry, astrology, perspective, and cosmography.[62] In 1561, after a period of curriculum reform, the statutes introduced a new three-year program for a bachelor's degree in mathematics and astrology.[63] In 1588, a lesser professorship devoted exclusively to mathematics was created in response to royal directives that sought to increase the number of persons in the Spanish realm knowledgeable in cosmographical science. Periodic changes to the university's mathematics, cosmography, and astronomy curriculum responded to an increasing preoccupation on the part of Philip II and his advisers that the kingdom lacked sufficient numbers of men capable of addressing the growing questions surrounding the geography of the New World and its navigation. A royal directive in 1593 explained the reason for underwriting the post: "so that enough able persons can be trained in this university to teach there and to have them also at the sea ports as well as in other places, because it is so necessary and navigation depends on it."[64] Educational efforts were not limited to the universities; in the following

60. Flórez Miguel, García Castillo, and Albares Albares, *Humanismo científico*, 47.

61. San Bartolomé, one of the principal colleges at the University of Salamanca, was home to many of the mathematics professors and their students. Ana María Carabias Torres, "Los conocimientos de cosmografía en Castilla en la época del Tratado de Tordesillas," in *El Tratado de Tordesillas y su época*, ed. Luis Antonio Ribot García (Madrid: Junta de Castilla y León, 1995), 965.

62. Francisco Javier Alejo Montes, *La Universidad de Salamanca bajo Felipe II: 1575–1598* (Burgos: Editiorial Aldecoa, 1998), 197–99.

63. For a copy of the statutes, see E. Bustos Tovar, "La introducción de las teorías de Copérnico en la Universidad de Salamanca," *Real Academia de Ciencias Exactas, Físicas y Naturales* 67–68 (1973): 243–44, 49–50n37.

64. "Se críen personas suficientes y hábiles para leer en esta Universidad y para tenerlos así mismo en los puertos de mar como en otra cualquier parte, por ser tan necesario y porque de ello depende la navegación." Provición Real, March 26, 1593, AUS 2870 Documentos Reales (1568–1600), in Alejo Montes, *Universidad de Salamanca*, 197–98.

chapters we will examine the curriculum of two new institutions formed to address these issues, the Academia Real Matemática in Madrid and the Cátedra de Cosmografía at the Casa de la Contratación in Seville.

When the professorship of mathematics and astrology at the University of Salamanca became vacant in 1576 and no suitable candidate solicited the position, university regents considered two candidates with markedly different backgrounds and expertise. One candidate, Jerónimo Muñoz (c. 1515–92), was an astronomer at the University of Valencia as well as professor of Hebrew and theology.[65] The other was Rodrigo Zamorano, a cosmographer working at the Casa de la Contratación in Seville. At the time he was teaching the principles of navigation to pilots destined for the Indies fleet and had recently written a well-received translation of Euclid's *Elements* (Seville, 1576). The two men were at opposite ends of the spectrum of cosmographical practice. Zamorano's work reflected the imperatives of the port city, the gateway to the New World. They were centered on astronomical navigation, preparing sea charts and teaching cosmography to pilots and sailors. Muñoz's interests were principally astronomical, not for the sake of mastering the position of celestial bodies in the hope of finding location at sea but directed at understanding the movements of heavenly bodies and correlating these with his own astronomical observations. Not surprisingly, the regents chose one of their own, but not before Muñoz negotiated a significant salary increase.

It is evident from a number of Muñoz's works that he understood natural philosophical speculation as concomitant to this role as astronomer. For example, extant manuscripts reveal he engaged in the natural philosophical debates surrounding the nature of the heavens, the mechanics of vision, and classical atomism. We can presume he discussed these topics with his students, since some of Muñoz's works survive as copies made by students.[66] His reputation as a keen astronomical observer and theoretician was noted by Tycho Brahe, who referred to Muñoz's observations on the comet of 1572 in his own work on the subject. As historian Navarro Brotóns points out, Muñoz's cosmology deviated markedly from Aristotelian views in some key aspects. He believed in the corruptibility of the heavens and argued against the existence of celestial spheres, two positions he felt could be sufficiently supported solely on the weight of mathematical and observational evidence.

65. Biographical notes on Muñoz from Víctor Navarro Brotóns and Enrique Rodríguez Galdeano, *Matemáticas, cosmología y humanismo en la España del siglo XVI: Los comentarios al segundo libro de la historia natural de Plinio de Jerónimo Muñoz* (Valencia: Instituto de Estudios Documentales e Históricos sobre la Ciencia, 1998), 19–29.

66. For an inventory of Muñoz's work, see ibid., 31–34.

He discussed Copernicus's heliocentric theory in his classes, both in Valencia and in Salamanca,[67] but refuted it using Ptolemaic arguments and Theon's commentary on the *Almagest*. Muñoz rejected the heliocentric thesis but still preferred to use Reinhold's Copernican Prutenic tables because they correlated well with his astronomical observations. These unorthodox positions did not seem to hinder his chances for getting the post at Salamanca, although they might have gained him some enmity in other circles.[68]

In the university milieu Muñoz was able to cultivate the applied aspects of cosmography, as well as indulge his interest in the theoretical aspects of astronomy. Like his central European counterparts, Muñoz sought to increase his social standing and gain intellectual independence by remaining in the institutional framework afforded by the university.[69] As we will see in the chapters that follow, by the late sixteenth century the utilitarian focus associated with cosmography's institutionalization at the royal court, the Casa de la Contratación, and the Council of Indies would have limited Muñoz's practice to solely those aspects that yielded utilitarian products had he worked in these settings.

Some of Muñoz's classroom lectures on cosmography dating from his years at the University of Valencia survive and help re-create the intellectual context of a late sixteenth-century cosmographer. His *Astrologicarum et Geographicarum institutionum libri sex* consist of his lectures organized around the Ptolemaic astronomical and geographical corpus.[70] In the

67. Salamanca was the only European university during the sixteenth century that listed Copernicus's *De revolutionibus* in its curriculum. Historians concur that if the book was discussed at all, the text was approached from an instrumentalist perspective. For more on the teaching of Copernicus at Salamanca, see Bustos Tovar, "Introducción de las teorías de Copérnico en la Universidad de Salamanca," and Manuel Fernández Álvarez, *Copérnico y su huella en la Salamanca del Barroco* (Salamanca: Universidad de Salamanca, 1974). By the late sixteenth century, the Copernican theory was openly discussed in Spanish cosmographical circles. Mariano Esteban Piñeiro, "La primera versión castellana de *De Revolutionibus Orbium Caelestium*: Juan Cedillo Diaz (1620–1625)," *Asclepio* 43, no. 1 (1991).

68. Editor's introduction to Jerónimo Muñoz, *Jerónimo Muñoz. Introducción a la astronomía y la geografía*, ed. Víctor Navarro Brotóns, Colleció Oberta (Valencia: Consell Valencià de Cultura, 2004), 25–29.

69. Muñoz fits neatly into the characterization Robert Westman made of the role of the astronomer in the later half of the sixteenth century: Westman, "Astronomer's Role in the Sixteenth Century," 127–33.

70. Muñoz's classroom material dated circa 1570 survives as manuscript copies made by two students. A set is in the Vatican Apostolic Library, Ms. VL 6998 titled *Astrologicarum et Geographicarum institutionum libri sex*, and another in the Bayerische

preface to the course of lectures, Muñoz explained that in the search for the causes of things and the *celorum archana* (secrets of the heavens), he favored the laconic style proper of the mathematician rather than to try to entice listeners with flowery discourses and verbal flourishes.[71] Indeed, the lectures are a straightforward course in cosmography, favoring the practical applications of the discipline and the use of mathematical proofs to settle theoretical debates but also at times touching upon the more pressing cosmological questions of the time.

The lectures were punctuated with references to classical authors, but the professor incorporated the work of modern astronomers into his class material and frequently corrected ancient authorities when modern observations or mathematical proofs challenged the classical canon. He also urged his students to use their personal observations and mathematical proofs to correct the classical canon. On aspects of practice, he adopted the approaches used in Apian's *Cosmography* and on Gemma Frisius's *De principiis astronomiae*. For his discussion of gnomics and methods for determining latitude, he sided with Oronce Finé. In fact, Muñoz referred to Frisius and Finé as his former preceptors.

Frisius's method of mapping using triangulation was among the techniques Muñoz taught his students. The Muñoz lecture notes show an example of this method employed in the vicinity of Valencia. Based on this and other similarities, Salavert and Navarro have shown conclusively that Abraham Ortelius's map of Valencia in his 1584 *Theatrum Orbis Terrarum* was based on Muñoz's surveys of the region.[72] The sixth book of his *Astrologicarum et Geographicarum* gives detailed directions for constructing terrestrial and celestial globes, as well as maps using at least seven different projections plus a trapezoidal projection he used in his map of the Iberian Peninsula. Muñoz repeatedly warned his students about relying on latitude and longitude values—even those given by mathematicians—that had not been verified or whose mathematical basis was uncertain. He cautioned them against inventing a new geography: "You should try most of all not to place anything unexplored on this type of description, for it is preferable that we indicate with imprecise lines this unknown region and leave it in obscurity, than to spread errors among students with false descriptions.

Staatsbibliothek, Clm 10674. For an expert edition and translation into Spanish of the version at the Vatican, see Navarro Brotón's edition, cited in note 68.

71. Muñoz, *Introducción*, 79, 244.

72. V. Navarro and Vicente L. Salavert, "Muñoz y la geografía descriptiva: La descripción de España," in Muñoz, *Introducción*, 40–56.

We lament that even learned men, tempted by money and fame, commit these errors."[73]

Elsewhere in his classroom notes, Muñoz gave an example of the type of geographical description that should accompany these maps. It is, as Muñoz's historians have characterized it, the eclectic work of an erudite humanist. His incomplete geographical description of Spain[74] utilizes philological and historical tools to decipher the geographical descriptions of Pliny, Mela, Ptolemy, Strabo, Antoninus Pius, and others, while his personal knowledge of eastern Spain is occasionally evident in the text.

Although Muñoz died in 1592, the new academic guidelines promulgated at the University of Salamanca in 1594 continued to reflect his influence. These reforms extended the mathematics and astrology program to four years. Now the first year was wholly dedicated to learning fundamental skills in mathematics and geometry to be used in the second year to master astronomy. In addition to the *Almagest*, the students read the works of modern astronomers. The approved reading list for teaching how to prepare astronomical ephemeredes now included Christopher Clavius, Regiomontanus's or Reinhold's tables of the sun, and Peuerbach's theory of the sun. In the second half of the second year, students still could elect to read Copernicus, supplemented with the Prutenic tables of Erasmus Reinhold. Enticing as it may be for the historian of science to imagine students electing to read Copernicus, students instead often preferred to study gnomics, the art of making sundials. The third year was devoted to Ptolemy's *Geography* and Apian's *Cosmography*. The curriculum also covered cartographic topics and several treatises on the use of measurement instruments, such as the astrolabe and astronomical rings, as well as the *Planisferio* by Juan de Rojas (Paris, 1551). The third year might also have included lectures on navigation—that is, study of the theoretical aspects of astronomical navigation, not practical nautical instruction. The fourth year was devoted to astrology using Ptolemy's *Tetrabiblos* and the works of Alcabisus.[75]

73. "Se ha de procurar ante todo que no coloquemos nada inexplorado en descripciones de esta naturaleza, pues es preferible que señalemos con líneas imprecisas que esa región es desconocida para nosotros, y dejarla en la oscuridad, que difundir errores entre los estudiantes con descripciones falsas, porque lamentamos que se cometan errores incluso por hombres sabios por el ansia de lucro y por la ambición de dar a conocer su nombre por todas partes." Muñoz, *Introducción*, 188.

74. Chapter 16 of the *Astrologicarum et Geographicarum* manuscript. Muñoz, *Introducción*, 214–26, 320–30.

75. Alejo Montes, *Universidad de Salamanca*, 200–201.

The program at Salamanca proved flexible enough to incorporate into the pedagogical program the latest mathematical tools of the cosmographical trade. In doing so, it was slightly ahead of other European institutions. A review of humanistic cosmographical studies in Jesuit institutions in France during the sixteenth century reveals a curriculum not unlike the one at Salamanca.[76] Although the theoretical framework remained Ptolemaic and largely Aristotelian, the instruments and data used to practice cosmography were wholly "modern" with a strong emphasis on the mathematical aspects of the discipline. Cosmographical education at Spanish university was on a par with, if not slightly ahead of, that at other European universities.[77] By the 1590s it had also begun to reflect the divergent paths cosmographical practice would take by the end of the century between the mathematical and the descriptive aspects of practice. Whereas the university curriculum reflected the canon of ideal practice, the intellectual boundaries of the discipline were also defined throughout the sixteenth century in a very different context—the sea.

Cosmography and the Sea: Mathematical Rationalism and Navigation Books

The curiosity of the European audience for whom Apian and Frisius wrote paled in comparison to the cosmographical fervor among explorers, businessmen, and government officials in the city of Seville. In Seville, accounts of the newly discovered lands circulated feverishly, challenging in equal measure what the learned and the illiterate thought about the world. In the courtyard of the royal Alcazar, the halls of the Casa de la Contratación, and the alleys of the sailor's neighborhood of Triana, men with practical concerns also tackled the cosmographical problems posed by the discovery and exploration of the new lands. Their attempts to lay out a conceptual scheme to explain the geography of the New World quickly abandoned any need to reconcile ancient authorities with the firsthand accounts arriving yearly with the return of the Indies fleet. As opposed to a cosmographer working in a country without access to the newly discovered territories, a cosmographer working in Seville, whether practicing privately or in the service of the Casa de la Contratación, was driven by a set of imperatives

76. Dainville, *Géographie des humanistes*, 3–45.

77. The Spanish astronomy and astrology curriculum was not unlike that of Italian universities. Grendler notes that professors at Bologna began emphasizing practical mathematics, geography, and mapmaking in the 1570s. Paul F. Grendler, *The Universities of the Italian Renaissance* (Baltimore: Johns Hopkins University Press, 2002), 419–22.

determined by Seville's role as the commercial center of trade with the Indies. These imperatives went far beyond intellectual curiosity. After all, commercial exploitation of the New World hinged on determining safe sailing routes and developing nautical techniques for transoceanic navigation. Cosmographers tackled these problems with zeal and in the process created a uniquely Spanish cosmographical genre—one with deep Portuguese roots: the navigation manual.

Part cosmographical treatise, part practical instruction manual, the navigation manual sought to codify and teach a set of principles vital for navigation. In differing measures, authors of such manuals turned in search of answers to disciplines that had been almost exclusively cultivated in university or courtly settings: cosmography and astronomy. By the mid-sixteenth century, the navigation manual synthesized the fundamentals of astronomical navigation in an expository narrative style that made the techniques approachable.

The principles of astronomical navigation had been well understood in university circles since the thirteenth century but had found limited acceptance among experienced sailors. Columbus, Magellan, and other great explorers made use of rudimentary astronomical methods during their voyages but relied on other navigation techniques, such as dead reckoning using a compass, to guide their ships. Sailors piloting ships during the initial years of the *carrera de Indias* learned their craft by experience, through careful observation of the sea and sky and with only the compass and portolan charts as technological tools. Mariners had relied on sightings of the North Star since ancient times to determine the four cardinal points at night. The period of astronomical navigation began per se with the introduction in the mid-fifteenth century of two nautical instruments for determining latitude: the Jacob's staff for measuring the elevation of the North Star above the horizon and the astrolabe for determining the height of the sun at noon. Both methods required making numerical adjustments to the values indicated by the sighting instrument.[78] These techniques for determining a ship's latitude

78. These corrections included the procedure known as the *Regimiento de Norte* or Regiment of the North, which accounted for the approximate three degrees of eccentricity between the North Star and the celestial pole. For daytime measurements, the observed elevation of the sun at noon had to be adjusted with the use of solar declination tables. The tables listed solar declination for every day of the year, and the pertinent factor had to be either added to or subtracted from the observed elevation of the sun, depending on the hemisphere and time of the year of the observation. Manuel Sellés, *Instrumentos de navegación: Del Mediterráneo al Pacífico* (Barcelona: Lunwerg Editores, 1994), 43–52. For a comprehensive introduction to the topic in English, see E. G. R. Taylor,

enjoy a theoretical coherency that lends itself to straightforward explications. But when these sight instruments were used at sea, the difficulties of using them quickly became apparent. Taking accurate readings while on a ship on the high seas was no small feat, and the inherently subjective reading of the astronomical measurements displayed by the instruments made them seem unreliable.

The navigation manual sought to establish the art of navigation on a theoretical footing consistent with established principles of natural philosophy. Authors drew upon the principles of the sphere as synthesized by Sacrobosco, the fundamentals of Ptolemaic cartography, and, in varying measures, practical nautical experience to produce this immensely popular genre. The genre, remaining faithful to its cosmographical roots, could also contain a geographical description of the world, with particular emphasis on the Indies and the Orient. The Spanish navigation manual also owed some of its content to the Portuguese tradition of *roteiros*. Since the 1480s the Portuguese had used astrolabes to calculate latitude by measuring the altitude of the sun in their trips to African Guiana, but this was far from standard practice. *Roteiros* included declination tables and instructions on how to use an astrolabe to measure the elevation of celestial bodies.[79]

The first navigation manual printed in Spain was written by Martín Fernández de Enciso and titled *Summa de geographia que . . . trata largamente del arte de marear* (Seville, 1519).[80] A lawyer by education, Enciso spent the years between 1508 and 1511 in America. As he stated in the dedication letter to Charles V, he wrote the book so the young monarch could learn easily about the newly discovered lands in his realm but also with the hope that the book would be useful to pilots and sailors sent by the king to discover new lands. A broad intended audience, to be sure!

Enciso's book was written in Castilian and used a cosmographical treatise to introduce a discussion of navigation techniques, a model that would be adopted by most navigation manuals to come. It began with the fundamentals of the sphere borrowed from Sacrobosco, emphasizing

---

The Haven-Finding Art: A History of Navigation from Odysseus to Captain Cook (London: Hollis and Carter, 1956).

79. For more on Portuguese celestial navigation, refer to the works of Luís de Albuquerque, in particular *Astronomical Navigation* (Lisbon: Comissão Nacional para as Comemorações dos Descobrimentos Portugueses, 1988).

80. An amended version was printed in Alcala in 1530 and again in Seville in 1546. It was translated into English in 1578. Luisa Martín Merás, *Cartografía marítima hispana*, Colección ciencia y mar (Barcelona: Lunwerg Editiores, 1993), 136.

the principles behind latitude and longitude coordinates. Throughout the book Enciso unapologetically punctuated his exposition with personal observations along with principles of natural philosophy. He explained how to determine latitude using the North Star or by observing the height of the sun and correcting the values using solar declination tables. Bowing to the cosmographical tradition that had preceded him, Enciso followed the chapters with a geographic description of the world along the stylistic lines of Pomponius Mela. But he deviated from the classical model in a key and fundamental way: he did away with the traditional division of the continents into three (Europe, Asia, and Africa) and instead divides the world into two halves: the Orient and the Occident, corresponding to the Old World and the New World, the known and the unknown. The Oriental half encompassed the three old continents, while all the newly discovered lands were in the Occident. For the description of the Orient he repeated well-known medieval accounts, but his description of the New World is an exceptional firsthand account.[81] He described the coasts of the New World in such detail that the reader feels a sailor could recognize the shoreline from onboard a ship. Significant geographical features, such as rivers or promontories, are described and their latitudes noted. All the lands are located following the tradition of the rutter[82] or pilot's book, with their position noted relative to a previously established reference point, using either degrees of latitude or compass bearings and distance. The geographical descriptions are not limited to the coast but also discuss people and territories further inland, particularly those that can be reached by river. The description of each territory typically contains a survey of the land's natural attributes, whether or not there was gold, and is often punctuated with anecdotes about the customs of the native peoples.

81. Mariano Cuesta Domingo describes Enciso's geographic description of America as the first "map in prose" of America: Mariano Cuesta Domingo, "'Tierra nueva e cielo nuevo,' navegación, geografía y mundo nuevo," *Boletín de la Real Sociedad Geográfica* 128 (1992): 27.

82. A rutter or *derrota* is a verbal description of an itinerary at sea. It gives sailing directions between two points by specifying the direction and distance to be traveled; it also provides descriptions of landmarks (often in the form of landscape drawings), currents, and tides. Rutters used compass bearings to specify direction; later they relied on latitude coordinates and, to the extent that they were available, longitude bearings. The word *rutter* is a linguistic reminder of the migration northward of navigation technology during the early modern era. It is an English corruption of the French *routier*, which in turn comes from the Portuguese word *roteiro*. J. B. Hewson, *A History of the Practice of Navigation* (Glasgow: Brown, 1951), 16–17.

By the 1550s the navigation manual had matured into books that almost exclusively discussed nautical topics and the cosmographical principles that pertained to navigation. A theoretical section instructing readers on the fundamentals of the sphere generally served as an introduction and prepared the reader for subsequent discussion of Ptolemaic cosmography. These preliminaries were followed by more practical instruction, such as how to use maritime charts to set a course at sea, how to take altitude measurements using quadrants or astrolabes, and how to apply solar declination tables to adjust values.

Inspired by Apian and Frisius, Spanish authors also used these manuals as a way to introduce instruments of their own invention, popularize new techniques, and discuss navigation problems. One such problem was the deviation of the compass needle from north or magnetic declination. The problem had vexed sailors since Columbus noted the phenomenon during his first voyage. The erratic behavior of the compass needle was known as the *nordestear* and the *noruestear*.[83] In long transoceanic voyages, the phenomenon added a new dimension of difficulty in charting a course at sea, since the deviation from north seemed to vary according to one's meridian. Navigation manuals became a forum for advocating different ways of coping with the problem, either by introducing instruments that "corrected" the compass reading or by trying to use the needle's deviation as an indicator of longitude. When traveling from east to west, as in the route taken by Spanish galleons sailing to the Indies, the compass needle deviated in a manner that suggested the deviation was proportional to longitudinal distance. This apparent correlation was used by generations of Spanish cosmographers in attempts to construct an instrument that would determine longitude at sea.

In Spain, as in the rest of Europe, the two most widely read navigation manuals of the sixteenth century were Pedro de Medina's *Arte de navegar* (1545) and Martín Cortés's *Breve compendio de la sphera y de la arte de navegar* (1551). Pedro de Medina (1494–1567) was a prominent figure in Sevilian cosmographical circles, taking part in an acrimonious debate between cosmographers and pilots who advocated different navigational

83. *Nordestear* referred to how much the needle deviated to the east of true north and *noruestear* to how much it deviated to the west of true north. Portuguese cosmographer Francisco Falero, working at the Casa de la Contratación in Seville, was the first author to discuss in print magnetic declination and its effect on the compass. Francisco Falero, *El tratado de la esphera y del arte de marear* (1535), facsimile ed., ed. Ministerio de Defensa and Ministerio de Agricultura (Borriana, Spain: Ediciones Histórico Artísticas, 1989), 29.

and cartographic methods. His book enjoyed wide diffusion, with twenty, mostly French, editions (figure 1.4).[84] The English, however, seemed to have favored Cortés's book, which appeared in at least six English translations with full attribution between 1561 and 1630.[85]

Save for the *Breve compendio*, we have no other work from the pen of Martín Cortés (ca. 1507–82).[86] His motivation for writing the *Breve compendio*, however, is clear. He explained in the dedication and prologue that he wrote the book out of frustration with the ignorance of pilots and their reluctance to learn the principles of astronomical navigation that could turn their dangerous craft into a more certain science.[87] Faithful to the genre, Cortés's book began with a theoretical section where he discussed the sphere and the composition of the earth according to Aristotle. This is followed by a discussion of movements of the sun and moon and their effect on earth. Imitating Apian, he included some *volvelles* as calculation aids. But whereas the introductory theoretical sections are peppered with classical references, in the book's final section classical citations fall by the wayside and the text becomes a straightforward instructional manual. Cortés addressed technical aspects of the construction of navigation instruments and the uses of sighting instruments and of sundials, including a clever one mounted on gimbals to compensate for the ship's motion. Cortés's explanation of the different methods for calculating latitude is one of the clearest of the genre. He was well aware that the mapping of the newly discovered lands had not been done as well as it could have had the task been done by "learned cosmographers and experts in the art of navigation."[88]

84. For the publishing history of Medina's text, see Marie Ange Etayo-Piñol, "Medina y Cortés o el aprendizaje de las técnicas de navegación en Europa en el siglo XV," *Revista de Historia Naval* 16, no. 64 (1998): 43, note 3. Medina never held an official post at the Casa de Contratación, although in 1552 he wrote a summary of the *Arte de navegar*, the *Regimiento de navegación*, which he intended for the newly established school of navigation. José María López Piñero, *El arte de navegar en la España del Renacimiento*, 2nd ed. (Barcelona: Editorial Labor, 1986), 162.

85. Lopez Piñero finds ten editions, while Navarro Brotóns lists six. López Piñero, *Ciencia y técnica*, 202, and Navarro Brotóns, *Bibliographía physico-mathemática*, 118–20.

86. Martín Cortés, *Breve compendio de la sphera y de la arte de navegar* (Sevilla: Casa de Antón Álvarez, 1551; reprint, Madrid: Editiorial Naval, 1990), 33–34, and Francisco Javier González González, "Martín Cortés de Albácar, Cádiz y el *Breve compendio de la sphera y de la arte de navegar* (1551)," *Gades* 22 (1997): 311–26.

87. Cortés, *Breve compendio*, f. 4v, 7r.

88. "Y no sería inconveniente, antes cosa justa y muy acertada (para quitar tantos errores de los cuales se sigue tanta confusión y peligros) que Vuestra Majestad mandase a doctos cosmógrafos y expertos en el arte de navegar que verifiquen las alturas de

Arte de nauegar
en que se contienen todas las Reglas, Declara
ciones, Secretos, y Auisos, q̃ a la buena naue-
gació son necessarios, y se deuẽ saber, hecha por
el maestro Pedro de Medina. Dirigida al sere
nissimo y muy esclarescido señor, don Phelipe
principe de España, y de las dos Sicilias. &c.
Con preuilegio imperial

Figure 1.4. Title page from the navigation manual by Pedro de Medina, *Arte de navegar* (Valladolid: Francisco Fernández de Córdoba, 1545). This item is reproduced by permission of *The Huntington Library, San Marino, California.*

Cortés and Medina insisted they had pedagogical reasons for writing the navigation manuals. Nonetheless, it is hard to imagine that these books were within the average pilot's reach. For although pilots held a privileged position in the ship's hierarchy (third only to the ship's master and captain), most were illiterate and notoriously undisciplined. Captains, although responsible for the military defense of the ship, were often completely ignorant of navigational matters, their appointment being an honorific title given to the highest-ranking gentleman onboard. The responsibility of guiding

polo que tienen los puertos, cabos, islas y pueblos marítimos: y asimismo describiesen verdaderamente las costas de la tierras, especialmente de la navegación de las Indias Occidentales o Mundo Nuevo" (Cortés, *Breve compendio*, f. 68).

a ship at sea fell on the pilot. A successful pilot, one with a storehouse of personal experience, who could recognize coastlines, predict storms, remember the location of treacherous shoals, and calculate the speed of the ship by keen observation of wind and water, was assured a career in the *carrera de Indias* throughout the sixteenth century.

Existing means of calculating position at sea, despite the methods touted by academic astronomers and cosmographical writers, remained notoriously inaccurate throughout the sixteenth century. Navigation instruments of the time, by design and because of available manufacturing methods, had an inherently large margin of error. Compounded by the difficulty of carrying out precise observation aboard a rocking ship, contemporary technology made the only method for finding position at sea, measuring the altitude of the sun or of the North Star to determine latitude, highly suspect. But when things went terribly wrong and a pilot failed to bring the ship to safe harbor, it was not uncommon for the blame to fall on his lack of skill in astronomical navigation. He was scorned because he had failed to use the principles explained so clearly in navigation manuals, and his illiteracy was often blamed for the disaster. In this climate of mistrust, pilots preferred to keep the results of their astronomical observations secret, because these often turned out to be unreliable or sometimes plain wrong. Some passengers—perhaps after having read Medina's navigation manual—thought that simply observing a pilot at work or mimicking his observations was enough to learn how to navigate.[89]

Despite authors' assertions that their books were meant for pilots, navigation manuals of the midcentury were written to formalize the principles of navigation, but hardly with real hope that practicing pilots would benefit from the books. So why write such books? Medina's and Cortés's books were likely written with a keen eye toward addressing a serious problem overtaking Spain's ventures overseas. Pablo E. Pérez-Mallaina argues that reliance on astronomical methods for navigation came about not because of any inherent technological superiority of astronomical navigation methods over traditional methods but because of a critical shortage of pilots in the Spanish empire with the experience necessary to guide ships on an Atlantic or—later in the century—a Pacific passage.[90] When the colonizing and

89. Pablo Emilio Pérez-Mallaina Bueno, *Spain's Men of the Sea: Daily Life on the Indies Fleets in the Sixteenth Century*, trans. Carla Rahn Phillips (Baltimore: Johns Hopkins University Press, 1998), 83–86.

90. Pérez-Mallaina explains that the need to outfit the great fleets with sufficient pilots exacerbated the problem and imposed on these men of the sea very different demands from those that faced explorers. It was one thing to stumble upon a unknown

exploitation phase of the American enterprise escalated after the Mexican conquest and the circumnavigation of the globe, the resulting increase in transatlantic expeditions created a critical shortage of pilots with sufficient experience to navigate in the open ocean.

When the shortage of experienced pilots threatened the imperial program and efficient exploitation of the newfound resources, government officials at the Casa de la Contratación and the Council of Indies devised a way they hoped would shorten pilot apprenticeships and produce pilots trained in astronomical navigation. To achieve this, they resorted to the same methods used for their own education—the classroom and the book. They also resorted to the only source of codified knowledge available to the early modern Western world regarding how to find one's place on the surface of the earth—cosmography.

The navigation manual is an example of the practices of a traditional craft assuming the trappings of a science as it migrated from the precinct of the trades to the exclusive district of the arts. Artillery and mining manuals of the mid-sixteenth century underwent a similar migration.[91] Authors of navigation, artillery, and mining technical manuals worked comfortably within the tradition of Aristotelian natural philosophy. Their familiarity with its language and forms of argumentation allowed them to frame their explanations in a manner similar to that used to legitimize a natural philosophical exposition. Authors of navigation manuals discussed matters intended to be reduced to practice; therefore they needed tools, instruments and maps, which the Ptolemaic corpus provided. When attempting to explain the craft of navigation, mid-sixteenth-century writers turned to well-established astronomical methods and practices associated with finding the position of an object in the celestial sphere, in the process recasting the craft of navigation in cosmographical language.

The navigation manual gave businessmen, government bureaucrats, and would-be adventurers a way of learning about the seafaring art; the

island during a voyage of exploration and declare it "found" or discovered; it was entirely another to return to the same spot year after year! Pablo Emilio Pérez-Mallaina Bueno, "Los libros de náutica españoles del siglo XVI y su influencia en el descubrimiento y conquista de los océanos," in *Ciencia, vida y espacio en Iberoamérica*, ed. José Luis Peset (Madrid: CSIC, 1989), 469.

91. Two examples of these types of texts are Niccolo Tartaglia's *New Science* (1537) and Agricola's *De re metallica* (1556). Although developing as a genre half a century after navigation manuals, Spanish artillery manuals rivaled navigation manuals in popularity. For a study of Spanish artillery manuals, see Jorge Vigón, *Historia de la artillería española*, vol. 1 (Madrid: Instituto Jerónimo Zurita, 1947).

books served as means of learning about where a venture was headed and what might be found on a distant shore. Not coincidentally, a similar surge in interest in astronomical navigation in late Elizabethan England answered similar motivations: a need to conserve profits by achieving certainty in navigation, a general excitement about overseas exploration, and a political will to expand beyond traditional borders.[92] Through the frequent association between astronomy and navigation, the writers of navigation manuals implied that the new art was grounded in the principles of the ancient and lofty science of astronomy, a science that in the hands of a few exceptional astronomers was capable of predicting with certainty the position of stars in the immutable heavens. The instruments and techniques used for astronomical observations, the astrolabe and quadrant, were the same as those used in navigation—why would they not prove to be just as accurate for finding the position of a ship at sea? Thus by subsuming into navigation the attributes of predictability and certainty long associated with astronomy, authors of navigation manuals gave their anxious readers what they most sought—assurances.

These types of books, as Pamela Long has explained, served to transform the mechanical arts into "discursive disciplines structured by principles, whether mathematical or other, presented in writing."[93] This transformation did not happen without contention. The genesis of the Spanish navigation manual reveals a tense relationship between cosmographers and pilots. Alison Sandman has located the production of these navigation texts in the context of a struggle between experienced ship's pilots (who rejected astronomical navigation) and a group she defines as "theory proponents" that included many cosmographers. During the first half of the sixteenth century, these camps wrestled for control over the navigation methods advocated by the Casa de la Contratación. Sandman argues that by the time Philip II ascended to the throne in 1556, theory proponents had consolidated their position as authorities, not only on geographical matters

92. John W. Shirley, "Science and Navigation," in *Science and the Arts in the Renaissance*, ed. John W. Shirley and F. David Hoeniger (London: Folger Books, 1985), 75–76. Bennett locates the production of these types of texts, in his case the treatises of Apian and Frisius, in a context of an expanding audience of practical men interested in mathematical solutions to the problems of surveying, navigation, military arts, and architecture: J. A. Bennett, *The Divided Circle: A History of Instruments for Astronomy, Navigation, and Surveying* (Oxford: Phaidon, 1987), 22–23. Ash explain the role of mathematicians as "expert mediators" in *Power, Knowledge, and Expertise in Elizabethan England*, 138–42.

93. Pamela O. Long, *Openness, Secrecy, and Authorship: Technical Arts and the Culture of Knowledge from Antiquity to the Renaissance* (Baltimore: Johns Hopkins Press, 2001), 176.

but also on navigation.[94] As we will see in chapter 7, debates about the merits of pilot's experience versus theoretical knowledge continued well after cosmographical practices had been institutionalized and officially endorsed by the Crown.

The interest in defining a set of principles that would place the practice of navigation on a more precise and certain foundation should be interpreted as part of a process of mathematizing the dimensions of space and time that was well on its way by the early fifteenth century.[95] Cosmography and its corollary, mathematical cartography, are emblematic of a transcendental shift in the Western perception of reality, from a view that emphasized qualitative properties prevalent in the Middle Ages to a quantitative view characteristic of the Renaissance and modernity.[96] The quantification of time and space, facilitated during the fifteenth and sixteenth centuries by new mathematical techniques, changed expectations regarding the kinds of information that could be known with certainty. If a science could have a mathematical foundation rather than a metaphysical or merely an empirical one, the age was inclined to give the art the attribute "certain." Spanish cosmographers were in the midst of this intellectual transformation and championed the capability of their science, arguing that its astronomical and mathematical foundations made theirs one of the certain sciences.

As Anthony Grafton points out in what is perhaps one of the biggest understatements of the historiography of the discovery, "The encounter between

94. The debates centered on a series of lawsuits between the cosmographers Alonso de Chaves, Pedro de Medina, and Pedro Mexía on one side and Diego Gutiérrez, master instrument maker at the Casa de la Contratación, supported by Sebastian Cabot, the pilot major, on the other. The principal bone of contention was whether pilots should be allowed to use navigation charts drawn with two scales, one of them correcting for the effect of magnetic declination on the compass needle. The bottom line with the lawsuits, however, was competition over the lucrative instrument trade in Seville and the monopoly that Gutiérrez, backed by Cabot, maintained on the trade to the exclusion of the other cosmographers. For more on these debates, see Ursula Lamb, *Cosmographers and Pilots of the Spanish Maritime Empire* (Aldershot, UK: Variorum/Ashgate, 1995), 3:40–57; Pedro de Medina, *Libro de las grandezas y cosas memorables de España* (Madrid: CSIC, 1944), xi–xix; and Alison D. Sandman, "Cosmographers vs. Pilots," PhD diss., University of Wisconsin, 2001, 283–88.

95. Alistair C. Crombie, "Science and the Arts in the Renaissance: The Search for Truth and Certainty, Old and New," in *Science and the Arts in the Renaissance*, ed. John W. Shirley and F. David Hoeniger (London: Folger Books, 1985), 19.

96. As Alfred Crosby explains succinctly, while building on the work of Thomas

Europe and the Americas juxtaposed a vast number of inconvenient facts with the elegant theories embodied in previously authoritative books."[97] If the discovery of the New World showed that the ancient authorities had erred in matters of fact, was Ptolemaic methodology still a valid approach? Spanish cosmographers grappled with this question and frequently disagreed on the best methodology to follow. Soon Spain's leading humanists were applying the tools they had mastered, linguistics and classical erudition, while authors of navigation manuals and some pilots at the Casa de la Contratación used the mathematical tools afforded by cosmography to address the technological problems of navigation. All the while the cosmographical opus, with its descriptive geography and maps, continued to evolve and to integrate into the discipline an epistemology that privileged eyewitness accounts and empiricism.

By the late sixteenth century, however, it became difficult to produce a Renaissance cosmography that conformed to the rules and standards associated with the discipline, particularly when the cosmographical discipline was asked to produce a useful description of the New World. Spanish cosmographers adopted the Ptolemaic model and associated methodology, not only for its elegant way of portraying the earth but also because it proved to be a valuable resource in the quest of empire. "Knowing" for sixteenth-century Spanish royal cosmographers went beyond satisfying personal curiosity. It meant organizing and presenting information about the new discoveries in a manner that served the empire effectively and provided utilitarian results. Thus they appropriated classical cosmographical models of practice and representation but suffused the discipline with a pragmatic and practical approach—witness the navigation manual—that ran contrary to Continental practice. As gatekeepers of firsthand cosmographical knowledge of the newly discovered lands, Spanish royal cosmographers were the first to wrestle with how to incorporate "a vast number of inconvenient facts" into a European understanding of the world. The following chapters explore how cosmographers working for the Spanish monarch fashioned a new conceptual framework that incorporated the reality of the New World.

Kuhn, David Landes, and Samuel Edgerton: Alfred W. Crosby, *The Measure of Reality: Quantification and Western Society, 1250–1600* (Cambridge: Cambridge University Press, 1997), 132.

97. Anthony Grafton, introduction to *New World, Ancient Texts: The Power of Tradition and the Shock of Discovery*, ed. Anthony Grafton, April Shelford, and Nancy Siraisi (Cambridge: Belknap Press of Harvard University Press, 1992), 5.

TWO

Cosmographical Styles at the *Casa*, *Consejo*, and *Corte*

For the pragmatic Philip II, a man keen on specifics and with a personal interest in geography, the New World with its natural wonders, its vastness, and its enigmatic peoples had to be accurately described, not only to satisfy his personal curiosity but also to assist in the administration of the largest empire the world had ever known. In addition, a good dose of imperial vanity and a Renaissance penchant for visual representations of monarchical power and prestige motivated the king to order that his territories be portrayed in maps, their natural resources cataloged and described, and potential sources of wealth surveyed.[1] During the reign of Philip II (r. 1556–98), the men commissioned to write the cosmography of the new overseas territories worked during a period of rapid and tremendous territorial expansion. New territorial claims, such as Miguel López de Legazpi's conquest of the Philippines between 1565 and 1571, and new political situations, such as the unification of the Spanish and Portuguese thrones in 1581, presented royal cosmographers with an endless and ever-changing stream of questions and problems.

By the second half of the sixteenth century, the systematic gathering of cosmographical information about the New World was institutionalized under government control in three principal sites engaged in cosmographical knowledge production. Professional cosmographers working at the Casa de la Contratación and the Council of Indies had access to a steady stream of oral and written sources of geographical information. At the Casa, cosmographers personally interviewed pilots arriving from the

1. Richard L. Kagan, "Philip II and the Geographers," in *Spanish Cities of the Golden Age: The Views of Anton van den Wyngaerde*, ed. Richard L. Kagan (Berkeley: University of California Press, 1989), 49–50.

Indies. From what they learned they pieced together the geography of the coastlines and wrestled with ways of conquering its navigation. The only stars in the sky that mattered were the ones that helped guide a ship at sea or locate the desired destination on a nautical chart. Cosmography at the Casa became navigation's handmaiden. For the Council of Indies, the problem was administration and oversight. How could councilmen in Madrid, thousands of miles from the Indies, know enough about this far-flung empire to effectively advise the monarch and legislate? Cosmographers at the Council relied on Renaissance cosmography's traditional taxonomy to organize the vast amount of geographic, ethnographic, natural, and historical data, using the empire's bureaucratic machinery to collect necessary information about the New World.

Cosmographical production, however, was not limited to the Casa de la Contratación and the Council of Indies. Sometimes cosmographical concerns were addressed by ad hoc royal appointments or by a cadre of cosmographers who worked for the royal household. Individuals also recognized the value placed by the Spanish monarchy on cosmographical knowledge and eagerly submitted contributions, some in the form of *relaciones* (narrative accounts). Cosmographers with royal appointments and those practicing cosmography privately, as well as those from the academy, could be called upon to serve on special councils convened to address specific cosmographical questions, most of which had significant political implications.

As we have seen, Spanish cosmographers worked within an intellectual tradition shaped by the humanist revival of descriptive geography and the rediscovery of Ptolemaic cartography. The cosmographical practitioners studied in this chapter, Alonso de Santa Cruz (ca. 1505–67), Juan de Herrera (1530–97), and Rodrigo Zamorano (1542–20), tried to reconcile the established literary styles and methodologies characteristic of Renaissance cosmography with the demands of practicing their science in the fluid cosmographical environment of the century of discovery. During the midcentury, Renaissance cosmography and its practitioners came under increasing pressure—exigencies placed on the discipline by the discovery of the New World and the accompanying demands of empire. These underscored the shortcomings of the literary genre and mode of representation associated with cosmography, and they resulted in a reconceptualization of what constituted cosmographical knowledge. Distinctive styles of cosmographical practice emerged in response, molded in large part by the institutions of empire that cosmographers served, issues of personal agency, and, increasingly, a preference for either mathematical or didactic cosmographical tools to explicate the New World.

"Like Scattered Pieces of a Puzzle": Compiling Knowledge of the New World

The driving force behind cosmographical activity in late sixteenth-century Spain was a *need to know* the New World. Renaissance cosmography, with its comprehensive approach, already possessed the conceptual tools to answer many questions about the New World and to organize the information in a way that could effectively address a number of administrative, financial, military, religious, and political concerns of the expanding empire. Geographic information answered the obvious questions about the empire's location and the extent of its borders, while Ptolemaic cartography provided a way of representing this geography with mathematical precision. Hydrographic reconnaissance judged the navigability of coastal waters and rivers, the existence of suitable ports, and the most efficient rutters to follow to reach new lands. Descriptive narratives summarized ethnographic information about the native peoples. Chronicles recorded the history of Spanish conquest and discovery in the new lands, while inventories of natural resources, especially mineral, often determined the Crown's interest in conquering and colonizing a territory. Cosmographers, the one group of scientific practitioners in early modern Europe whose intellectual heritage prepared them to compile this kind of knowledge, were given the task of collecting, organizing, and providing this information to the Habsburgs's administrative machinery. The cosmographical body of information they collected was recognized as having domestic and international strategic value—but only if it could be used effectively and kept secret.

To comply with the need for precise cosmographical knowledge during this period of rapid territorial expansion, the cosmographer's first task was to incorporate the New World into a new universal cosmography. This necessitated the refashioning of the methodological framework associated with Renaissance cosmographical practice. The new framework retained the conceptual foundation of Renaissance cosmography discussed in the preceding chapter, but the century of discovery presented a series of challenges that fifteenth-century armchair cosmographers never dreamt of when they were rediscovering Ptolemy's *Geography* or poring over Mela's description of the ancient world in their libraries.

Incorporating recent discoveries into a new world picture was intrinsic to postdiscovery cosmography, but in Spain and especially within the institutions of empire, the resulting cosmographies also had to respond to a number of demands imposed by the context where this cosmographical knowledge was formulated. At the Council of Indies, cosmography was perceived as a body of knowledge that served to inform legal and administrative actions. Thus, the products of cosmographical inquire had to be *useful*. This utilitarian mandate also implied that the body of cosmographical knowledge had to be current, timely, and accurate. To satisfy these demands, the discipline of Renaissance cosmography had to undergo a significant reorientation that would increasingly remove it from its humanistic roots. It had to incorporate new methodologies facilitating the collection of information from all corners of the world, provide a set of interpretive principles that would permit vast amounts of information to be organized efficiently, and articulate this information in ways that were comprehensible to the cosmographer's clients.

The sources of this information were the hundreds, if not thousands, of accounts (*relaciones*) that flowed from the New World to the Old. Some of these accounts were narratives, often written in the form of histories, which found their way to publishing houses throughout Europe. Not just any account qualified as a proper history, as the archbishop of Zaragoza, censor of Francisco López de Gómara's *Historia general de las Indias* (Zaragoza, 1552), reminds us in the prologue to that book. The archbishop draws a strict line between a history and an account of recent events. After listing the names of the writers he considers *historiadores*, naming Peter Martyr, Hernán Cortés, Gonzalo Fernández de Oviedo, and Gómara, the archbishop remarked, "These writers have written many and substantive works about the Indies and printed their works. All the others that are out there published, write a little about their own [experiences], which is why they do not belong among the historians; if it were so, all the captains and pilots that recount of their voyages, of which there are many, would call themselves historians."[2]

Yet cosmographers relied on the accounts of these very "captains and pilots" that the archbishop found so unsuitable to provide the material for their cosmographical works. But these published narratives were only one type of account or *relación* among the thousands sent to Spain since the days following the discovery of the Indies. Only a small fraction of narratives about the New World appeared in print; most circulated as manuscript copies. The news-bearing manuscript featured prominently in a culture that considered the written medium integral to its social organization and

2. "Estos autores han escrito mucho de Indias y impreso sus obras, que son de substancia. Todos los demás que andan impresos escriben lo suyo y poco, por lo cual no entran en el número de historiadores; que, si tal fuese, todos los capitanes y pilotos que dan relación de sus entradas y navegaciones, los cuales son muchos, se dirían historiadores." *Historia general de las Indias*, in *Pórtico a la ciencia y a la técnica del Renacimiento*, ed. María Jesús Mancho Duque (Salamanca: Junta de Castilla y León, 2001), 316.

increasingly to its political life. As Fernando Bouza observes, the circulation of manuscript copies permitted the *relación* to retain the immediacy of an unmediated communication. In a cultural context where printed material was censored and required printing licenses, the manuscript implied a freer medium of expression.[3] As the sixteenth century progressed, moreover, news-bearing narratives of the New World evolved into a veritable literary genre.

Relaciones can be defined broadly as accounts of personal experiences in the New World. They encompass personal memoirs, letters, chronicles, replies to official questionnaires, and even personal interviews with travelers from distant lands.[4] By midcentury, the structure and content of a *relación*, whether addressed to the monarch or to one of the royal councils, whether part of an official report or personal correspondence, followed a standard format. A similar standardization took place in the correspondence between Jesuit superiors and missionaries. When reporting about Jesuit activity in a particular region, the missionary was advised to also describe its climate, the location of the lands in terms of latitude and longitude, the customs of the inhabitants, their means of livelihood—in short, all the geographical information thought to be needed for superiors administering the apostolic program. Despite this standardization, the letters written by Jesuit missionaries in the Orient became *les sources vives* for many early modern cosmographers.[5]

Reasons for writing a *relación* varied. Some used a *relación* in the form of a letter of merit as a way of informing the king of their service abroad with the expectation of just rewards. Others wrote to denounce injustices: cruelty to the Indians, lawlessness on the part of Spaniards, or perceived grievances in the distribution of Indians and land. Sometimes *relaciones* were written to comply with official requests for information, as was the case with pilots at the Casa de la Contratación, who upon returning from the Indies had to go before the pilot major and report any new discoveries. Some wrote out of genuine intellectual curiosity, as was the case of the sources used by Sevilian doctor and naturalist Nicolás Monardes (c. 1508–88). After the publication of the first part of his *Historial medicinal* (1565), about the medicinal properties of plants of the New World, correspondents

3. Fernando Bouza, *Corre manuscrito: Una historia cultural del Siglo de Oro* (Madrid: Marcial Pons, 2001), 49–67.

4. I prefer this broad definition to the narrow categories Mignolo applies to these texts. He subdivides the genre into *cartas relatorias* (account letters) and *relaciones*, with a third category for chronicles. Mignolo, "Cartas, crónicas y relaciones," 58–59.

5. Dainville, *Géographie des humanistes*, 103–13.

and visitors from the Indies eagerly brought him specimens and were happy to share their personal experiences with medicinal products in the Indies.[6]

Others found the epistolary format the best way to relate news about the Indies to the metropolis. Many authors expected these communications to circulate widely, and it was not uncommon for letters with different versions of the same event to "compete" in the manuscript marketplace. Official chroniclers and cosmographers freely subsumed the *nuevas* or news in these manuscripts into chronicles and historical narratives and avidly collected such materials.[7] These communications became primary sources of information for the cosmographies, descriptive geographies, and maps they prepared.

We cannot gloss over the degree to which gathering cosmographical information would have been disorienting to the early modern cosmographer. The cosmographer was above all a compiler of information. Yet unlike encyclopedia writers, cosmographers did not simply collect random facts. Like scattered pieces of a jigsaw puzzle, cosmographical information arrived from beyond the sea in a disorderly manner. The cosmographer had to assemble these pieces into a composition that resembled the classical models he had been taught to emulate. To make sense of the avalanche of information, they subjected it to a process of translation that went beyond the simple act of finding European analogies. As it became assimilated into the European context, it did so under a new conceptual framework that eschewed emblematic correspondences and references to classical sources.[8]

How did cosmographers use this information? The cosmographer's task, as Ptolemy had carefully described in the *Geography*, was to interpret, decipher, and translate myriad sources into a coherent description of the world. For cosmographers in the king's service, the task was also to comply with the requirements of the organization they served, which demanded that they reduce cosmographical knowledge into a format or mode of representation that could be used. Thus, cosmographers working at the Casa de la Contratación were entrusted with maintaining up-to-date

6. José Pardo Tomás, *Oviedo, Monardes, Hernández: El tesoro natural de América, colonialismo y ciencia en el siglo* XVI (Madrid: Nivola, 2002), 107.

7. Bouza, *Corre manuscrito*, 147.

8. William Ashworth argues that the "demise of the emblematic world view" occurred prior to Baconianism and independently of it. William B. Ashworth, "Natural History and the Emblematic World View," in *Reappraisals of the Scientific Revolution*, ed. David Lindberg and Robert Westman (Cambridge: Cambridge University Press, 1990), 324.

navigation charts and rutters to the Indies, while cosmographers at the Council of Indies had to keep members of the Council informed about the geography, natural resources, and peoples of the Indies. Likewise, this information took on markedly different forms: charts and instruments for pilots, descriptive geographies, chronicles, and maps for the Council. Preparing these products was far from easy. A cosmographer was expected to compose a cohesive picture of the New World following mathematical rules and standards of prose narration dictated by the cosmographical genre. For cosmographers during the first fifty years after the discovery, who had been trained in the humanistic tradition to abstract information from classical texts, the lack of a consistent format or even a uniform set of underlying assumptions among the variety of *relaciones* presented a unique challenge that their training had ill-prepared them to face.

The Tordesillas Question

Increasingly, however, it was the mathematical underpinning of cosmography, and particularly its contribution to cartography, that came to be perceived as a tool that could clarify geographical questions with heretofore unprecedented certainty. Locating landmasses on a sixteenth-century map now required accuracy beyond that of a rough spatial organization of continents on a piece of paper intended to orient a reader relative to a text or even to serve the demands of astronomical navigation. For the first time in Western history, precise geodesic measurements became politically important. In 1493, Pope Alexander VI decreed in the Bull *Inter Coetera* that lands discovered west of a line of demarcation passing one hundred leagues west of the Azores and encircling the globe would belong to Spain, while those lying east of the line would belong to Portugal. In the 1494 Treaty of Tordesillas between Spain and Portugal, the Portuguese negotiated to have the line of demarcation moved to 370 leagues west of the Cape Verde and Azores islands.[9]

The language of the treaty, however, was vague and therefore open to speculation. It did not specify which of the Azores or Cape Verde islands was to be used for calculating the 370 leagues, nor did it consider the difficulty of measuring leagues at sea.[10] It was not even clear whether these

9. For an in-depth study of the political implications of the treaty, see Luis Antonio Ribot García, ed., *El Tratado de Tordesillas y su época*, 2 vols. (Madrid: Junta de Castilla y León, 1995).

10. It is interesting to note that the first line of demarcation proposed by Ferdinand and Isabella was based on observations of the effects of magnetic declination on a

two sets of islands were *in the same meridian*. Equally problematic was accurately translating leagues into degrees of longitude for territories in different parallels when the earth's size was still a matter of conjecture. Spaniards preferred to define one degree of longitude as measured at the meridian of Toledo, or 17½ leagues.[11]

The Portuguese chose instead to measure the distance between meridians along the equator, resulting in more leagues per unit of longitude and thereby placing more territory under Portuguese domain. After Magellan discovered the Moluccas on behalf of the Portuguese in 1511 and then deftly reclaimed them in favor of Spain in 1521, the problem took on a new urgency. A conference convened by Charles V in Badajoz in 1524 between "deputy commissioners, jurists and cosmographers from both sides" again failed to reach agreement, and Spain continued to claim that the Moluccas fell 30° within its side of the line of demarcation. The debate cooled off after 1529, when Spain leased the Moluccas and the territories located 297½ leagues (or 17° corresponding to 17½ leagues per degree of longitude) west of the islands to Portugal, only to be rekindled in 1564, when the Spanish launched a mission headed by Miguel López de Legazpi to conquer and settle the Philippines.[12]

In this political climate, it was a strategic necessity to have in royal service cosmographers who could convincingly debate cosmographical questions and determine longitudinal locations with exactitude. Resolving the geographical "question" presented by the Treaty of Tordesillas gave Spanish cosmographers a chance to coalesce as a community of practitioners. The sometimes-contentious juntas to discuss the "Tordesillas Question" became a forum where cosmographers could critically assess prevailing methodologies and develop argumentation to support cosmographical facts.[13]

---

compass needle made by Christopher Columbus during his first voyage—the first time the phenomenon was observed by a Westerner. Ricardo Cerezo Martínez, "El meridiano y el ante meridiano de Tordesillas en la geografía, la náutica y la cartografía," *Revista de Indias* 54, no. 202 (1994): 511.

11. For a detailed contemporary discussion of these issues, see María del Carmen González Muñoz, introduction to Juan López de Velasco, *Geografía y descripción universal de las Indias*, Biblioteca de autores españoles 248 (Madrid: Ediciones Atlas, 1971), 3–5.

12. Eduardo Trueba and José Llavador, "Geografía conflictiva en la expanción marítima luso-española, siglo XVI," *Revista de Historia Naval* 15, no. 58 (1997): 23–25.

13. The opinions issued on the matter by some of Spain's best-known cosmographers of the time are collected in BAH, 9/4797, "Parecer sobre la demarcación." Original documents are in AGI, P-49, R.12.

Alonso de Santa Cruz and His Cosmographical Opus

The three centers of cosmographical production considered here, the Casa de la Contratación, the Council of Indies, and the royal court, shared a similar intellectual heritage in Renaissance cosmography. All three could also lay claim to the work of one of Spain's leading cosmographers, Alonso de Santa Cruz (1505–67).[14] In many ways his career and cosmographical production defined what was expected of a cosmographer in the king's service and in general of cosmographical practice in Spain for the latter half of the century. In Santa Cruz, the sometimes contentious arts of navigation and cosmography found a single voice. The son of a well-to-do Sevilian ship owner, he experienced early in his life the vicissitudes of early modern navigation. In 1526, he participated as a budding businessman in an enterprise under the direction of the pilot major of the Casa de la Contratación, Sebastian Cabot (1476–1557, son of John Cabot). The original plan was to travel to the newly discovered Moluccas and circumnavigate the globe. But poor navigation—and the region's rumored wealth—landed the expedition on the coast of South America, where the explorers barely redeemed the enterprise by successfully mapping and exploring the mouth of the La Plata River.[15] During the five years the expedition lasted, which included a return trip via Veracruz and the Bahamas, Santa Cruz gathered a wealth of information about the geography and natural history of the area, some of which Fernández de Oviedo cites in his *La historia general de las Indias* (1535).[16] The trip was a life-changing event for Santa Cruz, who wrote to Philip II almost thirty years later, "After this [trip] I turned to learning about the science of Astrology and Cosmography."[17] The experience convinced

14. For a collection of historical documents detailing Santa Cruz's career and biography, see the editors' introductions to Alonso Santa Cruz, *Crónica de los Reyes Católicos*, ed. Juan de Mata Carriazo (Sevilla: Publicaciones de la Escuela de Estudios Hispano-Americanos de Sevilla, 1951), xxxiii–li, and Alonso Santa Cruz, *Alonso de Santa Cruz y su obra cosmográfica*, ed. Mariano Cuesta Domingo, 2 vols. (Madrid: CSIC, 1983), 1:35–58.

15. Santa Cruz, *Obra cosmográfica*, 1:53–56.

16. Schäfer points out that Gonzalo Fernández de Oviedo (1478–1577) was the first person to carry out functions similar to those of the cosmographer-chronicler of the Council of Indies, although it was not an official appointment as such. Oviedo was paid for the expenses he incurred while compiling *La historia general de las Indias* and for the cost of printing the first part of his work. Schäfer, *Consejo Real y Supremo de las Indias*, 2:352–53.

17. "Despues de esto yo me di a saber las ciencias de Astrología y Cosmografía" (AGI, P-260, N. 2, R. 6). "Borrador y apuntaciones para el prólogo del libro intitulado 'Islario General,'<ts>" in Santa Cruz, *Obra cosmográfica*, 1:57.

him that mathematics applied to cartography and navigation led to increased certainty and safety at sea.

In 1536, Santa Cruz began his professional career as a cosmographer at the Casa de la Contratación, having earned a reputation as an innovative instrument maker.[18] He was paid from the coffers of the Casa de la Contratación, but he tried to keep himself at the margins of the administrative structure of the institution. Along with his appointment, Santa Cruz obtained a series of royal orders that suggest he was engaged in projects beyond those demanded of other cosmographers associated with the Casa. One order expressly instructed pilots to visit Santa Cruz at his home to give him personal accounts of their navigations—which they were already obligated to present to the pilot major of the Casa de la Contratación.[19] Another royal order asked the governor of Castilla del Oro (Panama) to chart his territory and send the information directly to Santa Cruz.[20] Santa Cruz earned a coveted position of *contino*, or king's attendant, and he used his proximity to the monarch to secure royal endorsement for his cosmographical projects. Although obligated as a *contino* to reside at court, over the years Santa Cruz received royal dispensations allowing him to remain working in Seville. While at court he engaged with Emperor Charles V in long conversations about astrology and cosmography, which gave him the opportunity of tutoring the future Philip II on the sciences.[21]

His cosmographical inquiries took him in 1545 to Lisbon, where he met with leading cosmographers, talked to persons who had traveled to African kingdoms, and interviewed pilots who regularly sailed the easterly route

18. In 1536, Santa Cruz presented to the Casa de Contratación an instrument that used the effect of magnetic declination on the compass needle to determine longitude, as well as a map that reflected deviations relative to longitude (Santa Cruz, *Obra cosmográfica*, 1:62). For more on this instrument, see Santa Cruz's *Libro de las longitudes*, chap. 5.

19. It took some wrangling by Santa Cruz for officials at the Casa to finally agree to send arriving pilots to visit him at home to give their accounts (Santa Cruz, *Crónica*, xxxviii). These orders, and another one instructing the pilot major of the Casa de Contratación, Sebastian Cabot, to consult with Santa Cruz whenever Cabot was inspecting navigational instruments and examining pilots, were not without controversy. Cabot resented the infringement on his authority and alleged that Santa Cruz had gotten the orders surreptitiously. José Toribio Medina, *El veneciano Sebastián Caboto al servicio de España y especialmente de su proyectado viaje á las Molucas por el Estrecho de Magallanes y al reconocimiento de la costa del continente hasta la gobernación de Pedrarias Dávila*, 2 vols. (Santiago de Chile: Imprenta y Encuadernación Universitaria, 1908), 345.

20. As quoted in Santa Cruz, *Crónica*, xxxv–xxxvi.

21. AGS, Estado-121, f. 1–22. Published in Medina, *El veneciano Sebastián Caboto*, 348–49.

to India and the Spice Islands. By the early 1550s, Santa Cruz returned to Seville and, in what was perhaps his most productive period, became a near recluse in his house by the royal Alcazar in Seville,[22] preferring the comfort of his home "for the contemplation and recreation for his studies and life."[23] His sojourn in his hometown did not last very long; he was back at court in 1554. In 1563, the king instructed Santa Cruz to reside at court as the king's personal adviser on cosmographical matters, even though he held the post of cosmógrafo mayor of the Casa de la Contratación.[24]

Despite his official association with the Casa de la Contratación—he continued to collect his salary there—Santa Cruz's activities took place at the margin of the administrative machinery of state. His work was essentially a personal pursuit, encouraged by an admiring monarch and rewarded by royal favors. Several times throughout his life, he was asked by the king to advise the Council of Indies on cosmographical matters[25] and to serve on several cosmographical conferences.[26] Nonetheless, his relationship with the Council was often rocky. In a petition in 1557, he complained to the future Philip II that men of science were needed but not welcomed at the Council of Indies.[27] Santa Cruz's observation was not wasted on the future king. Years later, when the Council sought a full-time cosmographer, they recommended a jurist, Juan López de Velasco, for the job.

The extent of Santa Cruz's scientific work would have remained unknown to modern scholars but for the fortuitous inventory of an old leather trunk containing a collection of his papers and rich cartographic collection—an inventory made when López de Velasco took possession of

22. Letter from Hernán Pérez to the emperor, August 22, 1549, in Santa Cruz, *Crónica*, xliv.

23. AGS, Estado-84, f. 86. Letter from Santa Cruz to the emperor, 1551, in Medina, *Sebastián Caboto*, 346–47.

24. M. I. Vicente Maroto, "Alonso de Santa Cruz e el oficio de Cosmógrafo Mayor de Consejo de Indias," *Mare Liberum* 10 (1995): 519.

25. In 1554, the Council asked Santa Cruz to prepare a report on different methods of calculating longitude at sea in order to evaluate some instruments made by Peter Apian promising to solve the problem of determining longitude. Santa Cruz wrote the *Libro de Longitudes* in response (Santa Cruz, *Obra cosmográfica*, 1:93–95).

26. He also advised the finance and war councils. Angeles Arribas Lázaro, "Unas cartas de Alonso de Santa Cruz," *Asclepio* 26–27 (1974–75): 259.

27. Despite his disdain for the current councilmembers' lack of interest in cosmography, Santa Cruz asked the king in the same petition to appoint him to the Council of Indies (Vicente Maroto, "Alonso de Santa Cruz," 517). For other letters in which he refers to members of the Council of Indies in similar terms, see Arribas Lázaro, "Unas cartas de Alonso de Santa Cruz," 259.

Santa Cruz's cosmographical material upon becoming the new cosmographer-chronicler of the Council of Indies.[28] All of these works, however, remained unpublished at the monarch's request.[29] The inventory lists over 338 maps, including two atlases: a lost 169-map atlas and the *Islario general* with 120 maps (plates 1 and 2). In addition to maps, the trunk contained chronicles, geographic descriptions, and several cosmographical and astrological treatises. These documents became the nucleus of a set of reference materials passed from cosmographers and chroniclers of the Council of Indies to their successors well into the seventeenth century. Some of the documents, however, were scattered or lost over the years. Only one world map drawn in 1542 survives; it is currently at the Royal Library in Stockholm[30] (see plate 3).

We know that some years after his death Santa Cruz's family asked for the return of some of his documents. In response to the request, a senior councilman of the Council of Indies, Benito López de Gamboa, selected from among the documents those needed at the Council and recommended to the king's secretary, Mateo Vázquez, that the king might enjoy having some in his office, including a "curious description of Spain," a "clever book of water and fire inventions and machines of war."[31] Other documents were never lost but were assigned incorrect attributions over the years.[32] Two

28. AGI, P-171, N. 1, R. 16, f. 1–10v. "Minuta del inventario de los papeles de la antigua gobernación de Nueva España y Perú, que quedaron por muerte de Alonso de Santa Cruz, cosmógrafo de Su Majestad," October 12, 1572. The inventory is published in Santa Cruz, *Obra cosmográfica*, 1:73–78.

29. AGS Estado-143, f. 84. Quoted in Vicente Maroto, "Alonso de Santa Cruz," 521n1.

30. Alonso de Santa Cruz, *Nova verior et integra totius orbis descriptio nunc primum in lucem edita per Alfonsum de Sancta Cruz Caesaris Charoli V archicosmographum, A.D. MDXLII.* "A very new and complete description of the whole world now first prepared by Alfonso de Santa Cruz Cosmographer Major of the Emperor Charles V, 1542." Royal Library of Stockholm, Map Room.

31. IVDJ, Envio 25, f. 223. Letter mentioned in Vicente Maroto, "Alonso de Santa Cruz," 523n36.

32. Recently, Millán de Benavides has used the first folio of the manuscript *Epítome de la conquista del Nuevo Reino de Granada* (AHN, Indias 27) to attribute the work to Santa Cruz and also to map the trajectory of Santa Cruz's documents through the hands of official cosmographers and chroniclers. The *Epítome* was among the documents in Santa Cruz's trunk, as noted on the first page in what is likely López de Velasco's hand and along with his unmistakable annotation "relación notable OJO," common in other documents known to have been in his possession. Another annotation, the word *cespedes*, is common among cosmographical documents that had been in possession of Council cosmographer Andrés García de Céspedes, Velasco's successor as cosmographer at the

works, the *Islario general* and *El astronómico real*, had their title pages defaced by Andrés García de Céspedes, who replaced Santa Cruz's name with his own when he presented the works to Philip III.[33]

The *Islario general*

For Santa Cruz, composing a new "general" geography and history in the fashion of Ptolemy was the culmination of a life's work. As a cosmographer, he expected to write a cohesive work that both adhered to the discipline's mathematical underpinnings and met the standards of prose narration dictated by the cosmographical genre. Meeting these dual requirements was far from easy. In the final paragraph of the *Libro de longitudes*, he explains the cosmographical part of the project:

> I think I will describe in maps something about all the parts of the world, putting in each province the cities, places, rivers, mountains and other notable things. I will do the same for the West Indies, now rediscovered, and where I have been. So that those who come after me, using what little I have done, can make more precise and larger geographies. About all of these I intend, likewise, to write on at length. [I will write] about the succession of kings and nobles that have lived in each province, as well as the customs and the relationship between different peoples. All of this will come to light, should God give me a long enough life.[34]

Council of Indies, and ended up in the Jesuit college of San Isidro. Other marginalia suggest that Antonio de Herrera, Velasco's successor as chronicler of the Council, also consulted the material. Carmen Millán de Benavides, *Epítome de la conquista del Nuevo Reino de Granada: La cosmografía Española de siglo XVI y el conocimiento por cuestionario* (Bogota: CEJA, 2001), 52–53.

33. Mariano Esteban Piñeiro, María I. Vicente Maroto, and Félix Gómez Crespo, "La recuperación del gran tratado científico de Alonso de Santa Cruz: El astronómico real," *Asclepio* 44 (1992).

34. "[D]e todas las cuales partes del mundo pienso describir algo en tablas poniendo en cada una de las provincias, ciudades, lugares, ríos, montes y otras cosas notables que hubieren, y lo mismo haré de las Indias Occidentales, ahora nuevamente descubiertas, en mucha parte de las cuales yo tengo estado, porque con la mediana noticia que yo de ellas pudiere dejar puedan, los que después de mi vinieren, hacer su geografía mucho major y con más precisión, de todo lo cual es preciso así mismo escribir algo, así de la sucesión de los reyes y señores que en cada provincia de ellas ha habido, como de las costumbres y contrataciones de las gentes las unas con las otras, todo lo cual saldrá puesto a la luz, dándome Dios vida para ello." *El libro de las longitudes*, in Santa Cruz, *Obra cosmográfica*, 1:273.

Other parts of the *General Geography and History* would describe the continents or *tierra firme*, where Santa Cruz intended to include a "map of all [the continents] with the general and particular history of each province."[35]

Santa Cruz's *Islario general* was a subset of this larger project.[36] The *Islario*, as Santa Cruz explains, was a geography of only "all the islands that are known and have been discovered to date," described "using painted figures and in writing, noting their locations and rutters to reach them, and if available, their histories and antiquities."[37] Santa Cruz borrows for the *Islario* the narrative structure of the classical itinerary, organizing the text as Pomponius Mela had in *De situ orbis*, with continental coastlines determining the text's narrative sequence.[38] To orient the reader to each island's location relative to adjacent continental coastlines, Santa Cruz includes eight regional maps showing "the ocean's shoreline of all the world's geography."[39] Each island, in turn, is depicted in 102 specific maps employing a square cylindrical projection.[40] The coastlines show the distinctive exaggerations of portolan charts: coves and promontories are emphasized to help identify land visually from onboard a ship or, as Santa Cruz intended them in the *Islario*, to orient the reader and text relative to a coastline.

Originally dedicated to Emperor Charles V, the four-part *Islario* repeatedly addresses the emperor, which suggests that the work was originally

35. "Y para la tierra firme la traza de toda ella, con la historia general y particular de cada provincia." In the dedication to the *Islario general*, in Santa Cruz, *Obra cosmográfica*, 1:282.

36. "[E]ste nuestro libro, que por ser su material de islas pusimos nombre islario, la cual obra quisimos tratar aparte de nuestra general Geografía e Historia." *Islario general*, in Santa Cruz, *Obra cosmográfica*, 1:330.

37. "[D]emostrándole por figuras pintadas y escritas todas las islas que hasta hoy son conocidas y descubiertas, con las distancias y derrotas por do se han de caminar para ellas y las historias que de cada una de ellas se pudiesen hallar, con sus antigüedades." Santa Cruz, *Obra cosmográfica*, 1:282.

38. Naudé lists only three known cosmographical works dedicated exclusively to islands written before Santa Cruz's work, including the *Isolario di Benedetto Bordone* (Venice, 1528). Françoise Naudé, *Reconnaissance du Nouveau Monde et cosmographie a la Renaissance* (Kassel, Germany: Edition Reichenberger, 1992), 43–44.

39. "[L]a costa del mar de toda la geografía del mundo." *Islario general*, in Santa Cruz, *Obra cosmográfica*, 1:312.

40. In this projection, all meridians and parallels are depicted as straight lines, with the parallels and meridians equally spaced and perpendicular to each other, and with each degree, *regardless of latitude*, equaling approximately 17½ leagues. When used to depict large expanses, this projection causes severe distortion when moving either north or south from the equator, but it yields reasonable results for smaller areas.

intended for his personal use.[41] The first two parts describe the islands along the European Atlantic coast and the Mediterranean, while the final two parts discuss the islands off the coasts of Africa, Asia, and America. The difference between the first two parts and the last two could not be more striking. The first half of the *Islario* conforms to the established thematic and structural demands of humanist cosmography and as such is a classic product of an armchair cosmographer at work. The second part of the *Islario*, however, is written using a new set of imperatives, with Santa Cruz abandoning the epistemological methods of classical cosmography to write a descriptive geography of the islands of the New World using recent discovery accounts and personal experience.

In the *Islario*'s first two parts, Santa Cruz relies heavily on classical sources, comparing what the various ones had to say about each island but rarely challenging classical accounts with contemporary information.[42] When discussing islands known since antiquity, he prefers to dwell on historical events and classical legends, content to leave these islands enveloped in the mist of legends and folklore. History, and not geography, takes precedence. The significance of islands known since antiquity is located in their classical context rather than their present reality. Santa Cruz relegates topographical descriptions to brief notes, preferring to dedicate large parts of his narrative to resolving toponymic inconsistencies among his sources. Yet when he describes places he surely knew firsthand, like the island of Cádiz and the islands on the river Guadalquivir, Santa Cruz escapes the literary constraints of geographical exegesis and lets his personal experiences dictate the text. For the Santa Cruz of the first half of the *Islario*, erudition was the mark of cosmographical virtuosity.

When in the *Islario*'s last two parts Santa Cruz's attention turns to the description of the islands off the coasts of Africa, Asia, and America, he is happy to challenge what—if anything—ancient authors said about

41. Although surviving versions of the *Islario general* were prepared as presentation copies dedicated to Philip II, historians believe that Santa Cruz wrote the bulk of the work in the late 1530s, completing most of it by 1541 when he was living in the court of Charles V. Naudé maintains these changes were the result of Santa Cruz's dedicating the work to a different monarch, with "retouches successives et très discrètes" before a new dedication to Philip II was added before 1559. Naudé, *Reconnaissance*, 21–29, 35.

42. He lists forty-six sources, of which only seven were roughly contemporary: Jacob Ziegler, Olaus Magnus, Peter Martyr, Gonzalo Fernández de Oviedo, Américo Vespucci, Oronce Finé, and Johanes Stöffler. *Islario general*, in Santa Cruz, *Obra cosmográfica*, 1:311.

them.[43] In this portion of the text, he points out errors in Pliny's and Ptolemy's descriptions of islands in the Indian Ocean, as in the case of the island of Hergana, remarking, "Nowadays, there is no such island because either they located it incorrectly or the sea has consumed it as it has many others in the Persian sea."[44] Santa Cruz does not generally name the sources of the modern information, although based on the discovery accounts he retells it is likely he used Portuguese sources for his description of the islands in the southeastern Atlantic, Indian Ocean, and South East Asian waters. As he moves east along the coast of India, classical references disappear, replaced by accounts of the recent Portuguese discoveries in the region. As Santa Cruz heads west from the Canary Islands, Ptolemy and Pliny also fade into the horizon. Peter Martyr, Gonzalo Fernández de Oviedo, and a number of pilots and explorers become his sources for descriptions of the Caribbean islands and those off the eastern coast of the American continent and the Strait of Magellan. Some sections, like the description written in the first person of the islands on the La Plata River—including a curious encounter with some strange apelike creatures—are clearly personal observations from his time with Cabot's expedition.[45]

In the third and fourth parts of the *Islario* Santa Cruz organizes the description of recently discovered islands to address a wholly different set of questions. After briefly describing the island's position relative to the nearest mainland, he begins the narration of each island's history with the moment it is "found" by Western explorers. With the island's discovery and early years of colonization, he grants the island a historical existence. He then addresses the question of what is found there. He structures his answer first as a general geography of the entire island and then as a chorography of principal settlements. As part of the island's general description, he discusses significant geographical features, includes ethnographic accounts of the island's native population and their customs followed by a survey of the island's natural resources. For the *descripción particular* or chorography, he selects a number of principal settlements, locating each within the island's general geography and then noting their size, population, and significant features. Finally, he discusses the island's overall size and places the island within the Ptolemaic grid by giving its latitude, length of the longest day, and *klima* (a Ptolemaic designation corresponding to regions with

43. "[M]uchas de estas [islas] hay poco que examinar la concordia o discordia de los autores antiguos sobre ellas, pues los que de ellos han venido a nuestras manos, que son griegos y latinos, no tuvieron noticia salvo de pocas." Ibid., 2:195.

44. Ibid., 1:222.

45. Ibid., 2:361.

similar climates across latitudes). He never mentions longitude coordinates in the text, although his maps include a longitude scale.

To resolve discrepancies among his modern sources Santa Cruz resorts to the same methodology advocated by Ptolemy to sort out conflicting accounts; they are first compared and then implausible ones are dismissed. This time, however, a nameless arbiter enters the picture. To resolve conflicting geographical information about an island's location or size, he privileges "the most common opinion of those who sail there" over the written accounts of Oviedo and Peter Martyr.[46] His maps of Cuba and the Hispañola are both drawn according to what he considers commonly held notions among men familiar with the *carrera de Indias*. For more remote places, such as the Strait of Magellan, he relies on explorers' accounts: Ferdinand Magellan (1520), García Jofre de Loaysa (1526), and the armada led by the bishop of Plasencia (1540). A close look at Santa Cruz's cartography of the region shows how faithfully his maps followed the textual descriptions of the land he had collected from these explorers.

Take for example Plasencia's account of the exploration of the southern coast of the Strait of Magellan. Santa Cruz writes that after Plasencia's armada passed by a cape with many snowy peaks on the southern side of the strait, they found a great bay, with three other smaller bays lying in the shape of a cross within it. Toward the southern end of the large bay they observed many fires and thus named the land Tierra del Fuego. As in the case of his other island maps, Santa Cruz's map of the Strait of Magellan is a schematic tracing of the narrative account[47] (plate 4).

Santa Cruz faced a number of epistemological problems when selecting his sources. His most pressing concern was privileging his sources and organizing them based on a criterion that would establish their relative credibility. In the *Islario*, he approaches this problem by applying "all his strength and ingenuity" to searching for and inquiring into all relevant material. Eyewitnesses should preferably write these accounts, he states, adding apologetically that it is an inherently impossible task, given that not even the "famous ancient geographers" were able to collect all their

46. "[L]a opinion más común y recibida de todos los que navegan y pasean." Santa Cruz does this in the case of the island of Hispañola, dismissing Peter Martyr's and Oviedo's account of the island's size and preferring the opinion of a pilot, Andrés de Morales. He sides again with a pilot for Cuba's size. Ibid., 2:309, 20.

47. He marks the entrance to the Straits of Magellan in his map at longitude 45° (calculated from the Island of San Antón in the Cape Verde Islands), a surprisingly good estimate. (The entrance to the straits is 43° from San Antón.) His map indicates the entrance at almost the correct latitude of 52½°. Ibid., 2:368.

information from eyewitness sources. To supplement these accounts, he also interviewed and questioned "learned and expert persons" familiar with those lands. Lastly, he used what ancient and modern geographers had written "in general and in particular" about the islands.

He "diligently examined and consulted" all this material in order to write the "literal description" and to draw the "demonstration and painting" of all the islands now known to the people living in "our old and inhabited orb."[48] The author's nation of origin was another important criterion when Santa Cruz selected and privileged his sources. When it came to the geography of the New World, Santa Cruz considered only Spanish or Portuguese accounts as credible sources. All others, he assured the emperor, were simply writing about an imagined geography: "This is what has been discovered of this land and islands to this day by trustworthy authors ... what others write ... outside of Spain lacks merit and is done with temerity, because they assert things that can only be known through the experience and practice of Your Majesty's vassals and those of the serene king of Portugal and not through dreams or vain presumption, which should not be brought into writing Geography."[49]

From what has survived to this day of Alonso de Santa Cruz's cosmographical works, it is clear that he remained intent upon complying with the cosmographical methodology Ptolemy outlined in the *Geography*: examining sources carefully and systematically, presenting the material in the form of a narrative account describing historical, ethnographic, and natural features of the land, and using mathematical cartography to reconcile geographical coordinates.

48. "[E]sforzándonos en cumplir lo mandado de V. M. a todo lo que humanamente se ha podido hacer con nuestras fuerzas e ingenio, indagando y buscando con solicitud toda aquellas cosas que a este propósito hacen en parte vista y experiencia (que en él todo fuera imposible, pues ninguno de los famosos geógrafos antiguos tal hizo ni pudo) y parte solícita inquisición de personas sabias y expertas en mucho de ello y lo tercero la lección de escritores geógrafos antiguos y modernos, así generales como particulares, con diligencia examinados y conferidos y así exprimiendo junto con la descripción literal la demostración y pintura de todas las que hasta hoy se saben y tratan por los de este nuestro orbe antiguo y habitado." Ibid., 2:370.

49. "Esto es lo que de esta tierra e isla hasta el día de hoy está descubierto por autores ciertos y dignos de fé que son los que arriba habemos dicho, todo lo que demás ponen Oroncio y otros fuera de España carece de fe y es hecho no sin temeridad, pues aseveran lo que si algo se hubiese de saber había de ser por la experiencia y plática de los criados y vasallos de V. M. o del serenísimo rey de Portugal y no por sueños y presunciones vanas, que no se han de traer a consecuencia en tratación Geográfica." Ibid., 2:369.

The *Islario* is a work of transition, however; the attempt of a cosmographer to reconcile the established literary style, sources, and methodologies characteristic of Renaissance cosmography with the demands of practicing his science within a fluid cosmographical environment. Given this, it comes as no surprise that the *Islario general* is an imperfect text despite the more than twenty years Santa Cruz labored on the project. There are numerous inconsistencies between the different surviving copies of the *Islario*, as well as between geographical coordinates mentioned in the text and those shown in the regional and local maps, suggesting that Santa Cruz returned to the document many times to incorporate new material.[50] Practices that served a cosmographer in the late fifteenth century—a command of classical sources, textual comparisons, philological virtuosity, and Ptolemaic cartography—proved to be inadequate tools to formulate the cosmography of the New World. Santa Cruz had to rely on firsthand accounts of uneven quality, from the oral accounts of pilots to written accounts of men of letters like Fernández de Oviedo. Multiple sources describing the same territory were few, so he could not rely on textual comparisons to identify inaccuracies. Greek, Hebrew, and Latin were of little use for deciphering place-names; instead Santa Cruz would have needed to master Nahuatl, Tupi, and Arawak. And finally, Ptolemaic cartography, for all its mathematical elegance, demanded geodesic measurements of a precision that instruments of the time were unable to provide, particularly for longitudinal measurements.

Santa Cruz was also aware that his work had to conform to the literary exigencies of Renaissance cosmography and be, or claim to be, current, comprehensive, and complete. With this in mind, he could present the *Islario* only as a small component of a much larger project: his magnum opus, the *General Geography and History*. Despite having collected an impressive cartographic collection, having access to accounts of the latest discoveries, and counting on the monarch's support, he never considered the project completed: living in the heyday of discovery and exploration, Santa Cruz was working with geographical information that was constantly changing and growing in scope. Within this fluid medium, any attempt to present a static picture of the world was futile. Perhaps Santa Cruz realized this inherent conflict, and to his credit, he did not presume he could ever complete such a monumental task.

The genre associated with Renaissance cosmography proved unable to provide a mode of representation flexible enough for the demands placed on it by Santa Cruz's institutional patrons and Santa Cruz himself. After

50. Naudé, *Reconnaissance*, 100–102.

his death, the Council of Indies would take a quite different approach to cosmography, in large part motivated by the crises that had overtaken the genre. Rather than continue to view cosmography as a humanist's individual pursuit, it instituted a series of reforms that would turn cosmographical practice into a bureaucratic enterprise. Yet the bureaucratic approach the Council of Indies took was not the only path for cosmographical practice in the later half of the sixteenth century. At the royal court, cosmography, led by the dynamic Juan de Herrera, took on a decidedly experimental and mathematical style. For Herrera the way to settle cosmographical questions was to place the collection of necessary data in the hand of well-trained and capable individuals.

Experts to Explain the World: Juan de Herrera and the Expert Explorers

After Alonso de Santa Cruz's death in 1567 and for the duration of the reign of Philip II, cosmographical activity at court was under the direction of the architect of San Lorenzo de El Escorial, Juan de Herrera (1530–97).[51] Herrera acted as the king's personal adviser on cosmographical matters, as marginal notes on countless memoranda in the king's hand attest. The son of an impoverished *hidalgo*, Herrera had known that his future, like that of many proud nobles of reduced means, lay in the career of arms. In 1548, Herrera served in the royal guards during the future Philip II's tour of the domains he would inherit in Italy and Flanders. For the intelligent and observant Herrera, this was both an educational grand tour and a chance to gain entry into the future monarch's service.[52] By 1563, he was working as an assistant and draftsman to Philip's architect, Juan Bautista de Toledo.[53] After the architect's death, Herrera became the de facto architect of the many royal projects then under way, most impressive among them the monastery/palace of El Escorial.[54] His position near the monarch brought him in contact

51. M. I. Vicente Maroto and Mariano Esteban Piñeiro, *Aspectos de la ciencia aplicada en la España del Siglo de Oro*, 2nd ed. (Valladolid: Junta de Castilla y León, Sever-Cuesta, 2006), 8, and Catherine Wilkinson-Zerner, *Juan de Herrera: Architect to Philip II of Spain* (New Haven, Conn.: Yale University Press, 1993), 14.

52. Wilkinson-Zerner, *Juan de Herrera*, 3.

53. Ibid., 9, and Nicolás García Tapia and M. I. Vicente Maroto, "Juan de Herrera, un científico en la corte española," in *Instrumentos científicos del siglo XV: La corte española y la Escuela de Lovaina* (Madrid: Fundación Carlos de Amberes, 1997), 42, 45.

54. An appreciative Phillip II granted Herrera honors and financial rewards throughout his life. Herrera enjoyed a yearly income from grants and salaries of 1,050

with other humanists at court, among them Benito Arias Montano, with whom Herrera collaborated to assemble a world-class library at El Escorial. His rising fortunes also allowed him to put together an impressive personal collection of scientific and technical books and instruments.[55]

Herrera's cosmographical interests and mode of practice were different from Santa Cruz's. Whereas Santa Cruz focused on fashioning a humanistic cosmographical opus, Herrera seems to have solely favored the mathematical aspects of the discipline, focusing particularly on instrument design. Like Santa Cruz and most cosmographers of his era, however, he was also fascinated with the problem of measuring terrestrial longitude. The earliest evidence of Herrera's involvement in cosmographical matters dates from 1573, when he requested and was granted a royal license (an early form of patent) to sell some navigational instruments that he had invented and claimed were useful for determining latitude and longitude.[56] Although we do not know the precise nature of the instruments, Herrera's license suggests that one of the instruments measured the effect of magnetic declination on the compass and the relationship of declination to longitude. He also petitioned to have the instruments inspected and reviewed by the Council of Indies. Not only did he earn the Council's endorsement for the instruments' use on ships traveling to the Indies, but in addition, the Council, through its cosmographer-chronicler Juan López de Velasco, arranged for the instruments to undergo a trial at sea in the hands of Alonso Álvarez de Toledo, cosmographer of the armada, set to sail with the next fleet.[57] The trials sidestepped the Casa de la Contratación's long-standing monopoly on producing, reviewing, and certifying navigation instruments used in transoceanic travel. We do not know what reaction the cosmographers at the Casa had to Herrera's

---

ducats. He died a wealthy man in 1597. For an inventory and study of his possessions, see Luis Cervera Vera, *Inventario de los bienes de Juan de Herrera* (Valencia: Albatros, 1977).

55. Francisco J. Sánchez Cantón, *La librería de Juan de Herrera* (Madrid: CSIC, 1941).

56. AGI, IG-426. f. 275, "Privilegio a Juan de Herrera para instrumentos de navegación," December 13, 1573. Published in Luis Cervera Vera, "Instrumentos náuticos inventados por Juan de Herrera para determinar la longitud de un lugar," *Llull* 20, no. 38 (1997): 151–52.

57 . AGI, P-259, R. 58, "Conocimiento de Alonso Álvarez de Toledo, cosmógrafo de Su Majestad en la armada de los galeones, de los instrumentos de Juan de Herrera," January 8, 1574. Published in Cervera Vera, "Instrumentos náuticos," 152–53. The text of the memorandum does not mention that these were Herrera's instruments, but a note on the documents' reverse side mentions Herrera. Other historians have pointed out that this evidence is not definitive; see Vicente Maroto and Esteban Piñeiro, *Aspectos de la ciencia*, 387.

project, nor do we know the trial's outcome. Unfortunately, further historical records concerning these matters have not been located.

The years 1580 to 1582 would energize Herrera's interest in cosmography. When Philip II laid claim to the Portuguese crown in 1580, he asked his architect to accompany the royal court to Lisbon. Now with unfettered access to Portuguese maps, Herrera quickly realized that there were fundamental and alarming discrepancies between Spanish maps and these.[58] Reunification eased some of the political tensions that arose periodically about the line of demarcation,[59] but the cartographic material Herrera saw and that the Portuguese used to support claims that the Philippine islands fell on their side of the demarcation had to be quickly addressed and somehow reconciled with the Spanish view of the situation. On August 21, 1581, just two months after witnessing the triumphal entrance of Philip II to Lisbon, Juan de Herrera wrote to Juan López de Velasco, asking for Portuguese maps once in possession of another court cosmographer and sometime spy, Juan Bautista Gesio, to be sent forthwith to Lisbon. Gesio had taken the maps to Madrid in 1573 at the behest of Spain's Portuguese ambassador, Juan de Borja.[60] Herrera needed the maps to draw a new larger map, showing clearly the line of demarcation according to where the Spanish believed it to be. He intended to use this new map to amend modern Portuguese-made maps; "their modern maps are depraved," he said, adding with a wink, "because of what you know about the Portuguese."[61]

Herrera's cosmographical work while in Portugal went much further than simply suggesting the Portuguese amend their maps to conform to Spanish maps. He also coordinated an effort to establish in Lisbon controls similar to the ones used by the Casa de la Contratación in Seville to license pilots and inspect nautical instruments.[62] To this end a junta was held in 1583

58. Juan de Herrera, *Institución de la Academia Real Mathemática*, ed. José Simón Díaz y Luis Cervera Vera (Madrid: Instituto de Estudios Madrileños, 1995), 17–18.

59. Elliott, *Imperial Spain*, 271–75.

60. BL, Eg 2047, f. 325v, Antonio Gracían, "Copia de carta de Gracián al Presidente de Consejo de Indias," December 23, 1573, San Lorenzo. For discussion of Gesio's role as a spy for Spain, see Vicente Maroto and Esteban Piñeiro, *Aspectos de la ciencia*, 79–80.

61. "[P]orque la moderna [carta general] esta depravada, por lo que V. sabe de los Portugueses." Letter from Juan de Herrera to Juan López de Velasco, August 21, 1581. Published in Eugénio Llaguno y Amirola and Juan A. Ceán-Bermúdez, *Noticias de los arquitectos y arquitectura de España desde la restauración*, 4 vols. (Madrid: Imprenta Real, 1829; reprint, Madrid: Ediciones Turner, 1977), 2:358–59.

62. BN MS 5785, f. 179r–179v, José Luis Casado Soto, *Discursos de Bernardino de Escalante al Rey y sus ministros (1585–1605): Presentación, estudio y transcripción por José Luis Casado Soto* (Santander: Servicio de Publicaciones de la Universidad de Cantabria,

under the supervision of the viceroy of Portugal, Archduke Albert.[63] The junta resolved to send Portuguese pilots traveling sea routes to the Orient and Africa with two sets of maritime charts, one drawn based on Portuguese observations and one modeled after the master chart at the Casa de la Contratación, which reflected Spanish views. The pilots were instructed to make astronomical observations over a period of several years, at which point both sets of charts would be reconciled. This extraordinary project of cosmographical reconciliation was never carried out. After Herrera left Portugal in late 1583, the project apparently fell by the wayside and was not pursued, nor was the envisioned reform of Portuguese navigation charts ever completed.[64]

The sojourn in Portugal crystallized for Herrera the need for more individuals skilled in the mathematical arts who could be recruited to address not only cosmographical problems but technological issues as well.[65] He became the driving force behind the Royal Mathematics Academy in Madrid, an institution whose mission was to prepare young men to practice professions based on mathematics.[66] The academy's pedagogical program differed in fundamental ways from how mathematical-based disciplines were being taught at Spanish universities. Herrera's curriculum abandoned the traditional *quadrivium* and *trivium*. The course of study was designed instead to teach the skills and knowledge necessary for a particular set of

---

1995), 49, 109–10. See also M. I. Vicente Maroto, "El arte de navegar," in *Felipe II, la ciencia y la técnica*, ed. E. Martínez Ruiz (Madrid: Actas, 1999), 363–64.

63. Albert, a nephew of Philip II and brother of Rudolph II of Hungary, served as viceroy of Portugal from 1583 to 1593. He later married Philip's daughter Isabella and ruled the Low Countries from 1598 until his death in 1621. For more on Albert, see Werner Thomas and Luc Duerloo, eds., *Albert and Isabella, 1598–1621* (Louvain, Belgium: Brepols, 1998).

64. Bernardino de Escalante was responsible for preparing the charts and preparing the necessary instructions, but he returned to Spain with the king in 1583 and the project remained unfinished. Ricardo Cerezo Martínez, *La cartografía náutica española en los siglos XIV, XV y XVI* (Madrid: CSIC, 1994), 242.

65. Fernández de Navarrete noted Herrera's new focus in Martín Fernández de Navarrete, *Disertación sobre la historia de la náutica y de las ciencias matemáticas que han contribuido a sus progresos entre los españoles*, ed. Carlos Seco Serrano, Biblioteca de autores españoles 77 (Madrid: Ediciones Atlas, 1954–5), 77.

66. Historians Esteban and Vicente have investigated the origins and evolution of this institution in great detail and have published an indispensable collection of pertinent documentation. They maintain that the academy's principal objective was to train cosmographers. See Vicente Maroto and Esteban Piñeiro, *Aspectos de la ciencia*, 65–219.

professions—professions Herrera saw as vital to the Spanish monarchy. In 1584, to publicize the academy's plan of study, Herrera published its charter, the *Institución de la Academia Real Mathemática*.[67] The document informed prospective students—understood to be young hidalgos at court—of the academy's purpose and the kinds of professions that would be taught at the new institution. Plans for the academy began during Herrera's two-year stay in Portugal. Since all the courses were to be taught in Castilian, in 1581 Pedro Ambrosio de Onderíz, a student of classical languages, was sent to Lisbon to study mathematics and begin work on a series of translations of classic mathematical texts.[68] In late 1582, the king took Portuguese cosmographer Juan Bautista Labaña (or Lavanha) (1555–1624)[69] into his service, ordering him "to take care of and understand matters pertaining to cosmography, geography and topography."[70] He was also to "read mathematics in the manner and place where he is instructed to do so." On the same date, Onderíz was appointed as Labaña's assistant and formally instructed to translate the texts, principally Greek geometries, used at the academy.[71] Both men were under Herrera's supervision.

67. Herrera was granted a printing license on June 8, 1584. The short book is extremely rare and remained unstudied until 1995. Herrera, *Institución*, 24.

68. Vicente Maroto and Esteban Piñeiro, *Aspectos de la ciencia*, 74.

69. For biographical notes and a reprint of an experiment carried out by Labaña on the effects of magnetic declination on the compass needle, see Alfonso Ceballos-Escalera Gila, "Una navegación de Acapulco a Manila en 1611: El Cosmógrafo Mayor Juan Bautista Labaña, el inventor Luis de Fonseca Coutinho, y el problema de la desviación de la aguja," *Revista de la historia Naval* 17, no. 65 (1999).

70. APR, a. 1582, T. VI, f. 210, "Cédula de Felipe II desde Lisboa dirigida al pagador de las obras del Alcázar ordenando que pague, a partir del primero de enero de 1583, cuatrocientos ducados anuales a Juan Bautista Labaña por entender de cosmografía, geografía y topografía y leer matemáticas en la corte," December 25, 1582. Published in Vicente Maroto and Esteban Piñeiro, *Aspectos de la ciencia*, 116–17.

71. APR, a. 1582, T. VI, f. 211, "Cédula de Felipe II mandando al pagador del Alcázar que a partir de 1583 se le pague a Pedro Ambrosio de Ondériz 200 ducados por ayudar a Labaña a leer matemáticas en Palacio y traducir al castellano las obras necesarias," December 25, 1582. Published in Vicente Maroto and Esteban Piñeiro, *Aspectos de la ciencia*, 117–18. The books Onderíz translated feature prominently among subjects assigned at the Academia. By 1584 he had translated into Castilian books 11 and 12 of Euclid's *Elements*, as well as his *Optics*, Theodosius's *Spherics*, Archimedes's *On the Equilibrium of Planes*, and Apollonius's *Conics*. Only the *Optics* was printed (Euclid, *La perspectiva, y specularia de Euclides: Traduzidas en vulgar Castellano, y dirigidas a la* S.C.R.M. *del Rey don Phelippe nuestro Señor*, trans. Pedro Ambrosio Onderíz [Madrid: Imp. casa de la viuda de Alonso Gómez, 1585]). We can assume the others circulated as manuscript copies (Vicente Maroto and Esteban Piñeiro, *Aspectos de la ciencia*, 91). See also Peréz

Herrera conceived of mathematics as the foundation of all sciences. For him mathematics prepared the mind for learning not only about the material world but about the spiritual world as well. A lifelong follower of the Catalan philosopher Raymond Llull (ca. 1232–ca. 1315), Herrera adhered to the principal tenet of Llullist philosophy, that all knowledge stems from a single Art.[72] The Llullist Art, in addition to serving as the organizing principle for all forms of knowledge, functioned as a method of spiritual contemplation. In Herrera's *Institución*, mathematics become the equivalent of the Llullist Art, or as Herrera explains, "a discipline where all is one and manifests the true method and order of knowledge, preparing the understanding, so that lifted above all things material and sensible, it can elevate to contemplate the supernatural and intelligible."[73] In keeping with these principles, Herrera envisioned a curriculum that taught the fundamental elements of mathematics, arithmetic, and geometry as the basis for preparing students for a number of professions: arithmeticians, geometricians and surveyors, mechanics, astrologers, experts in gnomics, cosmographers and pilots, experts in perspective and optics, musicians, architects, painters, fortification builders, waterworks builders, and gunners. Interestingly, the reading list for the astrology course emphasized mathematics, instrumentation, and astronomy but did not include any interpretive texts such as Ptolemy's *Tetrabiblos*, a sign that Herrera's focus on the useful arts did not include prognostication.

In the *Institución* the curriculum to prepare cosmographers and pilots read as follows:

> Cosmography presupposes knowledge of the sphere and of the part of planetary theory that deals with eclipses, and as a subaltern of astrology, is aided by geometrical and mathematical principles. Anyone who chooses to make it his profession will be able to completely understand Ptolemy's *Geography*, which is where everything concerning this subject is accurately discussed. Also [the cosmographer] will know how to use terrestrial globes, and understand navigation charts and know how to make them, as well as the descriptions of the provinces, both general and particular.[74]

Pastor for a summary of the text: Cristóbal Pérez Pastor, *Bibliografía madrileña*, 3 vols. (Madrid: Tip. de los Huérfanos, 1891–1907), 1:111–12.

72. Cervera Vera has pointed out the importance of Herrera's Llullist philosophy in formulating the academy (introduction to Herrera, *Institución*, 50).

73. Herrera, *Institución*, 4v.

74. "La cosmografía presupone las noticia de la esfera, y de las teorías de planetas lo que trata de eclipses, y como subalterna a la astrología, ayudada de principios

Conspicuously absent from cosmographer's curriculum is the requirement to read Mela or Strabo, such an integral part of the university cosmographical education.

Students of the academy who trained in navigation, on the other hand, were expected to learn the theory of the spheres, understand navigation charts and how to locate landmasses on them, and become well versed in the use of the astrolabe and the Jacob's staff. The pilot should also learn how to make compasses and understand the effects of declination. To learn this art they were expected to read "some chapters of Ptolemy's *Geography* so they can learn how to make universal and local descriptions and navigation charts, and other things relevant to the navigation art."[75] Pilots, Herrera insisted, also needed to learn other things that "the prudent pilot only learns with experience [at sea], without which no one should use this title or be allowed to practice a trade so important for the Republic."[76]

The *Institución* records Herrera's interest in setting up a pedagogical institution for teaching future government officeholders the fundamentals of the technical arts. The aim was to produce men useful to the empire in the tradition of the "practical" (*práctico*) mathematicians and devoted to solving problems rather than to philosophical speculation. By locating the academy in the courtly context, Herrera managed to leverage the king's authority and enthusiasm for the endeavor. Moreover, by removing the instruction from the universities, Herrera had the flexibility to construct an educational project free from what he apparently saw as the universities' humanistic overreliance on classical authority and rigid statutes. He sought to educate the future problem solvers of the empire from among the hidalgo class that gravitated toward court in the hope of capitalizing on their privileged birth. Herrera's academy was in continuous operation

---

geométricos y aritméticos. Podrá quien de la hiciera profesión entender muy de raíz la Geografía de Ptolomeo que es donde cumplidamente se trata todo lo que pertenece a esta materia, saber los usos del globo terrestre, y entender las cartas de marear y saberlas hacer, y todas las descripciones de provincias así general como particular" (ibid., 13).

75. "Para lo cual se leerán algunos capítulos de la Geografía de Ptolomeo, el modo de hacer descripciones universales y particulares, y cartas de marear y algunas cosas tocantes a la arte navegatoria" (ibid., 13v–14).

76. "[Y] otras muchas cosas que con la experiencia el prudente Piloto alcanza, sin la cual no se le debe dar tal nombre ni dejar usar oficio tan importante a la República" (ibid., 13v).

from 1583 to 1604 and again, after a three-year hiatus, until 1625—the year its pedagogical mission was incorporated to the Jesuits' Colegio Imperial de San Isidro.[77]

Before his stay in Portugal, Herrera's cosmographical interests appear to have been narrowly defined by concerns about navigation and longitude. But his stay in Lisbon alerted him to the political and administrative relevance of cosmography. His close collaboration with the monarch and the trust the king had in his abilities allowed him to formulate a series of ambitious cosmographical projects designed to place Spanish cosmography on a factual basis—meaning one where facts were defined mathematically and empirically.

In late 1582, while still in Portugal, Herrera also set in motion a cosmographical project designed to determine conclusively the extent of the Spanish empire and to resolve once and for all its ambiguous boundaries. He chose for the project a young mathematician and astronomer, Jaime Juan, who was instructed to travel from Spain to Mexico and then on to the Philippines, all the while conducting astronomical observations. The purpose of these observations was to "fix" spatially key latitude and longitude coordinates.[78] Juan was to carry out these observations and measurements using instruments designed by Herrera and following procedures detailed in a comprehensive set of instructions.[79] He was to follow a trajectory that spanned the breadth of the Spanish overseas empire. He would depart

77. Mariano Esteban Piñeiro, "Los cosmógrafos del Rey," in *Madrid, ciencia y corte*, ed. Antonio Lafuente and Javier Moscoso (Madrid: CSIC, 1999), 128–33.

78. We know little of Juan's background. The date of his birth is unknown, but we know he was from Valencia and studied under Jerónimo Muñoz while the professor was still there (María Luisa Rodríguez-Sala, *El eclipse de Luna: Misión científica de Felipe II en Nueva España* [Huelva, Spain: Universidad de Huelva, 1998], 61). Schäfer suggests that Herrera introduced Jaime Juan to Philip II during their stay in Portugal but does not document his source for this information (Ernst Schäfer, "El cosmógrafo Jaime Juan," *Investigación y Progreso* 10 [1936]: 10). Jaime Juan's reputation as an astronomer comes from statements made by Herrera concerning a treatise written by "comendador" Jaime Juan Falcó y Segura concerning the mathematical problem of squaring the circle. The mathematical proof is at the AGS Mapas, Planos y Dibujos, VI-127, and AGS Guerra Antigua, Leg. 165, f. 250 and another in f. 249. Cited in García Tapia and Vicente Maroto, "Juan de Herrera," 43.

79. Although the project's directives bear López de Velasco's signature, Vicente and Esteban believe Herrera wrote the detailed instructions. AGS, GA-155, f. 150–151v, "Relación y apuntamiento enviado al Consejo de Indias sobre lo que debe hacer el cosmógrafo Jaime Juan," December 2, 1582. Published in Vicente Maroto and Esteban Piñeiro, *Aspectos de la ciencia*, 420–23.

from Seville en route to Mexico City, and along the way he was to make observations at sea and in all ports of debarkation. Once he was on the American mainland, the instructions specified that he was to note carefully his position on the east coast of Mexico, in Mexico City, and in the western Mexican port of Navidad. He was to continue these tasks during the Pacific crossing until he arrived at his final destination, Manila. The instructions make no allowances for exploring other territories or following a different route east. Nor was Juan encouraged to make observations other than those that yielded terrestrial coordinates.

Whenever he made landfall, Juan first had to observe the path of the sun and determine the true north-south line (*la meridiana del lugar*), marking the line on a large stone slab provided for this purpose. This observation was then to be compared with the orientation of a compass needle; the needle's deviation from true north, whether to the east or west (or the *nordestear* or *noroestear* of the needle), would be determined by measuring—using a special ruler provided for that purpose—its deviation from the north-south line. Juan was then to translate, "using the laws of sine and cosine," this linear measurement into degrees and minutes. These results were to be logged, noting the place where the observations were taken and the date. The next day, he was to place an instrument that "finds the meridian precisely" on the stone slab with the true north-south line marked on it and, at noon, use the instrument to measure the elevation of the sun. To determine latitude, the resulting measurement had to be manipulated by adding or subtracting solar declination following well-known rules known as the regiment of the sun.

The instruction even specified the wording Juan should employ to record the observations. In addition to making these observations, Juan was asked to take particular care to record the location of places where the compass needle appeared "fixed" (where it did not seem to deviate from true north or the *meridiano fijo de la aguja*). Once he had identified these "fixed" points, he had to calculate the distances between them, as well as the distance between these points and where other observations were made. Juan was also reminded to note the presence of nearby large mountains with magnetic stones or iron mines that might affect his compass readings.

Save for a brief description of the instruments Juan took on the voyage, we have little information on their design and construction. Two instruments in particular appear to have been of Herrera's invention. The first was to be used on land or sea to determine the time of day by using the elevation of the pole, and it could also be used to determine the "true meridian" or the deviation of the compass needle.[80] The reference to finding the "true meridian" suggests that the instrument's purpose was to determine longitude. The other one is described as a large, horizontal instrument that pilots and sailors could use at sea to determine how far away they had traveled from the location where the compass appeared fixed and consequently from their port of departure.[81] This last instrument was meant to calculate longitude at sea by using as reference coordinates the latitude and meridian where the compass needle appeared fixed. This type of instrument would be useful only if constructed using a mathematical model of the relationship between latitude, longitude, and compass deviation. Unfortunately, Herrera's instructions do not discuss the algorithm he used to design this instrument.

Herrera was well aware of the difficulty of getting pilots to use—and trust—astronomical navigation instruments. After all, cosmographers at the Casa had been vociferously complaining about this for decades. Juan was instructed to explain to the pilots and sailors onboard the benefits of using Herrera's instruments and to teach them to use them correctly. Juan was also to observe pilots while they worked, noting the instruments and methods they used, so that while at sea he could point out what they were doing wrong and "little by little lead them to understand the truth behind one thing or another." Herrera the pedagogue could not pass up the opportunity of inculcating in these rugged men of the sea some appreciation of the mathematical aspects of guiding a ship.

Herrera did not overlook the difficulties such a long voyage imposed on the traveler and made many provisions for Juan's comfort during the eight years the expedition was expected to last. Juan was granted an ample salary of four hundred ducats a year, to be paid throughout his trip rather than upon his return, as was customary.[82] He was to travel in the admiral's ship

80. "Un instrumento para poder por él saber en la mar y la tierra a cualquier hora del día la elevación del polo la hora que él [es?], hallar la meridiana o lo que norestea or noroestea la aguja." AGS, GA-155, f. 151v. Published in Vicente Maroto and Esteban Piñeiro, *Aspectos de la ciencia*, 422.

81. "Un instrumento grande horizontal con el cual los pilotos y marineros puedan en la mar, después de hallado por el instrumento antes de este lo que el aguja norestea o noroestea y la elevación del polo, lo que están apartados del meridiano fijo del aguja y, por el consiguiente, del meridiano del lugar donde partieron." AGS, GA-155, f. 151v. Published in Vicente Maroto and Esteban Piñeiro, *Aspectos de la ciencia*, 422.

82. AGI, Filipinas-339, L.1, f. 229–229v, "Orden de pago para Jaime Juan," May 5, 1583.

or flagship with two personal servants and a draftsman.[83] Letters were sent to the viceroys of the provinces he was to visit instructing them to provide for his comfort and treat him with the deference customary to a "criado del rey" (king's servant).[84] In addition, the viceroys had to provide him with an adequate number of porters to carry his instruments.

Although the project had the monarch's backing, it was submitted for evaluation and review to the Council of Indies, as was characteristic of Philip's administrative style.[85] The proposal sat at the Council for over a year. As part of the review process, Juan López de Velasco issued an opinion endorsing the project's objectives and added several other observations to Juan's already long list: Juan should record high and low tides along coasts and ports and notify the Council of the eclipse observations he intended to make, "so that they can be observed in other places as well, because if they cannot be compared with other observations little will come of his effort."[86] As to the instruments, López de Velasco appears perfectly satisfied with them, adding that if Juan de Herrera made them he did not need to inspect them. It appears, however, that he was not personally acquainted with Jaime Juan and recommended that he come to Madrid to be "examined." Upon submitting its final report to the king, the Council of Indies echoed Velasco's suggestion that Juan come to Madrid to be interviewed. They also believed that, while in Mexico, Juan should meet with Francisco Domínguez de Ocampo, a royal cosmographer sent to map Mexico as part of the Francisco Hernández expedition, and with Alonso Álvarez de Toledo, who was attached to an armada patrolling the Mexican coast.[87] Once he

83. AGI, Filipinas-339, L.1, f. 229, "Orden de acomodar a Jaime Juan en la flota de Nueva España," May 5, 1583; Aranjuez and AGI, Filipinas-339, L.1, f. 228v. "Licencia de pasajero a Jaime Juan, cosmógrafo," May 5, 1583; quoted in Schäfer, "El cosmógrafo Jaime Juan," 11.

84. AGI, Filipinas-339, L.1, f. 227v, "Orden al virrey de México de ayudar a Jaime Juan, cosmógrafo," May 5, 1583, and AGI, Filipinas-339, L.1, f. 228r, "Orden a la Audiencia de Manila de ayudar a Jaime Juan," May 5, 1583.

85. AGS, GA-133, f. 230, "Carta de Antonio de Eraso al Licenciado Gasca sobre la misión del cosmógrafo Jaime Juan en las Indias," December 2, 1582; quoted in Vicente Maroto and Esteban Piñeiro, *Aspectos de la ciencia*, 420.

86. AGI, IG-740, N. 103 (2), "Informe del Cosmógrafo-Cronista Mayor de Indias Juan López de Velasco con instrucciónes sobre lo que debería hacer Jaime Juan," February 2, 1583.

87. AGS, GA-155, f. 149–150. "Consulta del Consejo de Indias a Felipe II sobre la instrucción que se debe dar a Jaime Juan sobre la descripción de Nueva España y Filipinas," February 5, 1583; quoted in Vicente Maroto and Esteban Piñeiro, *Aspectos de la ciencia*, 424–25.

was in the Philippines, the Council said, Juan should collect the papers left after the death of missionary Fray Martín de Rada, an Augustinian who had traveled extensively through China and was known for his astronomical skills.[88] Upon receiving the Council's additional recommendations, the king had had enough of the Council's delays, especially given that it had called into question Jaime Juan's capabilities. In a short, blunt marginal note the king instructed the Council to incorporate Velasco's observations, settle all matters pending concerning Jaime Juan's salary, and do so quickly, so Juan could meet the fleet preparing to leave in the coming weeks.[89] The king's reply apparently did not speed matters up. On April 29 Herrera wrote to the king's secretary Antonio de Eraso asking him to "light a fire" under the matter (*poner calor a ello*) or "these gentlemen will never be cured of the longitudes and latitudes."[90] This time the Council responded quickly, and all the necessary paperwork was ready a week later.

Jaime Juan finally set off on his expedition in the late summer of 1583, but not before spending time in jail due to monetary disputes with officials at the Casa de la Contratación.[91] Luckily, the fleet's departure that year

88. Fray Martín de Rada (1533–78) accompanied Fray Andrés de Urdaneta in the 1565 Legazpi expedition to the Philippines. Urdaneta provided him with a large instrument "used to determine longitude" so that Rada could make astronomical observations. Rada computed his measurements using both the Alphonsine and Copernican tables (BAH, Colección Muñoz vol. 33, f. 137v–318, "Parecer sobre la demarcación"). In 1574, the viceroy of Mexico relayed to the king a "painting," likely a map, of the Islas del Poniente (Philippines) and some papers discussing latitude and longitude of these islands written by Martín de Rada (AGI, Mexico 19, N. 128, f. 5, "Carta del Virrey Conde de la Coruña, Martín Enriquez . . . informa sobre mapa de las islas del poniente y papeles de Martin de Rada," March 24, 1574). Subsequently, there was an attempt to collect Rada's papers, which included "many papers about observations and the natural history of the islands [Philippines]." Apparently, Rada had done the research in compliance with a royal order drafted years earlier by Juan Bautista Gesio (AGI, IG-740, N. 103, [4], "Orden al governador de Filipinas que recoja los papel de fray Martín de Rada y envien al Consejo de Indias," April 24, 1580).

89. AGS, GA-155, f. 155, "Contestación de Felipe II a la consulta del Consejo de Indias sobre la instrucción de Jaime Juan," February 24, 1583. The king agreed with the suggestions made by the Council concerning the eclipse observations and about collecting Rada's papers but did not address the matter of Juan meeting with Alonso Álvarez. Quoited in Vicente Maroto and Esteban Piñeiro, *Aspectos de la ciencia*, 425.

90. AGS, GA-144, f. 111, Juan de Herrera, "Carta de Juan de Herrera a Antonio de Eraso sobre viaje de Jaime Juan," April 29, 1583, Aranjuez.

91. AGI, IG-1956, L. 3, f. 200–200v, "Real Cédula al presidente y oficiales de la Casa de la Contratación para que suelten de la prisión a Jaime Juan," August 2, 1583, Madrid.

was delayed until late October, when he presumably joined the ships. During a layover in Havana he surveyed and drew a map of the city's vital port.[92] He arrived in Mexico City in time to observe and record the lunar eclipse of November 1584, but his trip was far from over. In early 1585, the archbishop viceroy of Mexico, Pedro Moya de Contreras, noted that he had observed the lunar eclipse of 1584 along with Jaime Juan and others, including the cosmographer of the Hernández expedition, Francisco Dominguez. He also noted that Juan was constructing a number of new instruments and was preparing to leave for the Philippines with the fleet led by Francisco Gali, a voyage that also was to involve exploring the California coast north of Mendocino. Fearing that Juan's observations of the effect of magnetic declination on the compass might be lost during the perilous voyage, the archbishop instructed him to make a clean copy, which he would send with the next fleet returning to Spain. The original letter has a marginal note indicating that upon arrival the documents should be given to Juan López de Velasco.[93]

Jaime Juan sailed on to the Philippines in 1585 as planned but died of fever before beginning observations in those islands.[94] Save for a copy of the eclipse observations made in Mexico City in 1584, which Juan sent to the Council of Indies, all his papers, including those concerning magnetic declination, remain unaccounted for.[95] According to a government official

92. AGI, Mexico 336B, R. 4, N. 179, Pedro de Moya y Contreras, "Carta del virrey Arzobispo Pedro de Moya y Contreras al rey," May 8, 1585, Mexico.

93. "A Jaime Juan, cosmógrafo, he dicho que se apreste para pasar a las Islas Filipinas en el navio que saldrá por marzo a cumplir el orden y mandato que trae de vuestra majestad para lo cual está haciendo ciertos instrumentos que dice ser necesarios, y por el riesgo que podrían correr sus trabajos de las observaciones del nordestear y noroesterar de las islas y tierras firme que ha andado le he dicho que las saque en limpio porque las tiene en borrador, para que yo las envíe a vuestra majestad en la flota y así lo hará, de las cuales me dió una memoria que va con ésta." AGI, Mexico 336B, R. 4, N. 176, Pedro de Moya y Contreras, "Carta del virrey Arzobispo Pedro de Moya y Contreras al rey . . . trata de tablas de Francisco Domínguez y eclipse," January 22, 1585, Mexico.

94. Santiago Vera, president of the *audiencia* of Manila, reported to the king that Jaime Juan died "of fevers" (*calenturas*) just as he had begun to perform his duties in the Philippines. In his memorandum, Vera took care to also report the date and time of the lunar eclipse of March 23, 1587, the one Juan would have recorded. AGI, Filipinas-18A, R. 5, N. 31, f. 1–8, Santiago de Vera, "Carta de Santiago de Vera sobre la situación en Filipinas . . . anuncia muerte de Jaime Juan," June 26, 1587.

95. The May 8, 1585, letter by Moya to the king also mentioned that, along with Juan's declination observations, he was sending a number of descriptions of China and Francisco Dominguez's description and map of "Nuevo Mexico and Cibola"; all are now lost.

in the Philippines, Juan stipulated in a codicil to his will that his books and belongings were to be sold but the manuscripts and instruments should be given to government officials for safekeeping until the king determined what should become of them.[96] There are some indications that Jaime Juan's papers, along with information left in the Philippines by Fray Martín de Rada, were used by the *procurador* of the Philippines, Fernando de los Ríos Coronel, to prepare his 1597 proposal for a project to discover the Northwest Passage through the elusive Strait of Anian.[97]

Was Juan's expedition another attempt by Herrera to test his longitude-finding instruments? Perhaps. Surviving documents do not articulate the project's ultimate goal, nor do they state the reasons used to justify the expedition. The particular emphasis Herrera placed upon gathering precise geodesic measurements suggests that the purpose of the project was to settle once and for all the questions surrounding the empire's boundaries, or as Herrera expressed, no doubt tongue in cheek, "to cure the Council of its latitudes and longitudes." Juan's travel itinerary betrays this intention. His course would traverse the empire almost linearly from east to west, or as close to linearly as navigation and logistics permitted. Likewise, he was given a single mission: to collect latitudinal and longitudinal observations along the way. Herrera was convinced that accurate geographical coordinates could be established definitively if only the right instruments (his own, of course) were placed in the hands of a capable man and the difficulties of the voyage minimized. Architecture historian Catherine Wilkinson-Zerner has described Herrera's architectural style as "pragmatic utopianism," where the utopian element was defined by order and symmetry.[98] The expedition Herrera prepared for Jaime Juan was as pragmatically utopian as his architecture. A clearly defined expedition, following a precise itinerary and putting perfect instruments in the hands of the right man, would, in Herrera's pragmatic mind, yield definitive geodesic results, once and for all. I cannot characterize the tasks asked of Jaime Juan as part of Herrera's project as anything other than cosmographical

96. AGI, Filipinas-18A, R. 6, N. 36, Gaspar de Ayala, "Carta del Licenciado Gaspar de Ayala, fiscal de la Audiencia de Manila . . . muerte y papeles de Jaime Juan," June 20, 1588.

97. AGI, Filipinas-18B, R. 7, N. 68, Fernando de los Ríos Coronel, "Carta de Fernando de los Ríos Coronel dando cuenta del astrolabio que había inventado," June 27, 1597, Manila. Ríos Coronel earned a royal endorsement for this plan in 1613. See AGI, Filipinas-329, L. 2, f. 170–171v, "Propuesta de nueva ruta de Filipinas a Nueva España," June 20, 1613.

98. Wilkinson-Zerner, *Juan de Herrera*, 164–65.

experiments.[99] Juan de Herrera's style of cosmographical practice centered on experiments, mathematics, and instrumentation illustrates the beginning of the divergence of two intellectual currents in Renaissance cosmography, one descriptive and the other mathematical.

The preference for firsthand investigation was by no means exclusive to Herrera. In the late 1560s an ambitious project to compile a natural history of Mexico and Peru was placed in the hands of Francisco Hernández (ca. 1515–87). Little is known of the genesis of the project, but historian Jesús Bustamante associates it with a series of reforms that took place at the Council of Indies while it was under the direction of Juan de Ovando. Bustamante finds it likely that Herrera was involved in articulating the project's goals.[100] But whatever its origins, historians have come to frame the project as an early example of modern natural history, principally because Hernández's work begins the slow erosion of the reliance on Pliny's *Natural History* for stylistic guidance or taxonomy, replaced by a systematic and novel attempt to classify the natural history of the New World.[101]

99. I deliberately characterize Herrera's activities as "experiments" based largely on Dear's definition of *experiment* as an activity "involving a specific question about nature which the experimental outcome is designed to answer." For more on the not unproblematic differentiation between experience and experiment in early modern scientific practice, see Dear, *Discipline and Experience*, 11–31.

100. Jesús Bustamante García, "De la naturaleza y los naturales americanos en el siglo XVI: Algunas cuestiones críticas sobre la obra de Francisco Hernández," *Revista de Indias* 52 (1992): 306.

101. Dr. Francisco Hernández earned a prominent place among sixteenth-century natural historians. For some examples of recent scholarship, see Raquel Álvarez Peláez, "Etnografía e historia natural en los cuestionarios oficiales del siglo XVI," *Asclepio* 41, no. 2 (1989); Raquel Álvarez Peláez, "La historia natural en los siglos XVI y XVII," paper presented at the Jornadas sobre España y las Expediciones Científicas en América y Filipinas, Ateneo Científico, Literario y Artístico de Madrid, 1991; Jesús Bustamante García, "Francisco Hernández, Plinio del Nuevo Mundo: Tradición clásica, teoría nominal y sistema terminológico indígena en una obra renacentista," in *Entre dos mundos: Fronteras culturales y agentes mediadores*, ed. B. Ares Queija and S. Gruzinski Seville: Escuela de Estudios Hispano-Americanos, 1997); José María López Piñero and José Pardo Tomás, *La influencia de Francisco Hernández (1515–1587) en la constitución de la botánica y la materia médica modernas* (Valencia: Instituto de Estudios Documentales e Históricos sobre la Ciencia Universitat de València, CSIC, 1996); Pardo Tomás and López Terrada, *Primeras noticias*; Simon Varey, Rafael Chabrán, and Dora B. Weiner, eds., *Searching for the Secrets of Nature: The Life and Works of Dr. Francisco Hernández* (Stanford, Calif.: Stanford University Press, 2000); and Francisco Hernández, *The Mexican Treasury: The Writings of Dr. Francisco Hernández*, ed. Simon Varey (Stanford, Calif.: Stanford University Press, 2000).

In January 1570, Philip II instructed Hernández to travel to Mexico and Peru to study the natural history of the Americas. A cosmographer, Francisco Domínguez de Ocampo, whose job was to compile longitude and latitude observations for the provinces and principal towns of New Spain, accompanied him.[102] Hernández would study the plants of Mexico and then move on to Peru to do the same. To this end, the royal order specified that he was to "consult, wherever you go, all the doctors, medicine men, herbalists, Indians, and other persons with knowledge in these matters . . . about herbs, trees, and medicinal plants."[103] The instruction also placed considerable emphasis on the fact that Hernández's mission was to seek out medicinal plants, and not just to identify these by hearsay but to conduct trials to determine their usefulness—it was to be a field research expedition. For the Council of Indies, speaking on behalf of the monarch, the expedition's ultimate goal was utilitarian: to find among the seemingly infinite variety of plants in the New World those that might be of medical use.

Hernández, however, left for Mexico with vastly different expectations. He conceived of the work as a natural history and not simply a compilation of medical material—he was to serve his monarch as Aristotle had served Alexander the Great.[104] After seven years abroad, Hernández accompanied his manuscripts and botanical paintings back to Spain and urged the king not to delay publication of the work. Concerned that others might not understand how he had organized the material, Hernández wanted to edit the final work himself and to see it published in Latin, Spanish, and the native Nahuatl. Indeed, the enterprise proved too big for Hernández's original Eurocentric conceptual framework. While in Mexico he had to create a new taxonomy based largely on the quasi-binary Nahua plant names, discarding to a great extent Dioscorides' classificatory schemes.[105] Despite Hernández's request to edit the work, in 1580 the task of organizing and

102. Domínguez remained in Mexico at the end of Hernández's expedition, earning a living as cosmographer and making navigation instruments. Save for his observation of the eclipse in 1584, there are no other surviving examples of his cosmographical production, although he spent years describing and mapping the Mexican territory and even wrote a description of China. For a biographical study of Domínguez, see Rodríguez-Sala, *Eclipse de Luna*, 67–83. Most of what is known of Dominguez's life comes from his statements of service (*relación de meritos y sevicios*) at AGI, P-262, R. 9, and AGI, P-22, R. 11.

103. Royal Cédula, January 11, 1570. Published in Hernández, *Mexican Treasury*, 46.

104. Letter by Hernández to Philip II, November/December 1571. Published in Hernández, *Mexican Treasury*, 48–49.

105. Bustamante García, "Francisco Hernández, Plinio," 259–61.

publishing a brief compendium of his work that included only medicinal plants was given to the Italian Nardo Antonio Recchi. It was through Recchi's book that Europeans learned of Hernández's work.[106]

As was his colleague Nicolás Monardes, Francisco Hernández was swept into the utilitarian culture that surrounded the Spanish imperial program and thus was obligated to share the concern for utility that Alonso de Santa Cruz had faced when writing his cosmography. Hernández's project as articulated by Herrera was purely utilitarian in scope. Hernández was instructed to compose a pharmacopoeia of the New World, not the hybrid-Plinian natural history he ended up writing. He was accompanied to New Spain by a cosmographer, Francisco Domínguez, who was to map the territory so that once the medicinal plants were identified they could be located for later exploitation. That Hernández saw himself as an Aristotle or a Pliny rather than as the empiricist Herrera sought makes this first voyage of exploration of the modern era a fascinating example of how the dynamics of empire shaped the practice of a scientific discipline by directing its output toward utilitarian ends.

It is evident that Herrera understood that natural history and cosmography were disciplines that required the expertise that only an academically trained individual could attain, but he also recognized that these were disciplines best practiced in the field, far from the relative comfort of a study in Madrid or Seville. He thus recruited individuals—Hernández, Juan—who shared his vision and were willing to take on the challenges of early modern transoceanic travel. They saw collecting accurate information, be it about natural history or geodesy, as a pursuit demanding personal agency and specialization. The vicissitudes of oceanic travel, premature deaths, and humanity's natural unpredictability, factors that often conspired to derail even the best-made plans, were simply obstacles to be surmounted on the way to gaining an understanding of the New World.

Cosmography at the Casa: Pilots and Maps

Court-sponsored cosmographical projects never rivaled the importance, quality, and volume of the work undertaken at the Casa de la Contratación or at the Council of Indies. With the establishment of the Casa de la Contratación in Seville in 1503, the city on the Guadalquivir River became the locus of navigation and exploration in Spain. The Casa's objective was to

106. José M. López Piñero and José Pardo Tomás, "The Contribution of Hernández to European botany and *materia medica*," in *Searching for the Secrets*, ed. Varey, Chabrán, and Weiner, 124.

determine and secure the personnel, techniques, and technologies that would allow for safe navigation to the Indies. Throughout most of the sixteenth century, cosmographers and pilots at the Casa de la Contratación experimented with various ways of collecting and presenting this kind of cosmographical information in a cohesive and useful way. Often employed in parallel with the mathematical methods advocated in navigation manuals, the descriptive rutter remained an effective way of introducing unknown sea routes and coasts to pilots. By the later half of the sixteenth century, the nautical charts prepared by the cartographers at the Casa de la Contratación drawn according to Ptolemaic rules became the accepted visual representation of the newly discovered geography.

The Casa was responsible for managing shipping and trade between Spain and the newly discovered lands. Soon after its formation, the Casa appointed a *piloto mayor* or principal pilot responsible for maintaining the *padrón real*, an official rutter for the trip to and from the Indies, the *carrera de Indias*. As early as 1519, the Casa de Contratación in Seville had specialists in instrument and mapmaking, and in 1523 it named its first specialist with the title of cosmographer.[107] They shared with the *piloto mayor* the responsibility for maintaining the *padrón real*. As the pace of exploration and settlement of the New World increased, so did the demand for pilots capable of guiding a ship safely across the Atlantic and, later in the century, the Pacific. By 1552, the Casa had a program in place to educate and examine pilots bound for the Indies, with the principal cosmographer at the Casa in charge of instruction.[108] The institutional directors of the Casa de Contratación saw classroom instruction—and astronomical navigation—as a way of shortening the years of apprenticeship necessary to learn the craft of piloting.

Cosmographical activity at the Casa de la Contratación coalesced around the figure of Rodrigo Zamorano (1542–1620) during the 1580s. A university-trained cosmographer, he occupied at one point or another each of the Casa's cosmographical posts during his almost forty years at the Sevilian institution. His interests extended beyond navigation to include

107. Pulido Rubio, *Piloto mayor*, 979–83. For more on the role of cosmographers at the Casa, see Pulido Rubio, 303–9; Lamb, *Cosmographers and Pilots*, 675–86 ; and Esteban Piñeiro, "Cosmógrafos del Rey."

108. Jerónimo de Chaves became the first professor of navigation at the Casa de Contratación (AGI, Contratación-5784, L.1, f. 95–95v, "Nombramiento de Jerónimo de Chaves como cosmógrafo y catedrático de Cosmografía de la Casa de la Contratación," December 5, 1552, Monzón). For examples of the examinations given to prospective pilots, see "De la Cosmografía. Examen de pilotos. De las cartas de marear. Instrumentos para navegar y del piloto mayor," DIU, 25:278–94.

astrology, astronomy, and natural history. Zamorano cultivated American plants in his garden in Seville and exchanged botanical specimens with Dutch naturalist Charles de L'Écluse (Clusius, 1526–1609).[109] Although he never published on the subject of natural history, he did write on astrology, mathematics, and navigation and published repeated editions of a chronology and almanac.

Zamorano's long career cannot be discussed here in detail, but a few glimpses of the practices he engaged in while at the Casa will suffice to show how he employed the theoretical constructs of Renaissance cosmography.[110] Zamorano was not content with simply teaching navigation to pilots and actively sought involvement in a number of cosmographical projects organized by the Council of Indies. In a 1582 petition letter to the king, Zamorano explained that in addition to lecturing on navigation at the Casa, he had taken on many other cosmographical duties that typically fell under the purview of the pilot major, such as inspecting nautical instruments and testing would-be pilots.[111] As was customary in a letter intended to gain a royal favor, Zamorano described his activities at the Casa as complementary to those taking place in other institutions. He pointed out that he collaborated with Juan López de Velasco, the cosmographer-chronicler of the Council of Indies, by observing a number of lunar eclipses for the purpose of determining longitude and for which he personally built, at his own expense, a number of instruments. He also reminded the king that he had just written a navigation manual where he "reduced navigation to general and easy art and rules" and where he included "amended" solar declination tables, thereby correcting a number of errors in the tables currently in use.[112] In his petition letter Zamorano presented himself as a man of science

109. Josep Lluís Barona and Xavier Gómez i Font, eds., *La correspondencia de Carolus Clusius con los científicos españoles* (Valencia: Seminari d'Estudis sobre la Ciència, 1998), 76–78, 125–26.

110. The Biblioteca Colombina y Capitular (BCC) contains most of the surviving documentation about Rodrigo Zamorano's life (BCC, vol. 38, f. 290–316, "Papeles sobre Rodrigo Zamorano"). For a biographical study, see Pulido Rubio, *Piloto mayor*, 369–711.

111. Zamorano wanted to be awarded the salary due the cosmographer of the Casa de Contratación in addition to the one he earned as a professor. AGI, P-262, R. 11, f. 1–3, "Rodrigo Zamorano cosmógrafo y catedrático de la Casa de la Contratación de Sevilla sobre se le acreciente el salario que tiene," May 14, 1582. Published in Mariano Esteban Piñeiro, "Los cosmógrafos al servicio de Felipe II," *Mare Liberum* 10 (1995): 529–30.

112. "[E]nmendé las tablas de la declinación del Sol y las demás cosas que en los Regimentos pasados andaban muy erradas por causa del movimiento del Sol y octavo cielo." AGI, P-262, R. 11, f. 1v.

and a cosmographer who gladly collaborated in projects not directly under his purview, capable of carrying out the inquiries needed to advance the cosmographical art.

In this same letter Zamorano also gives a very illustrative example of the kind of collaboration that ideally took place between the cosmographer of the Casa de la Contratación and navigators. It describes the cosmographer at work, trying to incorporate the latest geographical discoveries into the Casa's storehouse of knowledge. Zamorano met with Captain Pedro Sarmiento de Gamboa to discuss the geography of the Strait of Magellan and assist him in building instruments and preparing the nautical charts to be used during Sarmiento's 1581–83 return voyage to fortify and colonize the strait.[113]

Sarmiento had first explored the area in 1579–80, in a voyage motivated by Sir Francis Drake's incursion in the Pacific Ocean and his attacks on Spanish settlements along the South American coast. At that time Sarmiento's objective had been to intercept Drake at the Straits, but the Englishman chose instead to return via a Pacific crossing. Three years later, as Sarmiento's fleet prepared to depart on a new expedition, Zamorano conferred with Sarmiento and his pilot, Antonio Pablos, who had mapped the area during the previous expedition, to correct "great errors concerning longitude" in the rutter and charts to be used in the expedition. We also have a parallel account by Sarmiento about working with Zamorano to prepare the new maps. Together they examined "all the old and new maps and rutters from different chart makers and cosmographers to examine the longitude of two fixed places of different longitude . . . Seville and Lima and from them situated the coasts in their locations."[114] Both observed the lunar eclipse of 1578, Sarmiento while in Lima and Zamorano in Seville. Their

113. For more on Sarmiento's 1579–80 expedition, see Martín Fernández de Navarrete, *Colección de opúsculos del excmo. sr. d. Martín Fernández de Navarrete*, 2 vols. (Madrid: Impr. de la Viuda de Calero, 1848), 1:235–48. While in the Straits of Magellan, Sarmiento attempted to calculate longitude using the method of lunar distances or by "the full of the moon and the position of the sun at sunrise" (*por la llena de la luna y nacimiento del sol*). Pedro Sarmiento de Gamboa, *Derrotero al Estrecho de Magallanes (1580)*, ed. Juan Batista (Madrid: Historia, 1987), 177. For more on Zamorano's work with these explorers, see Cerezo Martínez, *Cartografía náutica española*, 239–43.

114. AGI, P-33, N. 3, R 27, Pedro Sarmiento de Gamboa, "Relación de lo sucedido a la Armada Real de S.M. en el viaje al Estrecho de Magallanes . . . el año 1581 hasta 1583." Published in Martín Fernández de Navarrete, *Colección de documentos y manuscriptos compilados por Martín Fernández de Navarrete*, ed. Museo Naval, 32 vols. (Nendeln, Liechtenstein: Kraus-Thomson, 1971), 20.1:212v–14v.

observations yielded a difference in longitude between the two cities of 74°, which they chose as the true longitudinal distance between the cities. (The distance is more on the order of 70°.)[115] The maps they examined, including two by Diego Gutierrez, showed Lima either 7° too far west or 3–4° closer to Seville (too far east). Although it is unclear from the memorandum which Gutierrez map Sarmiento refers to, one well-known Gutierrez map, the *Americae sive quartae orbis nova* (1562) published by Hieronymus Cock and believed to be based in the *Padrón real*, shows Lima 77° from Seville.

Both captain and cosmographer recognized the sensitive nature of the cosmographical material they had assembled. Zamorano was under the same secrecy constraints as the pilot major and the cosmographer of the Council of Indies. Although the revision of the cartographic material for Sarmiento's expedition had taken place under the express mandate of Philip II, the king warned officials at the Council of Indies to make sure consultations took place with due "secrecy and prudence." The new maps were to be kept in the "ark with three keys" at the Casa de la Contratación, where the *padrón real* was kept safely stored.[116]

Unfortunately, there are no surviving examples of Zamorano's cartography, which he practiced both privately and in his official capacity as cosmographer of the Casa de la Contratación.[117] He was also an instrument maker, having built all the astrolabes, Jacob staffs, compasses, and nautical charts for Sarmiento's departing armada. In addition to the Spanish translation of Euclid's *Elements* that had earned him consideration for the professorship at the University of Salamanca, Zamorano wrote a navigation manual, the *Compendio de la arte de navegar* (Seville, 1581; figure 2.1), and a chronology (Seville, 1585). The navigation manual followed the well-established format of earlier manuals popularized in Spain and Europe by

115. See chapter 6 for a detailed analysis of their computations.

116. "Ya se os escribió que enviase a los oficiales de Sevilla la descripción y patrón que trajo Sarmiento de las costas y navegación del Estrecho y así lo habréis, previniéndoles para que con todo secreto y recato y en presencia de Diego Florez hagan que el cosmógrafo tome la razón de todo ello, y ponga en las cartas haciendo solas aquellas que fueren necesarias para que esta armada las lleve, y sin quedarle ninguna otra se meta el patrón en la arca de tres llaves, y cuando vuelva esta armada se recobren las que llevan y se guarden." AGI, IG-739, N. 306, "Consulta del Consejo de Indias sobre expedición de Pedro Sarmiento de Gamboa al Estrecho de Magallanes," March 1, 1581, Madrid.

117. In the late eighteenth century, Fernández de Navarrete cited navigation maps that Zamorano might have published in quarto titled *Carta de marear* (Seville, 1579 and 1588). Martín Fernández de Navarrete, *Biblioteca marítima española*, 2 vols. (Madrid: Viuda de Calero, 1851; reprint, New York: Burt Franklin, 1968), 686.

Medina and Cortés. The short book, only sixty folios in quarto, covers the topics laid down by the 1552 statutes creating the *cátedra de cosmografía* at the Casa de la Contratación. Over the next thirteen years, the *Compendio* underwent six printings in Seville, with the last one appearing in 1591.[118]

By the later sixteenth century, cosmographical practice at the Casa functioned within narrow categories defined by an individual's post: the *catedrático* taught navigation to pilots, the cosmographer inspected instruments used in the *carrera de Indias*, and the pilot major updated the official rutter to the Indies. Cosmography within the walls of the Casa in Seville had a decidedly utilitarian bent. There was little room for the type of astrological and natural philosophical speculation in which Zamorano had engaged in his *Chronología*. There he discusses the meteorological and astrological effects of the nova of 1572 and expresses his opinion on the nature of this much-debated celestial event. He mentions that those observing the "comet" noted an absence of parallax that would suggest the comet was "generated" beyond eighth heaven, rather than in the sublunar sphere of fire as Aristotelian natural philosophers maintained. Citing Geronimo Cardano, with whom he seems to agree, Zamorano explains, "[In order to] preserve appearances and physical reasons it could not be understood that it be in any other region than in the eighth heaven; or we would have to concede the penetration of bodies and other things foreign to natural reason and mathematics." Zamorano sided with the controversial opinion of those who, like Jerónimo Muñoz, were willing to challenge the Aristotelian notion of the immutability of the heavens.[119] For a man who was interested in astronomical observations and who kept current with the latest European astronomical debates, the post at the Casa de la Contratación offered little opportunity or reward for astronomical and astrological investigation. This, as well as Zamorano's interest in botanical specimens from the New World, took place at the margins of his career at the Casa de la Contratación.

Santa Cruz, Herrera, and Zamorano belonged to three distinct generations of Spanish cosmographer-humanists, as Jesús Bustamante points

118. First published in 1581, the *Compendio del arte de navegar* was published again in 1582, 1585, 1586, 1588, and for the last time in 1591. Navarro Brotóns, *Bibliographía physico mathemática*, 324–26. Zamorano's book was later translated into English by Edward Wright in *Certaine Errors in Navigation* (London: printed by Felix Kingsto[n], 1610).

119. Rodrigo Zamorano, *Chronología y repertorio de la razón de los tiempos* (Sevilla: Imprenta de Francisco de Lyra, 1621), 203v.

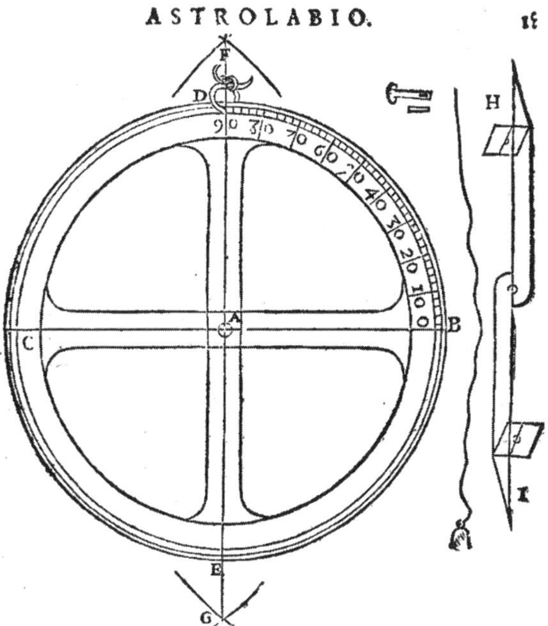

Y ajuſtada la regla ſobre el centro.A.ſeñaleſe la linea.B.A.C. que corte en dos partes iguales cada qual de los tres circulos.Y pueſta la vna punta del compas abierto ſegun la linea.B.C.enel punto.C. donde el Circulo mayor ſe corta con la linea.B.C.con la otra punta ſe ſeñalarà ſobre el pũto.D. vna parte de Circulo,y otra abaxo de.E.Y poniendo el compas aſsi abierto en el punto.B. ſeñalen ſe otras dos partes de Circulo, que cor-

D 4 ten

Figure 2.1. How to build an astrolabe, from Rodrigo Zamorano's *Compendio del arte de navegar* (Seville: Juan de León, 1588). This item is reproduced by permission of *The Huntington Library, San Marino, California.*

out.[120] From this generational perspective it would be expected that the problems of representation and epistemic dilemmas that faced the humanist Alonso de Santa Cruz would be recognized and addressed differently by the next generation. For Herrera and Zamorano, this implied the abandonment of the bookish epistemology and methodology characteristic of humanism and the embrace of an empirical episteme that required personal agency.

Our attention must now turn to the Council of Indies, as this institution became the locus of cosmographical activity in Spain during the closing

120. Bustamante García, "Círculos intelectuales y las empresas culturales de Felipe II," 44–45.

decades of the sixteenth century. The next chapter examines the Council's attempts to institutionalize the collection, compilation, and production of cosmographical knowledge. In contrast to the cosmographical activity at court, which took place largely through individual initiative, the Council adopted a bureaucratic approach that would yield new epistemological methods and new modes of representation in response to what its members perceived to be a deficiency in the discipline's ability to provide useful knowledge. Renaissance cosmography had to undergo a significant reorientation if it was to continue to serve as a viable vehicle for describing the reality of the New World.

THREE

Cosmography Codified

Cosmography as State Secret

Early in its history the Council of Indies recognized the strategic and sensitive nature of accounts describing the New World, whether they were historical, geographical, ethnographical, or natural historical. Therefore, for most of the sixteenth century the Council acted to safeguard this knowledge from eager and relentless enemies who challenged Spain's political and spiritual hegemony over these lands. The obvious enemies were foreign European nations operating beyond the Habsburg dynasty's sphere of influence, primarily France and England. State-sponsored piracy, as in the case of Francis Drake's attacks (1578–79), and the establishment of Protestant colonies, as in the French Huguenot colony in North Florida (1562–65), were constant reminders that both the material and the spiritual wealth of the Indies were in jeopardy. Internal enemies that threatened Catholic hegemony also abounded. Judaizers, purported Lutheran cells, and heretical cults mixing native and Catholic rituals all had to be guarded against. Plots by Spanish colonizers to break away from the monarchy, as those played out during the 1540s in Peru, were more than enough motivation to keep the economic particulars of each region secure.

The New World's cosmography, with its maps, geographical descriptions, and accompanying accounts that described a region's natural resources, had value not only because sources of wealth and trading routes had to be defended but because the Council of Indies was expressly chartered to administer and exploit resources on the Crown's behalf. The Council operated in the belief that the enemy—English, French, and later, Dutch—could not attack if they could not locate strategic ports or know the routes taken by the Spanish fleet in the *carrera de Indias* and the Manila galleons. The Casa de la Contratación had been instructed since 1510 to keep maps and rutters to the Indies under close guard, and by the 1560s the Council of Indies was following suit.

The containment of cosmographical information was achieved using various techniques. As Richard Kagan has pointed out, one was to impose authorship, publishing, and sales restrictions that effectively turned cartographic production, particularly nautical charts, into a state monopoly. He explains that "elsewhere in sixteenth century Europe, mapping was primarily a private affair.... The Spanish Habsburgs sought to make cartography a royal monopoly, an enterprise to be supervised and controlled by crown officials."[1] Another way to control cartographic production was to limit access to the primary sources that informed these maps and cosmographies: the rutters and logbooks of pilots running the *carrera de Indias*. Philip II often personally instructed the Council to collect and keep such material secret. In one instance, for example, while the Council was discussing establishing a permanent settlement in the Philippines, the king, referring to the original maps and rutter from the Legazpi expedition, instructed the Council to "look diligently for the papers and navigation charts there might be on this matter, put it all together, and keep it under close watch at the Council.... The originals should be sent to [the royal archive at] Simancas and true copies brought to the Council."[2]

In addition to trying to keep a close rein on cosmographical knowledge generated by the Casa de la Contratación and the Council of Indies, the Council tried to keep private parties from divulging this knowledge. Initial measures were aimed at containing politically sensitive material. Taking its procedural cues from the Inquisition, as early as 1527 the Council began issuing ad hoc bans on books that contained accounts of the conquest and colonization that it found troubling and that could cause political or

1. Richard L. Kagan, "Arcana Imperii: Mapas, sabiduría y poder en la corte de Felipe IV," in *El atlas de Rey Planeta: La descripción de España y de las costas y puertos de sus reinos de Pedro Texeira* (Madrid: Editoral Nerea, 2002), 60. For more on the economic and political linkages and the production of early modern maps in the context of commerce, see Jerry Brotton, *Trading Territories: Mapping the Early Modern World* (Ithaca, N.Y.: Cornell University Press, 1998), 151–60.

2. "[D]ígaseles que habrían de hacer diligencia en buscar los papeles y cartas de marear que hay sobre esto y juntarlo todo y tenerlo en el Consejo a buen recaudo y aún los originales se habrían de poner en Simancas y traer copias auténticas en el Consejo." AGI, IG-738, N. 82, f. 1, "Consulta de Consejo de Indias sobre población de Filipinas," July 5, 1566, Madrid.

religious controversy.[3] By the 1550s, the Council began acting proactively, requesting that all books that touched on the subject of the New World be submitted to the Council for review prior to publication. The Council's policy was thus directed at censoring accounts of the discovery and colonization containing material considered incendiary in the already volatile regions of the New World.

The same occurred for books that discussed problems associated with the governance of and spiritual ministry to the native populations. It was feared that public criticism or debate over imperial policies, such as the Crown's right to claim possession over all discovered territories or to allocate the native workforce to a Spanish colonizer (*encomienda*), could foster militant dissent. The Ginés de Sepúlveda–Las Casas debate made it evident to the Council that it had to systematically review books in order to censure debate on the "Indian problem." The debate took place in Valladolid between August 1550 and April 1551, in an attempt to address a grievance Sepúlveda had raised when his incendiary book *The Second Democrates* was deemed theologically unsound and therefore had not been granted a publication license.[4] For the Council of Indies, it highlighted the importance of acting proactively to moderate discourse on all matters concerning the Indies.

The censorship policy was formalized by a royal order issued on September 21, 1556,[5] prohibiting the printing and sale of any book "that dealt with the subject of the Indies" before it had been "seen and examined" by the Council of Indies. Local authorities were instructed to inform the Council about such books and could be directed to seize them. Printers

3. In this first instance, the banned book was Hernán Cortés's account of the conquest of Mexico. For a study of these regulations vis-à-vis those issued concerning religious matters by the Inquisition, see Fermín de los Reyes Gómez, *El libro en España y América: Legislación y censura (siglos XV–XVIII)* (Madrid: Arco/Libros, 2000), 178, 188–90.

4. Anthony Pagden points out that it was not a public debate. (It is famous now only because Spanish historians have pointed to it in defense of accusations that Spain did not care about what was happening to the Indians.) Anthony Pagden, introduction to Bartolomé de Las Casas, *A Short Account of the Destruction of the Indies* (London: Penguin Books, 1992), xxx.

5. AGI, IG-425, L. 23, f. 247–248, "Real Cédula requiriendo que todos los libros que se imprimieran o vendieran, sin expresa licencia, sobre cosas de Indias y los envíen el Consejo," September 21, 1556, Valladolid. This and other relevant orders appear in José Toribio Medina, *Historia de la imprenta en los antiguos dominios españoles de América y Oceanía*, 2 vols. (Santiago de Chile: Fondo Histórico y Bibliográfico José Toribio Medina, 1958), 1:5–10.

were prohibited from accepting a book for publication if the author did not present the appropriate license from the Council. The printer could receive hefty fines and have his presses confiscated if he failed to obey the order. The order apparently was not followed as faithfully as the Council had hoped—hardly an uncommon occurrence—and over the years the mandate was reissued several times.[6] The measure, however, came too late for Las Casas's A *Short Account of the Destruction of the Indies* (Seville, 1552; Antwerp, 1579). Even if the Council had acted sooner, it might have not foreseen that the brief text would have such devastating effect on Spain's reputation when it began to circulate in Protestant countries after the Antwerp edition appeared.

The reasons the Council of Indies alleged to justify censoring a book varied according to the nature of the work in question. The Council exerted its censorship privileges most heavily on religious works it deemed heretical or simply inappropriate for the newly converted. Historical works were examined to determine whether they contained versions of discovery and conquest that deviated from the official version—itself a work in progress—and that could cause offense or discord.[7] Authors were sometimes able to circumvent these restrictions by printing their works outside the kingdom of Castile or abroad, as López de Gómara and Girava did.[8] The restrictions were an almost insurmountable burden for American authors, particularly those who could not travel to Spain to lobby the Council of behalf of their work.

Secrecy also meant that many cosmographical works would never be published because the monarchy considered material in the work to be

6. Torre Revello has shown that censorship laws were blatantly ignored in the colonies, save perhaps those concerning books banned by the Inquisition. José Torre Revello, *El libro, la imprenta y el periodismo en América durante la dominación española* (Buenos Aires: Casa Jacobo Peuser, 1940), 243–45.

7. Juan Friede, "La censura española del siglo XVI y los libros de historia de América," *Revista de Historia de América* 47 (1959): 59.

8. Girava's books avoided scrutiny by Castilian censors since they were published in Milan and Venice. López de Gómara's *Historia de las Indias y la conquista de Mexico* was banned in late 1553, presumably because of its controversial account of the conquest of Mexico. AGI, P-170, R.1, *Cédula*, 1554, as cited in Schäfer, *Consejo de Indias*, 2:355n247. Gómara's book was originally published in Zaragoza (1552), Medina del Campo (1553), and before the ban was promulgated (or maybe it was ignored!) in Zaragoza (1554). In 1554, four editions appeared in Antwerp. It was not published again in Spain until the eighteenth century. For publication history of Gómara, see the introduction by José Luis de Rojas to Francisco López de Gómara, *La conquista de México*, Crónicas de América (Madrid: Dastin, 2000), 33.

important, valuable, and strategically sensitive. An important precedent was set concerning the secrecy policy governing cosmographical production when in 1563 Philip II denied Alonso de Santa Cruz permission to publish his works describing the Indies.[9] When during the 1570s a legal code came into effect that institutionalized cosmographical practice at the Council of Indies, these standing secrecy and censorship policies limiting the diffusion of historical, geographical, and natural historical information about the New World became the law. This legal code, the Ordenanzas de Indias (1571), however, also put in place what is unarguably the most ambitious program of knowledge collection in sixteenth-century Europe. It specified the methodology and epistemic criteria to be used to write the definitive—though secret—cosmography of the Indies.

Almost since its inception, members of the Council of Indies had realized that the efficient administration of the territories under their purview required *knowing* about these lands and their peoples. Thus, most markedly following the pacification of Mexico, the Crown, through its Council of Indies, routinely requested overseas civil and ecclesiastic authorities to prepare reports describing the geography, ethnography, and natural history of newly discovered and settled lands. These reports were requested periodically, every ten years or so, when the Council found itself with insufficient information to address a specific problem or administrative concern.[10] The Council, however, had no formal mechanism to process the cosmographical information yielded by these requests. *Relaciones* written to comply with the requests quickly lost their currency, and new requests for information had to be sent.

During the 1550s, the Council's members found themselves reluctantly relying on Alonso de Santa Cruz whenever they needed cosmographical advice, yet they grew increasingly dissatisfied with their access to timely information. In the following decade, the Council of Indies would react to the problem and initiate a series of reforms that would codify and define the parameters for cosmographical practice at the Council. These measures came in response to what they perceived as a recurring vacuum of knowledge about the New World but was, as I will argue, in effect a vacuum caused by the shortcomings of cosmography's traditional modes of textual presentation for handling so much new cosmographical information.

9. AGS Estado-143, f. 84, cited in Vicente Maroto, "Alonso de Santa Cruz," 521n1.

10. For more on these early questionnaires, see Raquel Álvarez Peláez, *La conquista de la naturaleza Americana* (Madrid: CSIC, 1993), 152–83, and Jesús Bustamante García, "El conocimiento como necesidad de estado: Las encuestas oficiales sobre Nueva España durante el reinado de Carlos V," *Revista de Indias* 60, no. 218 (2000).

In addressing the problem, they chose a style of practice diametrically opposed to that advocated by Juan de Herrera and closer in style to Alonso de Santa Cruz. Rather than engaging highly trained individuals to carry out cosmographical investigations firsthand, they elected to incorporate cosmographical practices directly into the bureaucratic fabric of the state and to design a new cosmographical opus in the form of *Books of Descriptions.*

Santa Cruz's Guidelines

In many ways the history of how the Council of Indies institutionalized cosmographical practice is inexorably tied to Alonso de Santa Cruz. Previously I discussed his work as an unofficial cosmographical adviser to the Council of Indies and how his cosmographical works were denied publication. In this section I will survey the methodological and epistemological approaches that Santa Cruz recommended explorers and cosmographers follow when composing the cosmography of the Indies.

In 1556, Santa Cruz enumerated the reasons as to why the king should consider a man like him—a cosmographer and chronicler—for a post as councilman in the Council of Indies, despite the reticence of other council members.[11] He met the Council's objections head on, dismissing them as the ignorant reaction of persons unaware of their own limitations and not without a good dose of arrogance. To buttress his position before the king, he enlisted the support of the Council's former president Luis Hurtado de Mendoza, marquis of Mondéjar. Hurtado observed on Santa Cruz's behalf that the "jurists" currently serving as councilmen "could not stand men who practiced a science other than their own . . . because they thought that as jurists they knew everything."[12] Santa Cruz was positioning himself as the much-needed resident expert on cosmographical matters at the Council.

Santa Cruz pointed out that, unlike most members of the Council, he had actually been to the Indies and had endeavored to learn about the

11. Santa Cruz relates that the Council responded: "[Y] respondiéronme que en el Consejo no tenía necesidad de saber las tales cosas, ni menos que en él no se trataba de ellas, y que no hablase más sobre ello." AGS, Estado-121, f. 23. Published in Medina, *Sebastián Caboto*, 1:350.

12. "[E]l me respondió que no habia cosa más necesaria en el Consejo que el tener conocimiento de las cosas que yo decía en mi petición; pero que tuviese por cierto que los juristas no podían ver hombres de otras ciencias que ellos no supiesen, más ver al enemigo, porque no sintiese sus faltas, y porque pensase que con ser juristas se lo sabían todo." AGS, Estado-121, f. 23. Published in Medina, *Sebastián Caboto*, 1:351.

Indians' way of life and their relationship with Christians in the New World. Should occasions arise when the Council had to resolve conflicts between Indians and Christians, he argued, he was capable of explaining possible causes of the conflict and giving an impartial opinion on the matter. The Council could also draw on his navigation experience to resolve lawsuits involving cases of possible pilot negligence. His cosmographical knowledge of the New World and his cartographic collection would likewise be helpful when the Council had to resolve disputes concerning land grants, territorial claims, or administrative divisions. Once in the Council, Santa Cruz suggested, he would request that viceroys and governors of the Indies prepare "a perfect geography and painting of each [territory] . . . so that once these arrived at the Council, I could study them and portray the provinces for which I now lack accounts."[13] The information necessary to compose these works, Santa Cruz reminded the king, would be gathered according to a set of guidelines he had given to the Marquis of Mondéjar years before. His presence in the Council, Santa Cruz also hoped, would stimulate other councilmen's curiosity about astrology and cosmography, or at least force them to show some interest so as to avoid public embarrassment (*cada uno se tendría por de menos valer si no lo hiciese*).

Santa Cruz maintained that cosmographical ignorance was not a problem limited to the Council of Indies; he complained that the Casa de la Contratación often remitted to the Council problems concerning cosmographical appointments, nautical instruments, pilots, and navigation charts that the Council was ill prepared to resolve. When the Council failed to resolve the matter, it sent the problem back to Seville, and "since officials at the Casa understand less than the ones in the Council, they gather information from sailing persons who serve their own interests and thus have

13. "Asimismo estando en el Consejo podría tener especial cuidado de que Vuestra Alteza enviase sus cartas á los Visorreyes y gobernadores que en aquellas partes están para que cada uno en su provincia hiciese la perfecta geografía y pintura de ella, poniendo, si fuese posible, los lugares principales en sus Alturas y los demás por sus derrotas y apartamientos de leguas y los montes, ríos y otras cosas notables, y poniendo por escrito las cosas notables de la tierra . . . por manera que viniendo las tales relaciones al dicho Consejo las podría tomar y precisar y poner en pintura las provincias de que yo no hubiese tenido relación, y lo que se enviase escrito guardarlo para hacer lo que tocase á la historia, y como todos los visorreyes é Audiencias y corregidores supiesen que había persona en el Consejo que lo procuraba y que a Vuestra Alteza hacían servicio, todos se desvelarían en enviar lo que acerca de lo dicho conviniese y todos los pasajeros procurarían traer lo que de allá se hiciese." December 14, 1556. Published in Medina, *Sebastián Caboto*, 1:348–49.

awarded salaries and done other things to Your Majesty's disservice."[14] An experienced cosmographer at the Council, Santa Cruz believed, could address these problems and advise the Council on the proper course of action. In his petition letter Santa Cruz does not articulate a desire to have final authority on cosmographical issues at the Casa, but as a member of the Council of Indies he would have been in a position to exert significant influence on the matter.

Santa Cruz envisioned a role that would not limit his duties to serving the cosmographical needs of the Council of Indies. He also saw himself standing next to the monarch, personally advising the king on cosmographical matters. The king, he suggested, could consult him on problems about the line of demarcation and Portuguese territorial claims in Asia. Or if, for example, the king had to make decisions about the location of settlements or needed to know the location of gold and silver mines or the boundaries of his territories, Santa Cruz could "place before His Highness a portrait of it with its description in writing to know the particularities of it all."[15] Lastly, Santa Cruz offered to write a history of the Indies, including maps of all the provinces in the "manner of Ptolemy." This, he added, would be to the perpetuation and good memory of the king. In his attempt to take on an official position at the Council of Indies, Santa Cruz had set down inadvertently the parameters followed three years after his death when the post of cosmographer-chronicler of the Council of Indies was officially instituted.

The contributions Santa Cruz made to the way cosmography was practiced in the Council of Indies went much further. In the set of guidelines he prepared for Hurtado de Mendoza, Santa Cruz articulated a sort of wish list of the kind of information necessary for writing a cosmography of the New World and for the collection of this information. The guidelines, known by the catalog title "About the discoveries in the Indies," reveal a Santa Cruz insatiable for cosmographical information about the New World.[16] For him,

14. "[Y] como los Oficiales de la dicha Casa entienden menos que los del Consejo, toman las informaciones de personas marineros que á ellos les parece para que se efectúe lo que ellos quieren, y así se han dado salarios y hecho otras cosas harto en deservicio de Vuestra Majestad." AGS, Estado-121, fol. 1–22. Published in Medina, *Sebastián Caboto*, 1:349.

15. "[P]odría poner á Vuestra Alteza delante la pintura de todo ello al propio y la escritura para saber otras particularidades sobre todo ello." AGS, Estado-121, fol. 1–22. Published in Medina, *Sebastián Caboto*, 1:349.

16. AGI, P-18, N. 16, R. 3, "Parecer sobre descubrimientos en las Indias." The document survives only as a draft copy, undated and unsigned but in Santa Cruz's handwriting, and with the telltale word "Céspedes" written above it, indicating that it formed part

the act of discovery involved more than chancing upon previously unknown territories and noting them accurately in a map; the moment of discovery also marked the point of departure for a series of geographic and ethnographic investigations essential to translating the new lands from the realm of the unknown to that of the known. For an early modern cosmographer, this translation involved composing a cosmographical description of the new lands and locating them on a map. In the document, Santa Cruz tried to distill into seventeen queries the information he considered essential for writing the cosmography of the discovered lands. He hoped that explorers and settlers would address the seventeen queries when they reported about new territories to the Crown and Council. In fact, he ultimately desired that the Council of Indies issue the guidelines as a royal order, thus making it obligatory for explorers and settlers to comply with the directives outlined in the document.

Santa Cruz began his guidelines with an administrative recommendation. He was uneasy about leaving the business of new discoveries in the hands of private individuals. Such individuals, he complained, rarely had enough capital to mount a successful expedition and were more likely to abuse the native population in their desperate search for profits. Santa Cruz preferred that the discovery enterprise be placed in the hands of a royal armada, under the direction of officers obligated to impose the laws the king had laid down expressly for the preservation of the Indians, as well as protect the king's interests. He also hoped that putting the responsibility of collecting cosmographical information in the hands of government officials would guarantee compliance.

The seventeen instructions began with a description of what Santa Cruz considered a discovery armada's ideal crew. It would consist of a captain who knew the "*principles* of the art of navigation" (his italics) and would not only serve as the final authority on board but also have a say regarding where the ship was heading—a lesson from his ill-fated expedition with Sebastian Cabot.[17] The pilot serving under his command should navigate using nautical instruments, among them one indicating how much the compass needle deviated from true north. Once on land, the first order

of the collection of papers belonging to the cosmographer of the Council of Indies. Published in Santa Cruz, *Obra cosmográfica*, 1:67–72. Sánchez Bella and Mata Carriazo place the document circa 1557, but Cuesta and Millán prefer to date the document in the 1540s. This earlier date corresponds to when the Marquis of Mondéjar was president of the Council of Indies, 1546–49.

17. Santa Cruz, *Obra cosmográfica*, 1:69.

of business was to determine the location of the newly discovered territory. Officers would set out to determine their location both astronomically (latitude) and relative to nearby territories. They were to inquire about the name of the land from the natives and also indicate the name given to it by Spaniards (*cómo se llama entre nosotros*). These officers would also conduct a geographical survey of the new territory. They were to describe the territory's natural features, noting the quality of the land and the locations of principal geographical features, such as mountains and rivers, as well as its climate and whether it was considered healthful by the native inhabitants.

Similar inquires were to be carried out to determine the natural resources of the new territory. Santa Cruz suggested that the discoverers address certain key questions. Are metals mined in the vicinity? Are there precious stones or pearls? The discoverers should investigate (*que procuren saber*) the animals found in the land, whether they were "like the ones we have here" or unknown ones that might be monstrous (*que sean monstruosos*). Likewise, discoverers and settlers should report on the types of plants the natives used for food and their harvest time, taking care to note which plants might have medicinal properties.

The most important observation among all those requested, Santa Cruz emphasized, was the recording of geographic locations as faithfully as possible. By this he meant the distances between the principal cities, significant geographical features, and neighboring provinces. These measurements and calculations should be carried out with the utmost care, because, he explained, it is the "principal thing we must know." In addition, he advised discoverers to draw a map "in the manner of navigation charts," using a compass rose and distance scale to record geographical coordinates.[18] In keeping with Renaissance cosmographical tradition, for Santa Cruz the textual description took precedence over the design of the map.

Santa Cruz's guidelines then turned to the description the peoples inhabiting this new land. The first matter to resolve was whether the territory was inhabited by "Gentiles or Moors." If inhabited by Gentiles, Santa Cruz instructed the discoverer "to try to know all their customs about their science or what they understand of the creation of the world and of the movement and makeup of heaven."[19] This inquiry should extend to their rites and religious beliefs: whether they knew of Christ, believed in heaven for the good

18. "[S]e tomarán unas hojas de papel y se pondrán en ellas los ocho vientos principales a manera de carta de mareas [sic] y puédese hacer un padrón de leguas para que lo que se asentare en ellas sea cierto" (ibid., 1:70).

19. "[P]rocurará saber todas sus costumbres acerca de su ciencia o lo que sienten de la creación del mundo e del movimineto y hechura del cielo" (ibid., 1:70–71).

and hell for the bad, and believed in the immortal soul. The explorers should also learn about the "nature" (*naturaleza*) or character of the inhabitants. Are they inclined toward war or rather toward trade? Santa Cruz's interest in the cultural context of the native peoples went beyond religion. He wanted to know whether they had "letters and science" and if among them were men given to study. Had they written chronicles, he exhorted the discoverers to have them translated into Castilian, "even if it costs money." Likewise, investigations should be carried out to learn about the native's customs: their manner of dress and of eating and drinking, marriage customs, and how they waged war. Lastly, the discoverers should learn about the native king and his household and, if at all possible, about the history of previous rulers.

Rather than demanding from discoverers a comprehensive cosmography, Santa Cruz was attempting to establish some order and consistency among *relaciones* about newly discovered lands by having them provide a subset of information he considered essential for writing the territory's cosmographical description. The document's seventeen clauses addressed three principal topical areas—the where, what, and who of discovery. He also made several methodological requests: that only qualified officers knowledgeable in astronomical navigation collect the geodesic information, that their observations be recorded on a map drawn to scale, and that they take care to indicate the direction and distance from adjacent territories. The most important information Santa Cruz sought was an accurate assessment of the new territory's geographical location and a description of its geographical features. Once these were achieved, the explorers should describe the land and its natural resources. This should include an eyewitness account of the discovery, a passage Santa Cruz always included in his cosmographical works about the New World. The description of the native inhabitants and their customs rounded off the cosmographical description.

Finally, Santa Cruz requested native historical accounts dating from before the arrival of the Spanish, as well as a review of the political and dynastic situation of the native states. These accounts would help solve a deficiency that Santa Cruz wrestled with in his *Islario general*, which he resolved by beginning the historical narrative of the New World at the moment of discovery by Europeans. In light of this document, it appears that Santa Cruz found the omission of prediscovery historical narrative in the *Islario* unsatisfactory and was seeking to remedy this shortcoming. Throughout the guidelines, Santa Cruz was very interested in identifying native individuals with sufficient *estudio* to provide this historical information; he urged discoverers to gather available books and even to figure out how to bring some of these learned persons to Spain, in the hope that once they learned Castilian they could clarify the books' contents.[20] When it came to pre-Spanish history, Santa Cruz clearly privileged native sources over secondhand accounts by Spanish explorers and understood that just as in his European context, a people's cosmology and science were measures of cultural sophistication.

Santa Cruz's cosmographical works, such as the *Islario general* and the *Epítome de la conquista del Nuevo Reino de Granada*, show he employed a thematic agenda similar to what he requested from the explorers. Santa Cruz's guidelines are the wish list of a cosmographer sitting in his study thousands of miles from the New World and envisioning how he wanted information from newly discovered territories to arrive at his doorstep. His "About the discoveries in the Indies" has come to be regarded as an important precedent to an ambitious project initiated by the Council of Indies in the 1570s to collect information through questionnaires and use it to prepare an all-encompassing cosmography of the Indies.[21] Santa Cruz's close association with the Council and the fact that his papers formed part of the collection of documents belonging to the first cosmographer-chronicler of the Council of Indies, Juan López de Velasco, suggest the document could have been used as a model for subsequent questionnaires. We must keep in mind, however, that Santa Cruz wrote the guidelines as a means of collecting information he intended for his personal use while working on his *General Geography and History*. The guidelines were not intended to yield *relaciones* that would be the Council's sources of information about new territories but accounts that would supplement a humanist cosmographer's reference library. In fact, Santa Cruz concluded the document with a plea to the president of the Council to issue the guidelines, "for it will serve God and His Majesty and I will receive from it much gratification."[22]

Santa Cruz's guidelines were never promulgated as a royal directive. His rocky relationship with the Council of Indies and his ambiguous

20. "[Y] si son hombres dados al estudio y pudieren haber algunos libros de ellos los habrán y cuesten lo que costaren y trabajarán como traer alguno de la Tierra que sepa leerlos porque aprendiendo nuestra lengua los pueda declarar" (ibid., 1:71).

21. Marcos Jiménez de la Espada identified the document as the definitive precedent to the questionnaire of the *Relaciones geográficas de Indias*, but Ismael Sánchez Bella and Raquel Álvarez Peláez suggest that the use of guidelines in the form of questions to collect geographical information about the Indies predates Santa Cruz's work. Ismael Sánchez Bella, "El *Título de las descripciones* del código de Ovando," in *Dos estudios sobre el código de Ovando* (Pamplona: Universidad de Navarra, 1987), 115.

22. "[H]ará mucho servicio a Dios y a su majestad y yo recibiré muy señaladas mercedes" (ibid., 1:72).

cosmographical appointments denied him an institutional base from which to launch the ambitious cosmographical project. Although it seems that later in life he recognized the need for institutional support, he operated most comfortably within a model of Renaissance cosmography that approached cosmographical production as an individual's task. It consisted of collecting firsthand accounts, cataloging them, cross-referencing them with classical narratives and resolving epistemological problems, and finally composing a comprehensive cosmography. He recognized, however, that with the support of a central authority the discovery enterprise could be administered in such a way as to make the task of collecting cosmographical information more efficient and certain. Before this could happen, though, the Council of Indies' view of what constituted cosmographical practice had to undergo a significant reorientation.

A Law to Define Cosmographical Practice

Not long after Santa Cruz's death, the methodological framework he had advocated for securing cosmographical information was assimilated into the Council of Indies. Nearly seventy-five years after the discovery of the New World and seventeen after the famous Las Casas–Ginés de Sepúlveda debate, where Las Casas reminded the Crown of its duty to safeguard the native population, the king continued receiving news of abuses committed by Spaniards on the native population and of shoddy viceregal administration in the New World. Troubled by these claims, as well as by the rising cost of administering the empire, the king ordered an audit (*visita*) of the Council of Indies in 1569.[23] To carry out what would become a four-year audit, the king chose Juan de Ovando y Godoy, a legal scholar and priest who had risen to prominence serving under Inquisitor General Cardinal Espinosa. Ovando is a prime example of the kind of man Philip II sought for the administration of his expanding empire. *Letrados* or men of letters like Ovando held many important positions that had formerly been awarded only to gentlemen of high birth. A native of Cáceres, Ovando was educated in the college of Saint Bartholomew in Salamanca, where he studied law and later taught. He was also a councilor for the Inquisition and *visitador* or auditor of the University of Alcala.[24]

Far from conducting a blame-finding mission, Ovando sought to determine what was at the root of what he perceived to be a systemic

23. Schäfer, *Consejo de Indias*, 1:136–39.

24. Stafford Poole, *Juan de Ovando: Governing the Spanish Empire in the Reign of Phillip II* (Norman: University of Oklahoma Press, 2004), 97, 197–99.

mismanagement of the spiritual, financial, and political governance of the Indies. He conducted countless interviews with members of the church, businessmen, and colonists from the Indies, as well as members of the governing bureaucracy. They generally replied with striking candor.[25] He also sent abroad repeated requests for information, many in the form of long lists of questions, ranging from reports on the geography of a particular region to reports concerning ecclesiastical jurisdiction or administrative organization.[26]

At the audit's conclusion, Ovando identified two problem areas which he believed accounted for many of the problems plaguing the administration of the Indies: a lack of clearly codified laws and ignorance in Spain about the New World. This last problem, he noted, was particularly pervasive among precisely the royal officials chartered with administering the new territories: the members of the Council of Indies. Information about the New World, Ovando noted, reached the Council in a haphazard manner, making it difficult for councilors to make educated decisions. The councilmen, Ovando opined, did not know enough about the New World to legislate effectively.

With the Crown's support, Ovando launched a comprehensive project aimed at compiling, organizing, and making available to members of the Council of Indies the information needed for the efficient administration of Spain's overseas territories, a corpus that would include a cosmographical description of the New World. A principal part of the project consisted in compiling, revising, and reissuing all existing laws, royal orders, and edicts issued since the discovery of the New World. This aspect of Ovando's project became known as the *Recopilación de leyes de Indias*. Sir Francis Bacon (1561–1626) advocated a similar project of legal reform in 1614 aimed at correcting the written record of English common law, a project that historians have associated with Bacon's subsequent interest in a reform of natural philosophy.[27]

Prior to Ovando's arrival at the Council of Indies, work had started on collecting the legal history of the Indies. Between the years 1563 and 1565, a young assistant by the name of Juan López de Velasco was scouring Council records for legal material and summarizing findings in a series of books. Once Ovando's audit began, these books became an intermediate step in

25. BL Add. 33983.

26. Sánchez Bella, "*Título de las descripciones*," 117–19.

27. Julian Martin, *Francis Bacon, the State, and the Reform of Natural Philosophy* (Cambridge: Cambridge University Press, 1992), 72, 108–9.

Table 3.1. Juan de Ovando's legal reforms of the Council of Indies

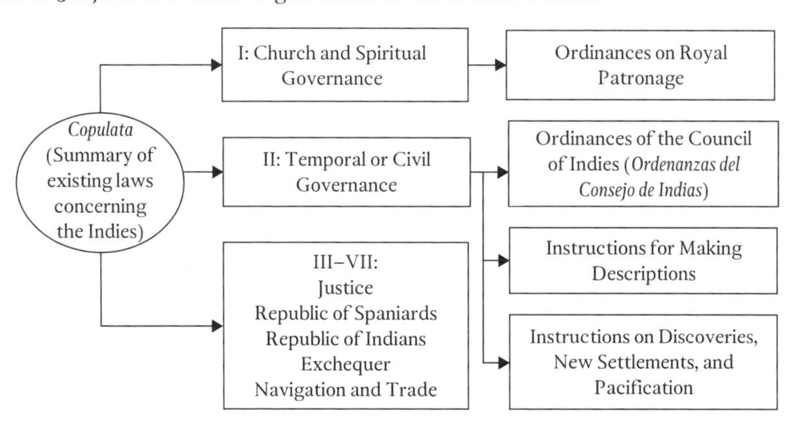

the plan for a new legal code of the Indies.[28] Using this compilation of existing laws, or *Copulata*, and adding to it other laws that came about from the suggestions made during the deposition phase of the *visita*, Ovando prepared a legal strategy that, in addition to yielding a unified and consistent legal code for the Indies, codified the administrative procedures used to govern the colonies. Ovando also incorporated a significant cosmographical program into this much larger project. At the completion of his reform project, Ovando would have produced a legal code consisting of seven parts or books, each covering a particular aspect of colonial administration: church and spiritual governance, temporal governance, justice, republic of Spaniards, of the Indians, of the royal exchequer, and finally navigation and trade[29] (table 3.1). After old laws had been identified, rewritten, and recorded in these books, they were to be submitted for review and amendment to the Council of Indies. Once approved by the Council, the book went to the king, and in some instances the pope, for final approval.

28. Under Ovando's supervision, the legal material Velasco had collected over the previous years was compiled into a single book. Only one copy of this work survives and was renamed by Peña Camara the *Copulata de leyes de Indias*. The original is in the BAH, códice 93. It was published in the DIU, vols. 20–25, under the misleading title *Gobernación espiritual y temporal de las Indias*. Jóse de la Peña Camara, "La copulata de leyes de Indias y las ordenanzas Ovandinas," *Revista de Indias* 6 (1941): 129.

29. Juan Manzano Manzano, "La visita de Ovando al Real Consejo de las Indias y el Código Ovandino," in *El Consejo de las Indias en el siglo XVI* (Valladolid: Universidad de Valladolid, Secretariado de Publicaciones, 1970), 119–21.

As of 1571, the project was past the compilation phase, but only the first book, on the church and spiritual governance (*De la gobernación espiritual*), had been completed, discussed in the Council, and submitted to the king.[30] Later that year, Ovando was appointed president of the Council of Indies, and his new duties most likely distracted him from personally working on the project. Ovando then changed strategy. Rather than waiting for entire books to be completed, he selected already finished sections that addressed areas that he considered in most urgent need of reform and shepherded these through the approval process. It appears that only the section from the first book on church and spiritual governance, titled *Ordenanzas del patronazgo real*, establishing the monarch's right of patronage over the church, was promulgated (June 1, 1574). Of the second book, on temporal or civil governance (*De la gobernación temporal*), only three sections were ever completed and submitted to the Council. These were the sections that dealt with the Council's day-to-day operation (*Ordenanzas del Consejo de Indias*);[31] a section describing how cosmographical information about the Indies should be collected (*Instrucciones para hacer las descripciones* or *Título de las descripciones*);[32] and lastly, a set of laws that dictated how new territories were to be discovered, settled, and pacified (*Ordenanzas de descubrimientos, nuevas poblaciones y pacificaciones*).[33] Of the other books Ovando foresaw as part of his reforms, only an outline—in Ovando's hand—of the fourth book on government of Spanish colonists (*De la república de los Españoles*[34]) and small fragments of the outline of book 5, on the government of the Indian

30. The first book in its original form never received royal approval. Several laws needed papal approval, and it was apparently never granted. Ibid., 122.

31. The section titled *Título del Consejo* was approved by the king in 1571 and appeared in print in 1585 under the title *Ordenanzas del Consejo*. Consejo de Indias, *Ordenanzas reales del Consejo de Indias: Gobernación y estado temporal* (Madrid: Casa de Francisco Sanchez, 1585). For a facsimile reproduction, see Antonio Muro Orejón, "Las ordenanzas de 1571 del Real y Supremo Consejo de las Indias: Reproducción facsimilar," *Anuario de Estudios Americanos* 14 (1957).

32. It is likely that the provisions listed in the *Título de las descripciones* were written and approved by the Council in 1571 and received royal approval on July 3, 1573. Juan Manzano Manzano, *Historia de las recopilaciones de Indias*, 2 vols. (Madrid: Ediciones Cultura Hispánica, 1950), 224–26. They were transcribed in the Council's general registry at AGI, IG-427, L. 29, f. 5v–66, "Ordenanzas para la formación del libro de las descripciones de Indias, 1573." Manuscript copies also exist at the BN, MS. 3017 and 3045. Sánchez Bella edited and published it in Sánchez Bella, "*Título de las descripciones*," 140–211.

33. AGI, IG-427, L. 29, f. 67–93v. This section was approved also in 1573 and published in DIA, 8:484–537 and 16:142–87.

34. Peña Camara, "Copulata de leyes de Indias," 141.

population (*De la república de los Indios*), survive.[35] Before his death in 1575, Ovando saw to completion only three parts of the legal reform: the laws concerning the spiritual governance of the Indies and its native population, the rules that brought order and accountability to the Council of Indies, and the guidelines for compiling cosmographical and historical information about the Indies.

Although the legal aspects of the whole project and its scope and complexity beset the Council for many years, the project that Ovando envisioned to remedy the Council's ignorance about the peoples, geography, and natural resources of the New World was truly magnificent. He recognized that the success of this project hinged on putting in place a conceptual structure that would organize efficiently all manner of information about the overseas territories and, more important, would function to keep this information accurate and current. This conceptual structure divided information into two categories: the moral and the natural. The moral encompassed matters considered "temporal" or changing and that were the result of human action in the territory described: government, religion, laws, cities, and the like. The natural had to do with "perpetual" aspects of the territory: land, sea, plants, and animals. As was usual with the historical genre at the time, the category "natural" also included description of the "natures and qualities" (*naturalezas y calidades*) of the native inhabitants.[36]

To catalog the natural, Ovando turned to Renaissance cosmography and its practitioners. Renaissance cosmography had a comprehensive taxonomy and a well-developed narrative genre for presenting information, and it even held the promise of a mathematically accurate cartography for determining geographical locations. Nonetheless, cosmography fell short in one critical area. The cosmographical genre had always held up as its crowning achievement the production of a magnum opus, an all-encompassing, encyclopedic exposition of all that was known about the natural world at a given time. (This was the model for the unfinished *Geografía general e historia* that Alonso de Santa Cruz labored all his life to create.) Three-quarters of a century of almost continuous discoveries and revelations about the world had shown that any such work was bound to become obsolete rapidly.

The body of cosmographical knowledge Ovando envisioned needed to be timely and current. The administrative needs of the empire could not wait for an individual to finish a work that could take a lifetime. To achieve

35. Manzano Manzano, "Visita de Ovando," 122.

36. Ponce Leiva, ed., *Relaciones histórico-geográficas*, xxi.

timeliness, Ovando conceived of a mechanism whereby cosmographical information was constantly, even automatically, updated as simply another function of the empire's record-keeping machinery. A lawyer by education, Ovando, not surprisingly, turned to the law to institute the mechanism for gathering, compiling, and making available to the members of the Council cosmographical information about the Indies.

With the king's approval in 1571 of the section of book 2 titled *Ordenanzas del Consejo de Indias*,[37] the laws regulating the office of the cosmographer-chronicler major of the Council of Indies were complete and formed a coherent body of directives. In fact, these rules remained in effect well into the seventeenth century.[38] In the case of the cosmographer-chronicler, the *Ordinances* specified the officer's principal responsibilities at the Council of Indies, with the *Título de las descripciones* or the *Instructions* describing the methodology he should use when compiling and cataloging information about the New World. The *Ordinances* and *Instructions* were intended to work together and complement each other. They spelled out, in a manner unprecedented in the intellectual history of Renaissance cosmography, the epistemic approach and methodologies that constituted cosmographical practice and in doing so broke definitively the ties that bound cosmography to humanism.

In the opening articles of the *Ordinances*, Ovando reminded the councilmen of their God-given responsibility to govern wisely and justly the new lands God placed in the monarch's care, a care the monarch had delegated to the Council of Indies. Ovando insisted that in order to carry out this divinely assigned responsibility, "nothing can be understood, or dealt with as it should, if the subject is not first known to the persons that need to inquire and make decisions about it."[39] He proposed making this knowledge available in the form of descriptions covering all that might come under the

37. I will use *Ordinances* as shorthand to refer in particular to the *Ordenanzas de Consejo de Indias* approved in 1571 and published in 1585.

38. Manzano Manzano, *Historia de las recopilaciones de Indias*, 1:175.

39. Article 3 reads: "Y porque ninguna cosa puede ser entendida, ni tratada, como debe, cuyo sujeto no fuere primero sabido de las personas que de ella, hubieren de conocer, y determinar. Ordenamos y mandamos que en los nuestros consejos de la Indias, con particular estudio y cuidado, procuren siempre tener hecha descripción y averiguación cumplida y cierta de todas las cosas del estado de las Indias, así de la tierra como de la mar, naturales y morales, perpetuas y temporales, eclesiásticas y seglares, pasadas y presente, y que por tiempo serán sobre que puede caer gobernación, o disposición de ley, segun la orden y forma del título de las descripciones, haciéndolas ejecutar continuamente con mucha diligencia y cuidado" (Consejo de Indias, *Ordinances*, 2).

Council's purview, that is, all matters related to the land and the sea, both "natural and moral, perpetual and temporal, ecclesiastic and lay, past and present." It was up to the cosmographer-chronicler major of the Council of Indies to catalog the natural aspects of this knowledge.

The last five articles of the *Ordinances*, 117 through 122, list the cosmographer-chronicler's responsibilities in detail. These articles make the cosmographer-chronicler responsible for compiling and maintaining the Council's books pertaining to cosmography, natural history, and the moral history of the Indies and instruct him to prepare the material so that councilmen may have ready access to it. The five articles were not meant to be comprehensive; the *Ordinances* worked in conjunction with the official's royal order of appointment. This document incorporated by reference the provisions of the *Ordinances* but could also list a number of other duties and responsibilities the king saw necessary to assign. The cosmographer-chronicler, like other officials at the Council, was primarily His Majesty's servant and could be called upon by the monarch to perform other tasks.

The cosmographer-chronicler did not work alone. He relied on the notary of the governance chamber to identify and make available to him the material needed to comply with his post's responsibilities. In article 75 of the *Ordinances*, the notary was instructed to keep in his care (*tenga a su cargo*) the *Books of Descriptions*. In this book the notary was to record all new descriptions as well as those written according to the guidelines laid down in the *Instructions*. The notary also had to forward to the cosmographer-chronicler "everything that came from the Indies concerning history and cosmography,"[40] so that the cosmographer-chronicler could "organize, correct, and verify" the material, as well as the maps that formed part of the description book. The notary served in a sense as the cosmographer's librarian, recording for future reference updated descriptions and keeping the cosmographer-chronicler abreast of new information as it arrived from the Indies. Performed on a continuing basis, the notary's responsibilities ensured that the cosmographer-chronicler had access to the latest cosmographical and historical information arriving from the Indies. Likewise, the notary was reminded of the utmost discretion and secrecy required by his

40. "El escribano de camara y governación del Consejo de Indias, tenga a su cargo el libro de las descripciones que ha de haber en el consejo, en el cual se asiente y ordene todo lo que se fuere describiendo de nuevo, e se hubiere de añadir en el, por la orden y forma que tenemos dadas en el título de las dichas descripciones, e así mismo tenga cuidado de dar al cronista cosmógrafo todo lo que viniere de Indias tocante a historia y cosmografía, para que lo ordene, ponga en forma, corrija y verifique las tablas del dicho libro" (ibid., 14).

post—after all, he would be privy to all exchanges between the king, the Council, and the colonies. The *Ordinances* instructed him to keep all sensitive material—including the cosmographical descriptions arriving from the Indies and those written by the cosmographer-chronicler major—securely locked away in the Council's "secret archive."[41]

The articles listing the cosmographer-chronicler's specific responsibilities begin with article 117 of the *Ordinances*, with an order to "make and organize" cosmographical tables (maps) of the Indies, taking care to record geographic coordinates "according to the geographical art," using both latitude and longitude as coordinates, as well as recording distances in leagues.[42] He was to record geographic features, "provinces, seas, islands, rivers, and mountains," both as written descriptions and as schematic maps (*en designo y pintura*). His sources of information were the "general and particular descriptions" that came from the Indies and the "accounts and notes" (*relaciones y apuntamientos*) provided by the notary of the governance chamber. The cosmographer-chronicler would then record this information in the "general" *Books of Descriptions* as instructed in the *Instructions*.

In the following article, number 118, the cosmographer-chronicler was given detailed instructions on the methods to use when collecting geographical coordinates for his general descriptions. For example, in order to determine longitude, he was to coordinate the observation of lunar eclipses in the Indies,[43] notifying local provincial authorities of the date and time of each upcoming lunar eclipse and providing them with the necessary instructions and instruments to make the observation. As eclipse observations yielded longitude coordinates, these were to be entered in the *Books of Descriptions*.

In addition to compiling descriptions of the new lands and positioning them correctly on the terrestrial orb, the cosmographer-chronicler had to

41. Ibid., 13.

42. The word tables (*tablas*) denoted both maps and written information arranged in columns and rows. Article 117 reads: "El cosmógrafo cronista que ha de haber entre los demás oficiales del consejo de las Indias, haga y ordene las tablas de la cosmografía de las Indias, asentando en ellas por su longitud, y latitud, y numero de leguas, según el arte de geografía, las provincias, mares, islas, ríos, y montes, y otros lugares que se hayan de poder en designo y pintura segun las descripciones generales y particulares que de aquellas partes se le entregaren, y las relaciones y apuntamientos que se le dieren por los escribanos de cámara de gobernación del dicho consejo, conforme a lo cual, y a lo que tenemos mandado en el título de las descripciones, prosiga lo que fuere a su cargo de hacer en el libro general de descripciones que ha de haber en el consejo" (ibid., 21).

43. Ibid., 21v.

record human activity in the territory described, in this case both native and Spanish. Article 119 instructed him to describe "with as great precision and as truthfully as possible, the customs, rites and antiquities [of the Indies], as well as incidents and memorable events."[44] Again he was instructed to use as his sources the "descriptions, histories, other *relaciones*, and replies to inquiries sent to the council." The cosmographer-chronicler was warned that the resulting history must be kept in the Council "so that it cannot be published or read, except what the council thinks should be made public." Just as the cosmographical description of the Indies was considered a state secret, so too was its history.

Similarly, the cosmographer-chronicler had to compile a natural history of the Indies, "because the natural things of the Indies should be known and understood." Rather than try to be comprehensive, article 120 advised him to summarize and continually gather (*recopile y vaya siempre coligiendo*) the natural history of the "herbs, plants, animals, birds, and fish and any other thing that is worthy of knowing in the provinces, islands, seas, and rivers."[45] Lastly, article 121 instructed the cosmographer-chronicler to compile navigation rutters to and among the Indies.[46] These should be based on travel accounts and rutters that pilots and sailors familiar with sailing routes to the Indies were directed to prepare. At first glance article 121 seems to call for duplication by instructing the cosmographer-chronicler of the Council of Indies to prepare rutters similar to those the pilot major at the Casa de la Contratación was under obligation to maintain—and indeed this was the

44. "Y porque la memoria de los hechos memorables y señalados que ha habido y hubiere en las Indias se conserve el cronista cosmógrafo de Indias, vaya siempre escribiendo la historia general de ellas con la mayor precisión y verdad que se pueda, de las costumbres, ritos, antigüedades, hechos y acontecimientos que se entendieren por las descripciones, historias, y otras relaciones, y averiguaciones que se enviaren a nos en el consejo: la cual historia esté en el, sin que de ella se pueda publicar ni dejar leer, mas que de aquello que a los del consejo pareciere que sea público" (ibid., 21v).

45. "Así mismo, porque las cosas naturales de las Indias sean sabidas y conocidas. El cronista cosmógrafo de Indias, recopile, y vaya siempre coligiendo la historia natural de las yerbas, plantas, animales, aves, y pescados, y otras cosas dignas de saberse, que en las provincias, islas, y mares, y los ríos de las Indias hubiere, según que lo pudiere hacer, por las descripciones y avisos que se enviaren de aquellas partes, y por las más diligencias que con autoridad nuestra, y orden de consejo se podrán hacer" (ibid.).

46. "Otrosí, el dicho cosmógrafo colija y recopile en libro, todas las derrotas, navegaciones, y viajes que hay de estos reinos a las partes de Indias, y en ellas de unas partes a otras, según le pudiere colegir de los derroteros y relaciones que los pilotos y marineros que navegan a las Indias, trajeron de los viajes que hicieren, como tenemos mandado" (ibid., 21v–22).

Council's intention. By giving the cosmographer-chronicler of the Council access to the same sources of information as those used by the pilot major to compile the *padrón real*, the Council was in effect establishing control over this vital aspect of overseas navigation. It would not be long before the Council was rebuking the Casa's cosmographers for not keeping the *padrón* current and for lapses in keeping it secret.

The last article concerning the duties of the cosmographer-chronicler takes care of some housekeeping and, more importantly, specifies the security measures to be taken to safeguard the cosmographer-chronicler's work. Article 122 refers back to article 75 and again instructs the notary of the governance chamber to provide the cosmographer-chronicler the papers he needs to comply with the responsibilities of his office. These, as well as the description he wrote based on them, had to be handled with secrecy, "without telling anyone about them or letting anyone see them."[47] Only with the Council's permission could they be shown to someone outside the Council of Indies. In a final note, and as a way to ensure that the notary fulfilled his duties, every year, before he could be paid the final third of his salary, he had to deposit all finished descriptions in the Council's "archive of secrets."[48] As we have seen, the spirit behind the secrecy provisions in the *Ordinances* was not new, although the specific instructions might have been. They were simply articulating secrecy policies concerning geographic information that dated back to the early days of the discovery.

When comparing the office of cosmographer-chronicler as specified in the *Ordinances* to the duties Santa Cruz listed in his petition of 1556 to become a member of the Council of Indies, it is evident Ovando and

47. "Y porque mejor pueda cumplir con lo que es su cargo el cronista cosmógrafo de Indias, mandamos a los escribanos de cámara de gobernación del consejo que le entreguen los papeles y escrituras que hubiere menester, dejando conocimiento del recibo de ellos, y volviendo a quien se los entregare, cuando se los pidan: los cuales, y las descripciones que fuere ordenando: guarde y tenga con secreto, sin las comunicar, ni dejar ver a nadie, sino solo a quien por el consejo se le mandare, y como las fuere acabando, las vaya poniendo en el archivo del secreto, cada año, antes que se le pague el ultimo tercio del salario que hubiere de haber" (ibid., 22).

48. We know from an inventory taken in 1766 that the cosmographer's work was indeed kept in the secret archive. It includes a section titled "Lista de papeles reservados juntos con los que suele remitir anualmente el cosmógrafo del consejo." BAH, 9/4855, f. 567–580, "Lista de los libros que existen en los armarios del archivo secreto del Consejo (de Indias) para su uso," August 29, 1766. For a bibliometric study of the secret archive as it stood in the early nineteenth century, see Margarita Gómez Gómez and Isabel González Ferrín, "El archivo secreto del Consejo de Indias y sus fondos bibliográficos," *Historia, Instituciones, Documentos* 19 (1992).

Santa Cruz shared the same opinion about the Council's need for cosmographical information. The *Ordinances*, however, rather than assigning these responsibilities on a councilman, created an official post, cosmographer-chronicler, to assume these tasks. Although the law did not explicitly mention the officeholder's qualifications, it implied that the individual must be capable of practicing cosmography according to the rules of the art. Creating the post through a legal order guaranteed a certain measure of permanence and continuity. But perhaps more significant, the law also professionalized the discipline by subsuming its practices, and its corresponding epistemological and methodological canon, into the post.

The *Ordinances* also ventures into cosmographical methodological issues, in particular singling out that lunar eclipse observations should be used to calculate longitude and that the cosmographer-chronicler should compile an official rutter using testimonies from pilots. In general, though, methodological questions were addressed in a separate document, the *Instructions*. The *Ordinances* do not explain why these two aspects were singled out. I speculate that these actions were such novel departures from the Council's traditional function, even those adopted by proxy from years of experience dealing with Santa Cruz, that Ovando considered them best included among the laws governing the Council's operation rather than relegating them to the *Instructions*.

Whereas the *Ordinances* specified the cosmographer-chronicler's responsibilities, the *Instructions* (*Título de las descripciones*) explained the mechanics of gathering, organizing, and compiling the general cosmography and hydrography of the Indies and the more specific geographic descriptions, natural histories, and historical chronicles.[49] It is important to keep in mind that the *Instructions* proposed a documentation program that went far beyond collecting this type of material. Most of the provisions established guidelines for how the Council of Indies, as well as the different governing units in the Indies, from viceroyalties to parishes, had to record, catalog, and maintain

49. The document currently exists as a sequential list of 135 articles or chapters. Internal references to previous articles in the document are numbered incorrectly. It seems that sometime between the time the document was first compiled and the time it was recorded in the register books, subordinate articles were confused with principal articles, so that now all appear sequentially numbered. A contemporary copy of the document survives as a transcription in the Council's *Libros Registros del Consejo* (AGI, IG-427, L. 29, f. 5v-66v and BN MS 3017). For quotations, I use the text of the *Instructions* as edited by Sánchez Bella, "*Título de las descripciones*," 140–211. The *Título* has also been published in Francisco de Solano, ed., *Cuestionarios para la formación de las relaciones geográficas de Indias, siglos XVI–XIX* (Madrid: CSIC, 1988), 16–74.

administrative, ecclesiastical, judiciary, and financial records. The law sought to create a vast recordkeeping system organized around a series of books recording the "state" (*estado*) of the Indies. The document's 135 articles range from orders instructing parish priests to keep a "Record Book of Souls" (*Padrón general de ánimas*) divided by neighborhoods, streets, and houses (*barrios, calles y casas*)—the equivalent to conducting a yearly census—to instructions on how provincial exchequers were to keep tax collection record books. The *Instructions* specify again and again that all these books were to be compiled after "investigations, descriptions and reports of everything concerning the state of the Indies" were carried out and prepared.[50] Preparing the record books was not meant to be a one-time exercise; they had to be maintained in perpetuity and continually updated (*perpetua y sucesivamente*), with yearly summaries and updates sent to the Council of Indies. Based on these summaries, the Council would compile and update its own *Books of Descriptions*, including those describing the cosmography, hydrography, geography, natural history, and moral history of the Indies.

The *Instructions* reflect Ovando's conviction that an orderly system for collecting and organizing information was essential for efficient colonial administration. Although a significant component of the program was aimed at cementing Spain's political control over the distant colonies, Ovando's project was also intended to assist local governance. For not only did the law put in place a system for gathering geographical information and making it available to the Council of Indies; it also urged local civil and ecclesiastic officials to prepare and maintain local geographical descriptions and maps for their own use.[51] *Local* geographical knowledge was seen as an important administrative aid. However, *general* geographical knowledge was another matter. The law made no provision for making available to local authorities the general descriptions and maps after these had been compiled by the ranking provincial authority or, ultimately, by the Council of Indies. The flow of information, as the document states repeatedly, was strictly one way. Each administrative unit was encouraged to "know" its own affairs and territory but prohibited from learning the whole.

The *Instructions* is organized into three main sections of increasing specificity (table 3.2). The first section, articles 2–13, names the officials responsible for complying with the law. The second section, articles 14–40, discusses the specific material to be "investigated, described, and

50. "[A]veriguaciones, descripciones y relaciones de todo el estado de las Indias" (article 1).

51. Article 88.

Table 3.2. Articles concerning the post of cosmographer-chronicler of the Council of Indies, 1571

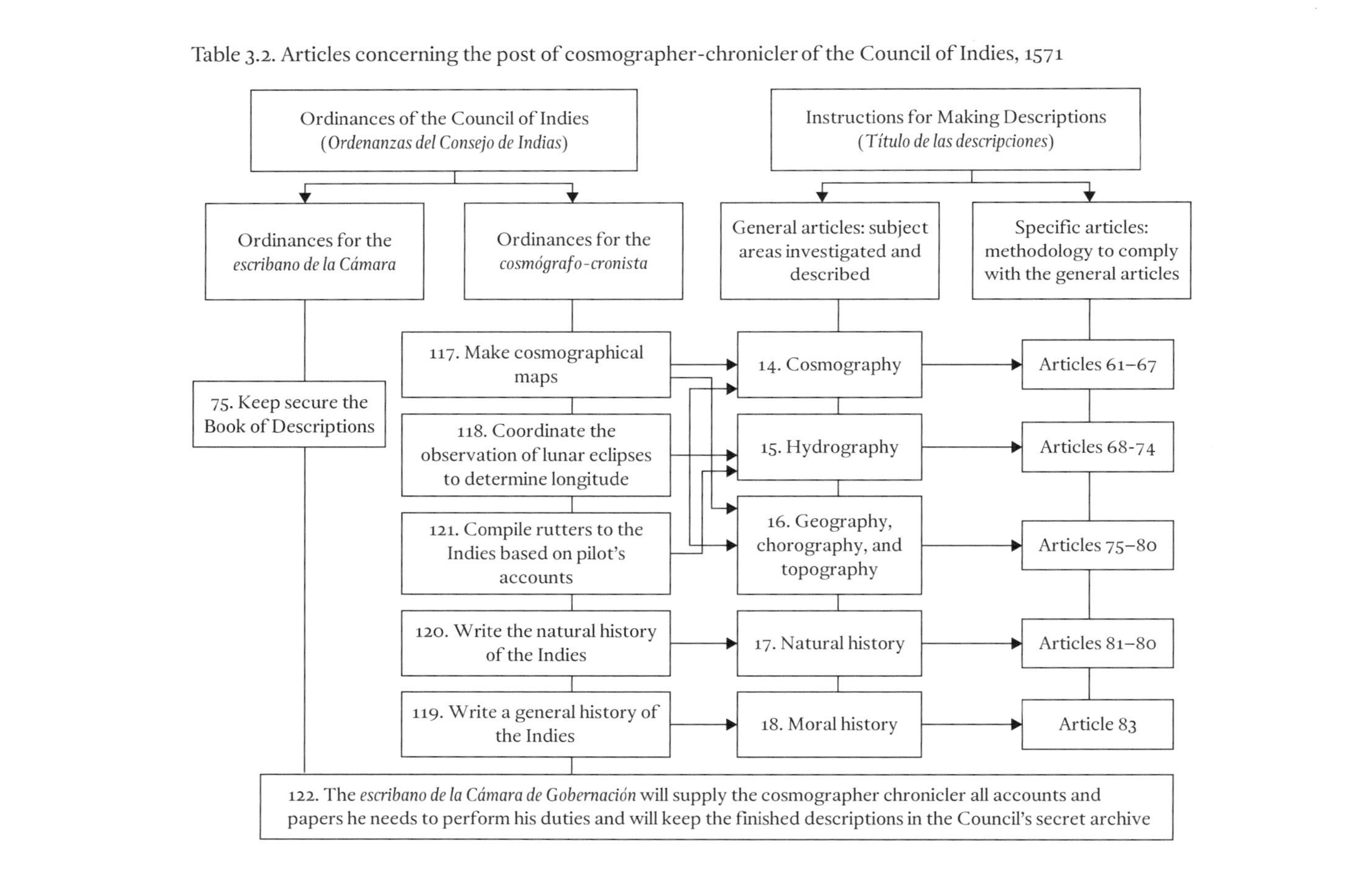

relayed" by these officials. And the third section, articles 41–132, discusses the "order and form" that must be followed when the requested reports and descriptions were prepared. Sometimes this last section amends the list of materials to be collected specified in the second group of articles. The articles in each of these sections fall into three principal areas: civil administration (*temporal*), ecclesiastic administrations (*espiritual*), and the cosmographical and historical description of the Indies. In general, the *Instructions* calls for a broad range of information to be collected at all levels of the civil and ecclesiastic administration, duly notarized, archived locally, and then sent from "inferior to superior" until it reached the Council of Indies in Spain. The law required most government and ecclesiastical officials to contribute to the documentation project. The parish priest and the provincial archbishop alike had to submit yearly reports and prepare description books, as did local caciques, viceroys, and officials at the Council of Indies.

The law did not limit this directive to persons enjoying official appointments. It also issued a blanket appeal to "any of our subjects and vassals," urging them to supply descriptions. The king, the *Instructions* specify, would consider these reports from individuals as a good service and compensate authors with "rewards and favors" (*gratificación y merced*).[52] Even the native population had to contribute to the descriptions, "using the *quipus* or paintings or by any means they might have to understand similar things."[53] By any measure, the breadth of the recordkeeping project was impressive. It generated books ranging from lists of the real estate possessions of the local archdiocese, to books recording civil posts, its officeholders, and their salaries—down to the most humble local official—to books recording the births and deaths among the Spanish and Indian population, down to a book, kept by the parish priest, listing the spiritual "status" of each Indian in his parish!

In Ovando's documentation scheme the books kept by the cosmographer-chronicler became the record book of the visible world, of all things "perpetual and temporal" on the land. These descriptions, which encompassed the land, its peoples, and its history, consisted of five books: a cosmography, a hydrography, a geography (including chorography and topography), a natural history, and a moral history. The first three books recorded "everything about the orb, land and sea" (*todo el orbe, mar y tierra*),

52. Article 13.

53. "[P]or los quipús o pinturas o por forma que ellos tienen para dar a entender cosas semejantes" (article 51).

with the first book being a general "cosmography of all the Indies" to be prepared by the cosmographer of the Council of Indies.[54] This general cosmography was composed using latitude and longitude to locate the Indies and its "principal" parts with respect to Spain and the universe. All cosmographers in the king's employ were requested to contribute to this general description and to channel these works to the royal cosmographer of the Council of Indies, who was responsible for compiling their contributions into one document, kept under lock and key, at the Council. Cosmographers were asked, "using a planisphere or globe," to divide and subdivide the Indies "according to the art of cosmography" into "plagues, climes, and parallels and meridians."[55] They were advised to use the customary divisions or climes when describing the new territory, "continuing these over the Indies"—in other words, placing the Indies in latitudinal relation with the Ptolemaic reference lines traditionally used to depict the European continent. They were also reminded to take special care to record all accidents of the land using latitude and longitude, a reminder that appears many times throughout the law.

In article 62, cosmographers were told to calculate longitude using the city of Toledo as the originating meridian and to number the parallels from east to west, rather than from west to east as the ancients had done. This reference point was chosen because it "is more natural and conforms to the discovery of the Indies."[56] All longitude calculations, the cosmographer was warned, had to be in agreement with the location of the concession line established by the pope. This was to be understood as 13° west of the Canary Islands and 23° west of Toledo and as relative to the line of demarcation with Portugal, considered to be located 29° west of the Canary Islands and 39° west of Toledo.[57] With article 62, the Spanish monarch had created

54. Article 14 and articles 61–67.

55. Article 61.

56. Article 62.

57. Ibid. "[E] otrosí, haciendo escala en el meridiano de la línea de la concesión que por Su Santidad nos está hecha de las Indias que es a doscientas y treinta e una leguas que hacen trece grados del meridiano de las Canarias y a veinte y tres grados del meridiano de Toledo. E otrosí, haciendo escala y parada en el meridiano de la línea de la demarcación e asiento e concordia que tenemos tomada con el serenísimo Rey de Portugal, que es quince grados, que son 266 leguas del de la concesión, y a veintinueve grados, que son 570 leguas de las Canarias, y a treinta y nueve grados que son 682 leguas del meridiano de Toledo." These distances are based on roughly 17.7 leagues per degree of longitude, except for the distance of 570 leagues between the Canary Islands and the line of demarcation, which is calculated at 19.65 l/° and might be a copyist error in the original.

a law requiring every map made by an official cosmographer to conform to the stated—albeit controversial—longitude of these politically sensitive coordinates.

Once the Indies, or more specifically "the part of the orb within our side of the demarcation," had been described in a general cosmography, article 63 asked that this territory be in turn "divided and subdivided" into "as many maps as there are principal parts." These maps not only should record the latitude and longitude of territories discovered so far but also were to be updated to record future discoveries. The territories should also be "extensively" described in writing "by degrees of latitude and longitude" to avoid mathematical mistakes.[58] Even the internal organization of the written text had to follow a particular order. Territories had to be described proceeding counterclockwise geographically, starting at the easternmost point, continuing to the northern and westernmost points, and ending with a description of the southern regions. Leaving little to chance, the law asks that the same territory be described for a second time using "the better known natural limits," meaning rivers, mountains, and other such natural features of the land rather than geographical coordinates.[59]

The law then addresses the specifics of how to determine the longitude and latitude of the Indies. It instructs the principal official of each administrative region to have the latitude (*altura*) of the principal places measured by a skilled person. The results should be kept in the local archive and also sent from "inferiors to superiors" until it reached the Council of Indies. To determine longitude, the law requires the diligent observation of lunar eclipses following the "order that our cosmographer major will send you or the ones given by cosmographers in each province." The law instructs all cosmographers to calculate eclipse dates and communicate this information to the local authorities in order for eclipse observations to be coordinated throughout the realm. Results from these observations were then sent to the Council, where the cosmographer used the observations to determine longitude and recorded the revised coordinates in the general description (*descripción general*). The exercise was to take place continuously until "the longitude of all the parts of the Indies has been taken."[60]

The second book of the "orb, land and sea" trilogy was a hydrography of the Indies. Again, all cosmographers in the king's employ were called upon to contribute hydrographical material, only this time the law expressly

58. "[P]or grados de longitud y latitud por escrito extensamente y no por suma, por la facilidad con que se causa error en los números" (article 63).

59. Articles 64–65.

60. All quotes from articles 66–67.

instructed the pilot major of the Casa de la Contratación and all other pilots and ship's captains sailing to the Indies to contribute descriptions as well. The hydrography was intended as a comprehensive description and a navigation guide of the bodies of water in the Indies, including the oceans and shorelines as well as rivers. The law requested descriptions of the tides, currents, and other accidents of the sea, also of prevailing weather conditions and seasonal changes, and how these might affect navigation. In sum, the general hydrography contained in addition to nautical charts "all the secrets that are known about the navigation of the sea and the *carrera de Indias*."[61]

The section of the law dealing with hydrography is particularly emphatic, insisting that the descriptions in the hydrography and the rutters had to be updated on a continual basis. The law warns that simply noting these geographical "accidents of the sea" in the rutter accompanying the general navigation chart (*padrón de la carta general*) did not suffice, "because the most important accidents cannot be described as well in [a chart] as they could in a book."[62] The law advises that it would be best to keep many of these details secret and confined to this book rather than including them on the official rutter or chart. Clearly, the Crown recognized the illegal dissemination of official maritime charts made by the Casa de la Contratación and believed that some secrets were best kept under lock and key at the Council's chambers.

The *Instructions* reaffirmed existing legislation ordering officials at the Casa de la Contratación to make and maintain the rutter and general chart of the Indies but amended the statute in place by instructing Casa officials to provide the Council of Indies with copies of the documents.[63] As new discoveries and investigations came to light, the cosmographer-chronicler of the Council of Indies was to "make and correct all the navigation charts given to pilots of the *carrera de Indias*." This should take place while the cosmographer-chronicler conducted a periodic audit (*visita*) of the Casa and after meeting with its cosmographers and pilots to collect their opinions.[64]

Cosmographers and pilots assigned to overseas posts were asked to write and describe the rutters of all their trips to the Indies and, upon returning to Seville, give a notarized copy of their log book to the

61. Article 15.

62. "[P]ues en ella no se pueden poner los accidentes mas necesarios que se deben saber como se podrán poner por escrito en un libro" (article 68).

63. Article 69.

64. Article 69. This section set the legal basis for the reform of the *padrón real* by cosmographers of the Council of Indies, but that was not carried out until 1596.

cosmographer and pilot major of the Casa de la Contratación.[65] These would be forwarded to the cosmographer at the Council. The article reminds pilots and ship's masters to carefully note the distance and direction traveled in every voyage in tables, as well as by "celestial observations and other instruments."[66] Pilots also had to swear that they would hand over to the Casa or the Council their original rutter and description book if they ever left the piloting profession or when they died. Furthermore, before a pilot could be certified to sail the *carrera de Indias* he had to swear he would not reveal "the secrets of navigation to any foreigner.[67]

The cosmography and hydrography books prepared by the cosmographer-chronicler at the Council in Madrid were conceived as all-encompassing descriptions; in the words used in the law, they were "general." The task of describing the particular geographical features of each territory, however, was delegated to each administrative unit. This meant composing a "geography, which provided a description of all the earth of the Indies, a chorography, which dealt with each region and province, and topography, which entailed a description of each particular place."[68] These geographies, chorographies and topographies were intended to be written by eyewitnesses and used locally.[69] Yet like the other series of reports, these descriptions were also to be sent every year to the cosmographer-chronicler. Based on this material, he then compiled a general geographical description that consisted of a "a book where he is to put maps and written descriptions, so that the first map contains the description of the mainland and adjacent islands, and then subdividing this into as many maps as there are principal regions."[70] The law does not attempt to specify how these maps were to

65. Article 71. In article 73 pilots were instructed that if they were not to return to Seville they were to leave a copy of the rutter with the local governor, who in turn should send it to the Council of Indies.

66. "Otrosí, vayan tomando la longitud de cada rumbo y derrota que llevaren por la singladura y por observaciones celestes y por otros instrumentos que para ello supieren y lo vayan todo poniendo muy precisamente en su libro y derrotero" (article 72). Alonso de Santa Cruz in the *Libro de Longitudes* defines sailing by *singladuras* as a way of knowing the distance a ship has traveled in an hour by noting the direction sailed, the type of ship, and the intensity of the wind (Santa Cruz, *Obra cosmográfica*, 143–44).

67. Article 74.

68. Article 75.

69. "[L]os que tuvieren noticia de ella por vista de ojos" (article 76).

70. "[L]ibro en que ponga las tablas en demostración y por escrito, de manera que en la primera table ponga toda la descripción de la tierra firme y continente con las islas adyacentes, la cual subdivida en tantas tablas cuantas fueren las regiones principales" (article 75).

be constructed, leaving the cosmographer to fashion them "according to the rules of his art."[71] The law cautions local authorities that if they did not have a cosmographer available to prepare the requested local maps, they were still obligated to describe all geographical features using latitude and longitude. They were authorized to use distances expressed in leagues only to specify locations in topographical descriptions. Each written description had to be accompanied by a topographical drawing (*padrón de demostración o pintura de su provincia*), with a copy sent to the Council of Indies.[72]

The hierarchical scheme adopted in the *Instructions*, particularly for the geographical sections, borrowed heavily from the organizing scheme and methodology advocated and used by Ptolemy in his *Geography*. Local geographical information was first collected, preferably from persons with firsthand knowledge, and was later organized and correlated by the geographer, in this case the cosmographer-chronicler of the Council of Indies. To compose the general cosmography and maps, the geographer worked from the specific to the general, until a complete vision of the world was carefully assembled. The *Instructions* also addressed a problem Ptolemy pointed out in the *Geography*—the inconsistencies inherent in firsthand accounts. To remedy this problem, the *Instructions* gave detailed directives about the material to be included, the way the information was to be collected and organized locally, and the persons responsible for doing so. These directives solved some of the problems Santa Cruz attempted to address in his guidelines regarding the nature of the *relaciones* available to the cosmographer. With the data collection and reporting parameters clearly specified, the reports, since they had been prepared with common underlying criteria, would be consistent and easier to correlate when it came time to compose the general cosmography.

The two final books under the purview of the cosmographer-chronicler, the natural history and moral history, were essentially conceived of as inventories: the natural history as a inventory of natural resources and the moral history as a listing of significant events in the region's history. The natural history was to be, in the law's words, "the description of natural things in the Indies that are in perpetuity and that rarely change . . . and that would be beneficial for those governing to know about."[73] The law emphasized the need to carefully record the description, and especially the utility, of the native animals, plants, and mineral resources, noting how these were used and how they could be better employed. It also requested a description of the native peoples, or rather, as the law states, of the "nations

71. Article 77. 72. Article 80. 73. Article 17.

of men that are [there] and their nature and qualities" (*las naciones de hombres que hay y las naturalezas y calidades de ellos*). Any person receiving a salary as chronicler and residing in a given territory was charged with writing this natural history. If a chronicler was not available, then the region's chief medical official (*protomédico*) or the appointed provincial doctor had to prepare the natural history. If neither was available, the task fell to the notary of the local governing authority. With respect to the natural history's "form and order," the compiler was instructed to keep a book divided into chapters by subject matter;[74] after writing what was currently known, he should leave the rest of the chapter "open" (*abierto*) so that as new things were discovered or learned they could be added.[75]

The fifth book was a moral history of things "contingent and variable" (*Historia moral contingente y variable*).[76] Perhaps not surprisingly, given the Council's concern about uncensored historical narratives, this task was not assigned to the local chronicler, even if there was one, but rather given to the notary of the local provincial council and to the notary of the governance chamber of the Council of Indies in Madrid.[77] Royal notaries, assumed to be "loyal, legal, and capable individuals," were authorized to collect material from local archives about the history of the discovery and conquest of each province. The history of the former Indian kingdoms and of the native inhabitants had to include accounts of the native's customs, rites, and types of government in significant detail.[78] The chronicle also had to record the effect on the natives of the Spanish presence and if they had been harmed by it. Finally, it asked for a record of the actions of Spaniards since the day they entered the region. Once again, this moral history was to be kept closely guarded and sent to the Council of Indies.[79] This is the only section of the instructions for the *Books of Descriptions* that asked the compiler to consult trustworthy persons (*personas fidedignas*), should material in the archives not prove sufficient.

While the cosmographical project outlined in the *Instructions* adopted some methodological and epistemic aspects from Ptolemy, the scope of the project defied anything envisioned by the ancient geographer or by the

74. Article 17 includes a comprehensive list.

75. Article 82.

76. Article 18.

77. Article 83. Recall that in article 75 of the *Ordenanzas* the Council's notary of the governance chamber was instructed to maintain the chronicle.

78. Article 18.

79. Article 83.

humanist scholars who had laid down the discipline's foundation. This was not a bookish attempt at synthesis but rather an immense, state-sponsored project that required significant administration. Ovando's project was a deliberate effort to design ways of collecting and organizing knowledge to serve a utilitarian purpose. Furthermore, it was designed so as to construct this body of knowledge upon a basis of "facts," or as the law states, "known and investigated things," an aspect discussed later in this chapter. As such, it is a very early example of the systems of classification of useful knowledge that would become instrumental in the development of modern science and best known through the work of Francis Bacon and his *The Advancement of Learning* (1605), *The New Atlantis* (1627), and *New Organon* (1620).

It comes as no surprise that the scheme outlined in the *Instructions* turned out to be too ambitious for distant territories to implement in its entirety. Although the information was forthcoming and there was an honest effort to comply with the regulations, the reports were uneven and inconsistent. In the case of the city of Quito, as Sánchez Bella has found, after the *Instructions* was promulgated in 1573 the regulations were not read before the local council until August-September 1574, and then were not sent to subordinate administrative units until almost two years later, and significantly shortened. The final report was complete and ready to be sent to Spain on the last day of 1576.[80]

Also in 1576, the Mexican viceroy, Archbishop Pedro Moya de Contreras, complained to the Council of Indies that complying with the *Instructions* was a "laborious business, especially since so many books had to be made at our expense."[81] Some local clerics, he said, had already prepared their reports "in their own way." He notified the king that he would continue to collect their reports with the hope of preparing a more polished version but would ensure it was done in a timely manner. The archbishop did not have high expectations for the requirement that local clerics compile moral histories of the native population. He feared that they would report "less than is necessary and in such a form that it will be difficult to understand what

80. Sánchez Bella, "*Título de las descripciones*," 124–29.

81. Letter from the archbishop of Mexico to the king, March 28, 1576. Published in Francisco del Paso y Troncoso, *Epistolario de Nueva España, 1505–1818*, 16 vols. (Mexico: Antigua Librería Robredo, 1939–42), 12:12–13.

they wrote" due to the multiplicity of nations, languages, and customs.[82] He promised to continue seeking persons to prepare the works, although he complained, "These people are not inclined to similar exercises." To expedite compliance, he proposed to have a history written by a local Franciscan, Bernardino de Sahagún, translated and sent.

The archbishop's criticisms hint at some fundamental problems with the project Ovando had designed. The *Instructions* assumed that persons with sufficient education and experience would be available (and willing) to prepare the requested material. As Moya points out, this expectation could not be met in the context of the rough reality of the New World. For all the care that Ovando took to define what should make up the "moral history" and the cosmography, it still framed these requests in a language that presupposed that respondents would know the structure, conventions, and methods associated with what was an elite humanistic genre. To a significant portion of Spanish settlers in the New World, this was essentially unintelligible. The lukewarm response to the *Instructions* and the many years it took to elicit a response from the overseas territories also contributed to the project's demise. By the time some of the reports reached Madrid in late 1576, the cosmographer-chronicler of the Council of Indies had superseded the original information-gathering methodology outlined in the *Instructions* in favor of questionnaires.

Legal Culture and Cosmographical Methodology

With the *Instructions*, Ovando put in place a law that established a vast documentation system intended to neatly record on an ongoing basis the reality of the New World—a reality as conceived by a legal scholar convinced, somewhat naively, that access to accurate and timely information would allow a reader in Spain to learn enough about the empire's distant domains to be able to legislate justly and wisely. The *Instructions* did not escape the influence of legal procedure in late sixteenth-century Spanish society, an emphasis on the law and legal protocol so widespread that historians refer to Spanish society as permeated by a "legal culture."[83] Evidence in the form of documents and records was the handmaiden of the legal culture within which Ovando conceived of his reforms. The law addressed the written

82. "[C]uando cada clérigo se ponga a inquirir algo será lo menos de lo necesario, y irá por tal término que será peor entender lo que escribieren" (in Paso y Troncoso, *Epistolario*, 12:13).

83. Richard L. Kagan, *Lawsuits and Litigants in Castile, 1500–1700* (Chapel Hill: University of North Carolina Press, 1981), 137–50.

works generated by cosmographers the same way it addressed fiscal reports written by accountants or books of parishioners compiled by priests: as another useful tool for managing the empire and thus too important to be left at the margins of the law.

The *Instructions* approached cosmography in a manner unprecedented in the discipline's history. Instead of simply specifying the recordkeeping system to follow, the law reached into the realm of practice and attempted to codify the methodologies associated with the discipline. The process of codification inherently required that the practices be framed in legal language. This implied that they first needed to be defined and the boundaries of the practice clearly delineated. The *Instructions*, however, was intended to remain faithful to the traditional practices associated with these disciplines. In writing it, Ovando respected the Ptolemaic methodology of proceeding from the specific to the general when composing geographical descriptions and maps, and he advocated, with a fervor that would have made the great astronomer proud, the use of latitude and longitude coordinates to determine location.

But the law also delved into aspects of practice perhaps best left to a cosmographer to decide. For example, in some instances the law established the epistemological basis for what constituted a cosmographical fact in an attempt to resolve via the letter of the law epistemological questions associated with cosmography. Recall as an example that the law instructed, that if available, eclipse observations were to be the preferred source of longitude information, with firsthand rutters based on distances and bearings as the alternate source. The project's overarching epistemological basis continued to be the implied creditworthiness of eyewitness (*por vista de ojos*) accounts structured in the form of local geographical, ethnographical, and natural historical descriptions. To establish the credibility of eyewitnesses, the law specified the individuals who had to compose and maintain these books: local government officials, priests, or other "learned persons" who were familiar with the area. In instances where the law did not specify a particular methodology to determine the facts, it resorted to a general criterion. Repeatedly, the law instructed that matters inscribed in the *Books of Descriptions* had to be "known and investigated things" (*cosas sabidas y averiguadas*).[84] This is the phrase used throughout the *Instructions* to indicate when something had reached the status of "fact."

In recent decades, historians have studied the influence of legal techniques in early modern Europe, particularly in England, in establishing the

84. Articles 63, 82, and 83 of the *Instructions*.

modern concept of "fact."[85] Barbara Shapiro and Julian Martin find that the epistemological philosophy of Francis Bacon, with its emphasis on witnessing and firsthand accounts as means of ascertaining "matters of fact," was profoundly influenced by his legal training.[86] This was clearly also the case nearly forty years earlier with Juan de Ovando and the *Instructions*.

Like Bacon, Ovando proposed a scheme for organizing knowledge that relied on eyewitnesses and notarized depositions. In the Spaniard's case, the scheme married Renaissance cosmography, the law, and the imperial bureaucracy. However, Ovando's emphasis was not in setting down the methodology to be followed for determining when something was "known" or how it was to be "investigated." These remained firmly empirical and nestled comfortably within the system of Aristotelian natural philosophy. Ovando's interest lay in translating procedural aspects of law—the collection of evidence, establishing the credibility of witnesses, recording testimonies—into a methodology that yielded the desired corpus of cosmographical, natural historical, and historical information.

Bacon would go that one crucial step further and define the epistemic and methodological how-to of fact construction. He addressed the questions Ovando left—save for a few cases—unanswered: how "things" were

85. The analogous word in Spanish to the English noun *fact* is *hecho*. The etymological root of the word, often spelled during the sixteenth century as *fecho*, is the Latin *factum* or "deed." Historians have noted that by the seventeenth century in England and France, the significance of *fact* had migrated beyond its traditional definition, "event," used principally to denote human action, to being used to denote natural phenomena. In Ovando's and López de Velasco's work, *hecho* retained its traditional definition, while it seems that the act of establishing when "something" became analogous to fact or *cosa sabida y averiguada* did not correspond to a distinct noun. For more on the concept of "fact" in early modern Europe, see Lorraine Daston, "Baconian Facts, Academic Civility, and the Prehistory of Objectivity," *Annals of Scholarship* 8 (1991); Lorraine Daston, "Strange Facts, Plain Facts, and the Texture of Scientific Experience in the Enlightenment," in *Proof and Persuasion: Essays on Authority, Objectivity, and Evidence*, ed. Suzanne Marchand and Elizabeth Lunbeck (Turnhout, Belgium: Brepols, 1996), 42–59; Martin, *Francis Bacon, the State, and the Reform of Natural Philosophy*, 72–80; and Barbara J. Shapiro, *A Culture of Fact: England 1550–1720* (Ithaca, N.Y.: Cornell University Press, 2000). On the role of social conventions in establishing the creditworthiness of witnesses when constructing knowledge claims in seventeenth-century England, see Steven Shapin, *A Social History of Truth: Civility and Science in Seventeenth-Century England* (Chicago: University of Chicago Press, 1994).

86. Shapiro mentions that she is unsure of the Spanish case and wonders whether the "fact collection" done by the cosmographer-chronicler of the Council of Indies used the same legal techniques as those used in England to establish legal "matters of fact" (Shapiro, *Culture of Fact*, 36, 106–7). Chapter 6, below, shows that this was indeed the case.

to be "known" and how they were to be "investigated." The truth is that although both statesmen shared a conviction regarding the value of knowledge in the governing of empire, they faced vastly different realities. Bacon could fancy a future time and place—a new Atlantis—where a reformed natural philosophy could be coaxed to yield knowledge useful for a budding empire. Ovando, in contrast, suffered from an embarrassment of riches. The nature and peoples of the New World thrust upon men like Juan de Ovando and Alonso de Santa Cruz a reality so complex, so monumental in scope, and so enigmatic that simply attempting to catalog the whole was ambitious enough.

Ovando's project was in keeping with a generalized and increasing dependence on documents that accompanied the growth of the Spanish empire during the sixteenth century and the prominent place *letrados* had in its administration. During the reign of Philip II there was a dramatic increase in the use of written consultations when councils or individuals needed to seek the king's opinion or decision on any of a wide range of matters. This was largely a result of the size and extent of the empire; but it also reflected the king's personal inclination toward written records and his preference for replying personally to these consultations.[87] Royal secretaries even advised petitioners to leave wide margins on their communications so that the king could personally annotate documents!

This mode of written consultation also established a historical record, which as long as it was stored to facilitate retrieval, could be consulted repeatedly. The archives established to store these documents served as a form of institutional memory, akin to a reference library, that could be marshaled to many ends.[88] In Ovando's reforms, cosmography became the classificatory scheme to organize the archives that described the human and natural elements of the New World. The *Instructions* created a written memory for the Council of Indies and in the process created a written image of the New World—an image constructed systematically from firsthand reports that were to be methodically recorded and archived in the Council chambers. For a jurist like Ovando, this was as close to truth—and as far from imagination—as you could get.

Ovando designed the law to yield specific products to inform the Council of Indies about the New World. The law sought from the cosmographer a cosmographical description and map of the Indies to be kept in the form

87. Bouza, *Corre manuscrito*, 261–66.

88. Bouza reminds us that archives are far from "innocent" keepers of paper. The decision regarding what is included and excluded from the archive is "constructed, modeled, eliminated, [and] used" (Bouza, *Corre manuscrito*, 286).

of a new type of cosmographical opus, the book of descriptions, with the compilation of its content an ongoing activity, woven into the administrative fabric of the empire by the hands of capable bureaucrats. Therefore, the *Instructions* defined cosmographical practice by methods that achieved this specific mode of representing cosmographical knowledge. In doing so, the *Instructions* implicitly recognized that the literary genre associated with Renaissance cosmography had failed to keep pace with the needs of a body charged with governing an ever-expanding empire. The information the Council so eagerly sought would now be readily available in the Council's chambers. By design, the *Books of Descriptions* would contain material that was current, neatly cataloged by jurisdiction, and summarized yearly by the cosmographer-chronicler. Rather than waiting what seemed a lifetime (and sometimes was) for a comprehensive cosmographical opus, the Council would have access to a constantly updated body of knowledge about the natural world, the geography, and the peoples of the Indies.

Ovando and the Council of Indies failed, however, to recognize a danger that lurked in the efficient schema of the *Books of Descriptions*. If the *Instructions* was interpreted broadly, likely what Ovando had envisioned, it would have given the cosmographer-chronicler an efficient mechanism for collecting and cataloging cosmographical information in a timely manner; otherwise, it had the potential of simply functioning as a bureaucrat's to-do list.

When the first cosmographer-chronicler major of the Council of Indies took office as a result of the legal reforms spearheaded by Juan de Ovando, cosmography in Spain entered an institutional phase that at once consolidated the profession's authority and challenged traditional modes of practice. Obligated to comply with the procedures and methods specified in the *Instructions*, the cosmographer-chronicler practiced his discipline within a strictly codified intellectual arena. The extent of this codification was unprecedented in the tradition that had yielded Renaissance cosmography. Would the measures taken in the *Instructions* to define precisely the bounds of cosmographical practice in the Council of Indies yield a conceptual framework that could encompass the New World? Or would they prove to be restrictive and stifle cosmographical practice? The following chapters explore the career of the first cosmographer-chronicler of the Council of Indies and attempt to answer these questions.

FOUR

The Cosmographer-Chronicler of the Council of the Indies

When the new ordinances of the Council of Indies went into effect in 1571, the task of managing cosmographical information requested by the ordinances fell to Juan López de Velasco. Velasco would serve for nearly twenty years as the official gatekeeper of cosmographical knowledge about the New World and compose the Spanish empire's first official—though secret—cosmography of the Indies. He deftly balanced bureaucratic authority and the prevailing economy of favors to secure valuable information about the New World, and in so doing he became the locus of an information exchange network that produced, collected, and compiled cosmographical knowledge. As a consummate bureaucrat (or person engaged in the *oficio de papeles*), he struggled during his tenure to balance the scientific and secretive aspects of his official post with his humanistic sensibility. As the first official cosmographer of the Indies, he unwittingly defined what constituted "official" cosmographical practice in the institutional environment of the Council of Indies. This made his work the target of attacks from others who conceived of cosmography as constituting a wholly different set of practices from those outlined in the *Instructions* and *Ordinances*. Velasco's life and social context—the lens through which he viewed and expressed his intellectual sensitivities—are impossible to disengage from the complex patronage system in which he operated and are an essential introduction to the cosmographer's work.

The Empire, Patronage, and the Humanist: Juan López de Velasco, 1571–90

Juan López de Velasco (ca. 1530–98) was born in Vinuesa (Soria) to Juan López Carrasco and Catalina Velasco.[1] He had two brothers, Pedro and Francisco, and two sisters, Catalina and a sister who preceded him in death. Both brothers emigrated to the New World; Pedro, a cleric, left for Quito in the 1560s, while Francisco died in Nueva Granada before 1598.[2] It is unclear where or even whether Velasco attended a university. The López de Velasco family's scant patrimony has led historians to assume that financial resources would have permitted Velasco only to attend a university close to his hometown. Yet inquiries at the University of Alcala and the Colegio-Universidad de Sigüenza have not produced any documentary evidence that Velasco attended either of these schools. A clue to his educational background may lie in Velasco's correspondence with Francisco Cervantes de Salazar, who taught at the Universidad de Osuna between 1547 and 1550.[3] Velasco refers to him as teacher (*maestro*) and in one letter apologizes for not having kept up his Latin studies. Unfortunately, the student record books for the university's early years no longer exist.

By 1563 Velasco was working at the Council of Indies preparing legal summaries, and by 1569 he was at Juan de Ovando's side as the future president compiled into a unified legal code hundreds of laws and directives

1. The following studies reconstruct aspects of Velasco's life: José Antonio Pérez-Rioja, "Un insigne visontino del siglo XVI: Juan López de Velasco," *Celtiberia* 8, no. 15 (1958); Manuel Miguélez, *Catálogo de los códices españoles de la Biblioteca del Escorial*, 2 vols. (Madrid: Imprenta heletica, 1917); and the introduction to López de Velasco, *Geografía y descripción*, v–xlviii. Additional material was gleaned from Velasco's personal correspondence with government officials at the British Library, Instituto de Valencia de Don Juan, and the Biblioteca Zabálburu. His will is at AHPM, Gonzalo Fernández, tomo 1638, f. 15–24, "Testamento de Juan López de Velasco," May 1, 1598, and was published by Pérez-Rioja.

2. AGI, Quito-24, N. 9, and AGI, Quito-1, f. 43, "Sobre hacer merced a Pedro López de Velasco . . . ," June 3, 1595. Francisco went to Nueva Granada in 1574 as assistant to the local magistrate. AGI, Pasajeros, L.5, E. 3383, October 25, 1574.

3. Francisco Cervantes de Salazar (ca. 1514–75), the author of the *Crónica de Nueva España*, was from Toledo, studied at the Universidad de Salamanca, and later taught at the Universidad de Osuna. He founded the Colegio de la Viscaínas in Mexico City and taught canon law and rhetoric for many years at the University of Mexico. Salazar was an ambitious man, and many of the letters in this collection are about his efforts seeking appointments. Francisco Cervantes de Salazar, *Crónica de la Nueva España*, ed. Agustín Millares Carlo, Biblioteca de autores españoles 44 (Madrid: Atlas, 1971), 13–23, 92–95.

issued since the discovery of America.[4] Among the depositions Ovando collected as part of the audit is one by Velasco written in 1567, in which he mentions his more than four years of association with the Council of Indies as secretary to Juan Sarmiento and Francisco Tello de Sandoval during their terms as Council president.[5] The deposition reveals a Velasco concerned about the Council's hasty management of matters related to the administration of the Indies and complaining that councilors spend too much time resolving legal disputes and meeting with businessmen, presumably pitching business ventures overseas. He argued in favor of requiring members of the Council to have lived in the Indies, contending that otherwise they inherently lacked the experience necessary to carry out expeditiously the duties of the office. The young secretary recognized, as Alonso de Santa Cruz had years before, that firsthand knowledge of the New World was an important requirement for anyone expected to legislate over the distant lands, although we do not have evidence Velasco himself spent any time overseas. It is difficult to judge whether Velasco's deposition had any bearing on the formulation of the *Instructions.* He did take part in the *junta magna* that deliberated on the findings of Ovando's *visita*, although in what capacity is unclear.[6]

While he was Juan de Ovando's assistant, Velasco began collecting geographical descriptions of the New World, many of which were *relaciones* written in response to specific requests for information issued by Ovando during the audit. A number of these documents survive and can be identified by an annotation on the top left-hand margin in Velasco's distinctive handwriting describing them as "accounts from time of the audit" (*relaciones del tiempo de la visita*).[7] For example, along with Velasco's deposition is

4. AGI, IG-425, L. 24, f. 157, "Pago de 200 reales a Juan López de Velasco por trasladar un libro de las cosas del oficio del Consejo," September 15, 1563. IG-425, L. 24, f. 195, "400 reales por trabajo de sacar resumen de ciertas cosas contenidas en los libros de Indias," 1564. Again in 1565 another 40,000 maravedís for "lo que ha trabajado . . . en provisiones y cédulas que se han despachado desde el descubrimiento" (IG-425, L. 24, f. 241). On December 1569, he was remunerated 300 ducats for services during Juan de Ovando's audit of the Council of Indies (IG-426, L. 25, f. 39–39v).

5. BL, Add 33983, f. 294–295, Juan López de Velasco, "Depositions before Juan de Ovando," 1567–68. Juan Sarmiento was president of the Council of Indies from January 1563 to March 1564, while Don Francisco Tello presided from April 1565 until 1567. Schäfer, *Consejo de Indias*, 1:334.

6. Poole, *Juan de Ovando*, 131.

7. Jiménez de la Espada was the first to note the significance of the annotation. Marcos Jiménez de la Espada, *Relaciones geográficas de Indias. Perú* (Madrid: Real Academia Española, 1881–97), reprint, Biblioteca de autores españoles 183–85 (Madrid: Ediciones Atlas, 1965), 183:48.

one by explorer and conquistador (*adelantado*) Pedro Menéndez de Avilés concerning his recent expedition to Florida to rid the territory of Huguenot colonists and pacify the Indian population.[8] Velasco contextualized for the Council of Indies Avilés's testimony with a brief outline explaining the historical precedents of the relationship between Florida's native population and the Spaniards.[9]

It was Velasco's handling of this kind of material that made him in Ovando's eyes a suitable candidate to fill the post of cosmographer-chronicler major of the Council of Indies. Sometime in 1570 or 1571, as it became clear that the new ordinances governing the Council of Indies would come into effect, Ovando proposed Velasco for the new post. His letter endorsing his protégé survives. It is a sparse, undated communication likely addressed to Cardinal Diego de Espinosa, then president of the royal council and Philip II's chief minister. Ovando mentions that Velasco had been collecting cosmographical material for some time and that if given the post he "will know how to do it well."[10] It is hardly a ringing endorsement of the young secretary's talents. The absence of any reference of Velasco's cosmographical expertise suggests that he did not practice the discipline with any degree of specialization. Furthermore, it suggests that Ovando did not think the specialized skills associated with cosmographical practice were very important given the way the office and practice were defined in the *Ordenanzas*.

Velasco's order of appointment was recorded on October 20, 1571.[11] It essentially reiterated the functions associated with the post as delineated

8. BL, Add 33983, f. 324, "Depositions before Juan de Ovando," 1567–68. Depositions by Pedro Menéndez de Avilés relating to the Indies, chiefly Florida and Terra-Nova, Madrid, March 28, 1568, with supplementary deposition dated April 14, 1568.

9. AGI, P-19, R. 23, "Relación que da Juan de Velasco cosmógrafo mayor de S. M. de lo sucedido desde el descubrimiento a la Florida desde el año de 14 hasta el de 65." The document is in Velasco's hand. Its catalog description is incorrect, since the document discusses events up to 1568.

10. "Suplica V.S. su Illust. sea servido que en el officio de cosmógrafo y coronista de las cosas de Indias se provea en Juan de Velasco porque lo sabría bien hazer y tiene hecho mucho tambien en los papeles de Indias que es necessario que ponga luego en ejecución." IVDJ, Envio 25, n. 528, Juan de Ovando, "Carta de Juan de Ovando pidiendo puesto de cosmógrafo para Juan López de Velasco," ca. 1571. Published in Jean-Pierre Berthe, "Juan López de Velasco (ca. 1530–1598), cronista y cosmógrafo mayor del Consejo de Indias: Su personalidad y su obra geográfica," *Relaciones* 19, no. 75 (1998): 150.

11. AGI, IG-874, "Provisión de Felipe II nombrando a Juan López de Velasco cronista-cosmógrafo mayor de las Indias. Juramento de éste y toma de posesión de su oficio. Asiento de título en los libros de la contaduría de la Hacienda de Indias," October 20, 1571. Published in Vicente Maroto and Esteban Piñeiro, *Aspectos de la ciencia*, 415–17.

in the *Ordinances* and the *Instructions.* The order gave him the title "Cosmographer and Chronicler Major of the states and kingdom of the Indies, islands and mainland of the Ocean Sea" and granted him all the honors, rewards, and immunities associated with the post. Specifically, the order instructed him to "make and compile" the general history, both moral and particular, of memorable events in the Indies, both past and present. Likewise, he was to record the land's natural history. The order also included the directive to "see and examine such histories as others have made." Although it does not use the word *censor*, it suggests that Velasco was to review these histories critically and determine whether any damage would come from their publication. As far as cosmography was concerned, he was to "organize, put into form, and practice" (*ordenar, poner en forma y ejercicio*), the cosmography of these lands and "as is, should, and can be done" by other chroniclers and cosmographers in the kingdom. In addition to the customary pledge to comply with the duties of his post, his oath of office included the pledge that when instructed to do so he would "keep secret" certain matters entrusted to him. Once the oath was in place, the new cosmographer-chronicler was to be given all the "histories, *relaciones*, communications, memorandums, letters, and other books and papers there are and that concern his duties."

Velasco was sworn in as the first cosmographer-chronicler major of the Council of Indies with a salary of 100,000 maravedís a year. (In comparison, Rodrigo Zamorano earned 60,000 maravedís a year as lecturer in cosmography at the Casa de la Contratación.) In 1572, Velasco's wages were effectively increased by 50 percent when he was granted an annual stipend of 50,000 maravedís (or 266 ducats) in addition to his salary.[12] The young man's rising fortunes did not go unnoticed. In a rare glimpse into the cosmographer's personal life, we learn that his former teacher, Francisco Cervantes de Salazar, encouraged his sister, Catalina de Sotomayor, to approach Velasco as a possible husband for one of her two daughters. Doña Catalina, though clearly amicable to the match, had to explain to her brother that it would require seven or eight thousand ducats to discuss these matters with Velasco. In addition, she explained, "he is little inclined toward marriage, since he is so well established and with such good prospects."[13] Velasco never married but maintained a lifelong friendship with Doña Catalina's daughters, bequeathing in his will to Doña María de

12. AGI, IG-426, L.25, f. 169–169v, March 24, 1572.

13. Letter from Catalina de Sotomayor to Cervantes de Salazar, Toledo, April 24, 1572. In Agustín Millares Carlo, ed., *Cartas recibidas de España para Francisco Cervantes de Salazar (1569–1575)* (Mexico: Antigua Librería Robredo, 1946), 82.

Peralta and Doña María de Espinosa some jewelry and religious pieces, "in appreciation for all the good will they have shown me and how much they have prayed to God on my behalf."[14]

Velasco had landed one of the few professional positions for a man of science that garnered a government salary in Spain. But when we compare his subsequent career to that of a peer, he seems to have drawn the short straw. While conducting a *visita* to the Universidad de Alcala in 1562–65, Juan de Ovando had employed the young Mateo Vázquez, a fellow Sevilian, as his secretary. In this case Ovando proved far more effective in finding his protégé a suitable position. On Ovando's urging, Vázquez became the personal secretary to Cardinal Diego de Espinosa, then president of the royal council and inquisitor general. After the cardinal's death in 1572, Vázquez began to assume a dominant position within the close-knit circle of royal secretaries; he earned the king's trust and served as his right-hand man for more than twenty years.[15] As secretary of the Cámara de Castilla, the principal dispenser of royal patronage, Vázquez would play a considerable role in the professional future of Juan López de Velasco.

As was customary among officeholders, over the next twenty years Velasco sought to increase his income by petitioning the Crown for special favors (*mercedes*), expenses (*ayuda de costa*), and other benefits, such as housing and promotions. He took a multipronged approach when asking for these favors. He petitioned the Council of Indies for rewards in his capacity as cosmographer, while often in parallel petitioning, through Mateo Vázquez, the king's chamber council (*consejo de cámara*) and the Council of Castile to reward other projects. Favors or *mercedes* in early modern Spain were not limited to rewards for the petitioner. In this economy of *mercedes*, a petitioner could also ask the Crown to extend favors to the petitioner's network of debtors and friends.

Velasco knew how to handle this currency of favors. One exchange involved Cervantes de Salazar, who from Mexico in 1570 was trying to gain a position in the recently established Mexican Holy Office. During this time Velasco was working on a commission from the Inquisition to expunge the *Lazarillo de Tormes*, a project that placed him in circle of the new inquisitor general of Mexico, Pedro Moya de Contreras. At Cervantes de Salazar's

14. AHPM, Gonzalo Fernandez, tomo 1638, f. 15–24, "Testamento de Juan López de Velasco," May 1, 1598, Madrid.

15. A. W. Lovett, "Philip II and Mateo Vázquez de Leca: The Government of Spain (1572–1592)" (Geneva: Librairie Droz, 1977), 13–15; Antonio Feros, "El viejo monarca y los nuevos favoritos: Los discursos sobre la privanza en el reinado de Felipe II," *Studia Historica: Historia Moderna* 17 (1997).

request, Velasco wrote to the archbishop recommending Salazar for the post of "consultant" for the new office.[16] Velasco also arranged for the prerequisite documents certifying Salazar's purity of blood (*probanza de limpieza de sangre*).[17] Salazar later sought the office of *maestrescuela* (chief judicial authority at the Universidad de Mexico), and again Velasco worked to advance his petition.

The relationship between Velasco and Salazar was certainly not one-sided. In 1573, Velasco wrote to Salazar, who had served since 1558 as official chronicler of Mexico City, thanking him effusively for his offer to place some historical material at the cosmographer's disposal. As Velasco explained, "[S]ince I am new in this office, I have become greedy for any [historical] material, any part of it and from anyone I will value very much, but so much more if it comes from your hand."[18] In return, Velasco promised to show Salazar's books—Salazar was a Latinist in addition to historian—to the "persons of letters at this Court," and Velasco lists the leading humanists of the time: Antonio Gracián, Zurita, Ceriolán, Covarrubias, and of course, the maximum exponent of Spanish humanism, Benito Arias Montano. Velasco held on to the historical material Salazar gave him, probably the only existing manuscript of the *Crónica de Nueva España*, until 1575, when an indefatigable Doña Catalina convinced him to return the material.[19]

Velasco's first significant reward came after he completed his first work as cosmographer of the Council of Indies, the *Geografía y descripción universal de las Indias*. The reward amounted to 400 ducats and a letter describing the work as "extensive and well-prepared" and stating that he had put in good order the geography and description of the Indies.[20] The Council

16. Letter from López de Velasco to Moya de Contreras, August 30, 1570, Madrid. In Millares Carlo, ed., *Cartas de España*, 57.

17. Letter from López de Velasco to Cervantes de Salazar, June 24, 1571, Madrid. In Millares Carlo, ed., *Cartas de España*, 72–73.

18. "V.m. me ofrece de papeles y cosas concernientes a historia, de que ahora, como novel en el oficio, yo ando avaro de este material, que cualquiera parte de ella y de quienquiera que sea no dejaré de estimar, cuanto más lo que fuere de mano de V.m." Letter from López de Velasco to Cervantes de Salazar, June 10, 1573, Madrid. In Millares Carlo, ed., *Cartas de España*, 107–8.

19. Letter from Catalina de Sotomayor to Cervantes de Salazar, April 14, 1575, Toledo. In Millares Carlo, ed., *Cartas de España*, 125–26. The daughters of Doña Catalina later sold the manuscript to the Council of Indies for 40 ducats in 1597 (Cervantes de Salazar, *Crónica de la Nueva España*, 93).

20. AGI, IG-738, R. 17, f. 249, "Informe favorable del Consejo de Indias para que se concedan 400 ducados a López de Velasco por el libro de 'La geografía y descripción de las Indias,'" December 7, 1576. He was paid a month later as recorded in IG-426,

deemed it a notable work that could have easily been divided into many books. They requested some changes to the work and noted that Velasco had relied on documents recovered from the heirs of Alonso de Santa Cruz to prepare the new geography. Along with the reward came the king's tacit approval of Velasco's work. The cosmographer, however, was not entirely satisfied with the reward. He complained to Vázquez, "Although the money is quite a bit less that what I or others thought was merited by the [king's] approval—according to what I have been told—I remain consoled and honored and very happy that your mercy has seen to it that it be ordered."[21]

Although the pecuniary reward was somewhat meager in proportion to the scale of the work, the monarch's recognition yielded some intangible rewards as well. In a society where closeness and access to the king amounted to honor, Velasco was now recognized by the monarch for his contributions as cosmographer of Indies. A few months later, Velasco, taking advantage of what he perceived to be the king's good opinion of him, asked Vázquez to forward to the king a petition to be appointed the prince's tutor.[22] He pointed out that he had spent many years studying the Castilian language and had even written a book on the subject;[23] he said that he would be happy to serve without further monetary reward solely for the great privilege and honor of instructing the prince in the language. The petition apparently did not progress, and in subsequent communications the same year Velasco began to complain to Vázquez of housing and financial difficulties.[24] He also sought to leverage his favor with the king to advance two requests of a more personal nature. He asked the king to provide some monetary assistance to the sister of Antonio Gracián, the king's former secretary, who after her brother's death found herself in financial

---

L.26, f. 22v-23, "Orden de pago de 400 ducados a Juan López de Velasco," January 11, 1577.

21. "[Q]ue aunque el dinero a sido harto menos de lo que yo y aun otros pensarían con que la aprobación que hizo de él segun lo que me han dicho quedo consolado y honrado y muy contento con que vm se haya hallado el decretarlo por la confianza que tengo de que en cualquiera ocasión me ha de hacer." IVDJ, Envio 100, f. 311, Juan López de Velasco, "Carta a Mateo Vázquez sobre San Isidoro y por gratificación a su libro," October 14, 1576, Madrid.

22. IVDJ, Envio 100, f. 309, Juan López de Velasco, "Carta a Mateo Vázquez pidiendo plaza de maestro del Principe," February 8, 1577, Madrid.

23. He published it five years later. Juan López de Velasco, *Orthographía y pronunciación castellana* (Burgos: Felipe de Junta, 1582).

24. IVDJ, Envio 100, f. 310, Juan López de Velasco. "Resumen de dos memoriales de López de Velasco a Mateo Vázquez," June 28, 1577, Madrid.

difficulties.[25] The petition was well received by the king, and he granted her a generous sum of more than 2,000 ducats.[26] In another request, Velasco asked for a position of court page for a young man, son of a Juan de Medrano, who was fluent in Latin.[27] Again, his request was granted.[28] The details remain unclear as to how Velasco's housing situation was finally resolved in Madrid's notoriously tight housing market, but by August 1577 Velasco was thanking Vázquez for solving his housing problems.[29]

Velasco kept the bureaucratic wheels of fortune turning at the Council of Indies. Also in 1577, the Council granted him the offices of keeper of the seal and register for the recently created new province (*audiencia*) of Nuevo Reino de Granada.[30] The offices promised to supplement his wages by 300 ducats a year. The ad hoc grants continued to flow from the Council of Indies. In 1579 the Council of Indies awarded him half the income derived from a *regimiento* for the city of Santa Fé in Nueva Granada, estimated to be worth 200 ducats,[31] followed in 1581 by one-quarter of the revenues

25. IVDJ, Envio 100, f. 307–308v, Juan López de Velasco, "Carta de López de Velasco a Mateo Vázquez," June 14, 1577, Madrid.

26. IVDJ, Envio 100, f. 299–300v, Juan López de Velasco, "Carta de López de Velasco a Mateo Vázquez agradeciendole merced . . . ," July 12, 1577. The money for the gift was drawn from funds confiscated from Dr. Cuesta. Half of the 4,310 ducats went to Lcdo. Benito López de Gamboa, a member of the Council of Indies who had been acting president after Ovando's death, and the remainder to Gracian's sister, in care of her father.

27. IVDJ, Envio 100, f. 297, Juan López de Velasco, "Carta a Mateo Vázquez sobre varias cédulas incluyendo una de Indias y la descripción de España," July 16, 1577, Madrid.

28. IVDJ, Envio 100, f. 222, Juan López de Velasco, "Carta de López de Velasco a Mateo Vázquez sobre cédulas a enviar a los monasterios pidiendo libros de San Isidoro . . . ," October 28, 1577. Vázquez replied that Medrano should be taken in as a page. The young man wrote Velasco a grateful reply note in Latin. IVDJ, Envio 100, f. 221, "Ilustri Ioanni de Belasco a García de Medras gratiarum actio," November 10, 1577.

29. IVDJ, Envio 100, f. 292–293, Juan López de Velasco, "Carta a Mateo Vázquez sobre aposento, favores, libros de Don Jorge de Beteta dona al Escorial," August 21, 1577, Madrid.

30. AGI, Santa Fé-1, f. 10, "Si se podría dar el sello y registro de la audiencia del Nuevo Reino de Granada a Juan López de Velasco, cronista," August 8, 1577. Before granting the request, the king asked for an explanation for the reasons the offices were vacant, how much they were worth, and whether it would be best to keep them separate. The appointment was confirmed on November 3, 1578, as recorded in AGI, Santa Fé-145, N. 11.

31. AGI, Santa Fe-1, N. 23, "Merced a Juan López de Zubizarreta y a Juan López de Velasco de un regimiento de la ciudad de Santa Fé en el Nuevo Reino," March 30, 1579. In a later communication, the Council estimated the office was worth 200 ducats.

provided by the office of *procurador* de la Audiencia Real de los Reyes (Lima, Peru), with estimated earnings of 100 ducats.[32] The sizes of the awards coming from the Council of Indies were definitely on a downward trend, but Velasco was able to supplement his income with grants from the other royal councils.

In addition to his work at the Council of Indies, Velasco worked on a series of projects at court that placed him in a position to request entry into the circle of courtiers that served as "His Majesty's servants" or *continos*. He wrote a petition to that effect in 1578, surely following the lead of Alonso de Santa Cruz, who had been appointed *contino* by Charles V in the 1530s. Vázquez summarized Velasco's petition, explaining that Velasco was asking for the position of *contino* so he could hire an assistant to help him with the project of compiling his descriptions and also for his work at the library of El Escorial, obligations he had taken over after the untimely death of Antonio Gracián. The king replied in the letter's margin: "It is not convenient to grant him the position of *contino*, but if something else is proposed I may be consulted again."[33] The position of *contino* required a lifelong commitment on the part of the Crown, and throughout his reign Philip II—who declared the monarchy insolvent three times—was reticent to provide for these expensive posts. Soon thereafter, the Chamber Council granted Velasco the rents derived from the registry office of the *guarda mayores de los montes* of Madrid, Escalona, and Guadalajara.

It soon became apparent to Velasco that the registry of the *guarda mayores* would yield very little by way of remuneration, and he approached the king's Chamber Council in 1581 with another petition. This time he asked for the *escribanía mayor de rentas* and apparently was granted the office.[34] Velasco continued to harbor aspirations of gaining a position that would bring him closer to the court's inner circle. In August 1581, he asked Mateo Vázquez to again forward to the king his desire to teach the prince how to read and write Castilian. After reminding Vázquez that he is an "old and

32. AGI, IG-740, N. 91, f. 1, "Memorial del Consejo de Indias desaconsejando la concesión de nuevas gratificaciones a López de Velasco por su libro," September, 28, 1582.

33. "[L]o del asiento de contino no conviene pero si se ofreciere otra cosa se me podrá acordar." AHN, Cámara de Castilla, Consultas de Gracia, 4408, f. 14, "Consulta sobre petición de Juan López de Velasco de contino por trabajo en la librería de El Escorial," February 28, 1578.

34. The back of the letter indicates that Council granted the petition. AGS, Camara de Castilla-499, f. 507, "Petición de Juan López de Velasco para la plaza de escribanía mayor de rentas," May 4, 1581.

clean Christian . . . of competent age and in good health," he reiterated the arguments made in 1577. By now, however, he had published his *Orthographia*, and he referred to it to support his claim as a linguist of Castilian. He hinted that his pedagogical approach was based on the student's "imitation and example of the teacher rather than on concepts of the art" and offered once again to do the work without expecting any reward.[35] The king again dismissed his petition.[36]

In the following year of 1582, Velasco suffered another blow when the Council of Indies refused to grant him any additional rewards for his second geographical book, the *Sumario* or *Demarcación y división de las Indias*. In this instance, the Council objected to Velasco's new petition and wrote the king, summarizing all the *mercedes* Velasco had received as the Council's cosmographer-chronicler. The king replied in his own hand tersely: "Reply to him that he should be satisfied with what he has already received."[37] No doubt Velasco was disappointed, and his relationship with the Council of Indies suffered. The timing could not have been worse, given the important cosmographical projects under way. In 1583, Jaime Juan, under the auspices of Juan de Herrera, was readying his voyage to the New World, and the replies to the questionnaires to the Indies were beginning to bear fruit only to come into the hands of a disillusioned Velasco.

Velasco's dissatisfaction with his post as cosmographer-chronicler came to a head in 1584. He once again turned to Vázquez for relief. In a letter dated April 25, 1584, Velasco asked Vázquez to help him find another "more honorable and beneficial" occupation, perhaps a post that would both bring him

35. "[E]l salir los niños con ello consiente mas en la imitación y ejemplo de él que enseña que en los conceptos del arte." IVDJ, Envio 99, f. 316, Juan López de Velasco, "Carta de López de Velasco a Mateo Vázquez donde se ofrece a ser maestro del Príncipe," August 12, 1581. A similar letter is at BL, Add 28342, f. 384: Juan López de Velasco, "Carta de López de Velasco a Mateo Vázquez sobre su ortografía y a proposito de la enseñanza del príncipe," August 12, 1581.

36. By 1583 it became clear that the original rewards bestowed by the king's Chamber Council were of little monetary value, and Velasco presented another petition, this time also mentioning that he been handling the estate of Diego Hurtado de Mendoza at the Crown's behest. AHN, Camara de Castilla, Consultas de Gracia, 4409, f. 2, "Gratificación a Juan López de Velasco por trabajo en la librería de El Escorial y administración de bienes de Diego Hurtado de Mendoza," January 1, 1583.

37. "[S]e le puede responder que se contente con lo que se ha hecho con el." AGI, IG-740, N. 91, "Memorial del Consejo de Indias desaconsejando la concesión de nuevas gratificaciones a López de Velasco por su libro. Contestación al margen de Felipe II sobre que se recojan los libros y se pongan en lugar seguro," September, 28, 1582. There is a copy of the document in AGS, GA-137, f. 256.

honor and pay enough to sustain him. He began his letter by reminding Vázquez of his past career as secretary to three presidents of the Council of Indies. "For over twenty years or more I have been handling papers in His Majesty's service, since Lcdo. [Licenciado] Castro went to Peru to begin compiling the laws of the Indies."[38] Surely Vázquez recalled, Velasco continued, that during Juan de Ovando's audit of the Council, he personally organized and wrote the reports of all the general matters that resulted from it and that these were signed by the king. Understanding that his career lay in bureaucracy (*ejercicio de papeles*), he explained to Vázquez that over the years he had endeavored to study languages and sciences that could be practiced outside of the universities, not for personal gain but to be able to handle capably the papers with which he was entrusted.[39] Ovando had given him the post of cosmographer-chronicler major of the Council of Indies only because a more suitable office was not available at the time and "not because it is my occupation, but more to keep me busy while another suitable post became available."[40] Despite this, he felt, "I have in good conscience done all I could [as cosmographer] and much more than my predecessors."[41] He explained he had become very disillusioned with his current post and its future prospects, not only because he had been poorly rewarded but principally because the post was not in line with his "inclinations" nor with the direction of his studies.[42] He hoped Vázquez could find him a suitable post, one that would allow him to marry well and live without financial worries to better serve the king.[43] Reluctant to suggest a particular office or to have to justify himself in writing, he asked Vázquez to

38. "A veinte años o mas que trato papeles del servicio de S.M. desde que el licenciado Castro lléndose al Perú comenzó la recopilación de las leyes de Indias." BL, Add 28345, f. 67, Juan López de Velasco, "Carta de López de Velasco a Mateo Vázquez sobre petición de otro oficio," April 25, 1584. Lope García de Castro was in Peru from 1564 to 1572.

39. "[C]on fin y deseo de ocuparme en algún ejercicio de papeles atendiendo siempre el estudio de lenguas y ciencias que fuera de universidades se pueden profesar no tanto por valerme de ellas como por hallarme con más suficiencia para lo que de papeles se me encargase" (ibid.).

40. "[M]ás por entretenimiento para esperar otro cargo que por justa ocupación" (ibid.).

41. "[P]or cumplir con la conciencia he hecho en el lo que se ha podido y más que todos mis predecesores" (ibid.).

42. "Se me ha gratificado mal . . . me tiene en desgana . . . el oficio no es conforme a mi inclinación ni al fin a que encaminan[?] mis estudios" (ibid.).

43. "[S]i me viere con alguna ocupación de buen nombre, casarme bien y pasar la vida mejor y con mas facultades para servir a S.M." (ibid.).

consult the particulars of his situation with Juan de Herrera. Herrera would explain matters further and act as an intermediary if Vázquez chose to raise his petition to the king. In closing, Velasco stated that only because he felt that the king had a "reasonable" opinion of him (*tiene de mi razonable concepto*) had he taken the step of presenting his situation to Vázquez.

We do not know how Vázquez handled Velasco's latest request. It was apparently not until almost two years later that Velasco made another petition, this time asking specifically to be taken on as the king's secretary. The request survives in a memorandum by Vázquez to the king concerning some open positions for royal secretaries. Vázquez wrote, "Juan López de Velasco, the cosmographer, has asked for the title of secretary, and [he] is very capable on matters concerning the Indies, but in the style and way of expressing himself in Your Majesty's affairs he needs to learn more."[44] Again, the outcome of the petition is not clear, because the king did not address the matter in his reply. But in a letter three months later, Velasco is much happier and very relieved. News had reached him of "the very big gratification [*merced*] that Father Mariana has told me about," and he adds elegantly, "Your virtuous character and noble condition oblige me to your service more than all the interests of the world."[45] Velasco may have been referring to the Jesuit Juan de Mariana (1536–1623), a influential historian and author of polemical books on economics and politics. The cosmographer also thanked Vázquez effusively for a gratification given to Antonio Gracián's mother. Not one to let a chance pass him by now that he was in his patron's goodwill, he reminded Vázquez in the same letter that Gracián's other brother also needed assistance.

Most likely Velasco's reward came in the form of an appointment to the Council of the Exchequer. Yet Velasco tested the Council of Indies' generosity one final time by requesting that it also appoint him to the body's accounting office (*contaduría*). But by 1588, the additional duties Velasco had taken on at the exchequer were interfering with his work at the Council of Indies. Upon reviewing the request, the Council found that he had not been fulfilling the duties of the post and therefore could hardly be expected to hold yet another. The Council of Indies forwarded the matter to the king. The matter came to the attention of the ever-present Vázquez, and he

44. "Juan López de Velasco el cosmógrafo ha pedido título de secretario, y es bien suficiente de noticia de lo de Indias, pero en el estilo, y manera de decir en despachos de V.M., haven[?] menester aprender más." Zab, Altamira, 142, D. 140, "Consulta al rey sobre petición de Juan López de Velasco," March 19, 1586, Aranjuez.

45. IVDJ, Envío 37, n. 46, Juan López de Velasco, "Carta de López de Velasco a Mateo Vázquez," June 16, 1586.

interjected a reply to the Council. He advised that they indeed should propose new persons for the post of cosmographer-chronicler but that as far as Velasco was concerned, he should be given a fair hearing to explain how he had complied with the duties of his post, "because it would be unjust to condemn him without hearing him."[46] When confronted by Vázquez with the Council's complaints, Velasco complained bitterly that he had asked for the *sello de ordenes* purely for financial reasons, so that he would not have to "tire my friends or look for a improper way to earn a living."[47] Vázquez, or rather the monarch, did not disappoint Velasco this time. In September 1588 Velasco was appointed royal secretary and specifically secretary of the Council of the Exchequer with a salary of 200,000 maravedís, twice his salary at the Council of Indies.[48] Velasco would remain at the exchequer until his death on May 3, 1598.

In parallel to his considerable duties at the Council of Indies and in what at times amounted to another full-time occupation—the one that gave him the most satisfaction—Velasco voluntarily took on a number of projects in tune with his humanistic sensibilities. We find him engaged in linguistic pursuits and gravitating toward a circle of other functionaries with humanistic inclinations, including the historian Ambrosio de Morales, the king's personal secretary Antonio Gracián, and Benito Arias Montano. Velasco worked alongside them collecting books in Castilian for the new library at El Escorial, the manifest expression of the Spanish Renaissance and humanism. It was a task that he clearly enjoyed and that was well suited to his

46. "Sera bien que me propongan personas de que resultara ver lo que convendría en lo de la ayuda de costa se oiga a Juan López de Velasco que muestre como ha cumplido la cualidad con que se le señaló porque no es justo condenarle sin oírle." Marginal note in Vázquez's hand and signed with his initials. IVDJ, Envío 23, caja 1, leg. 144, Hernando de la Vega y Fonseca, "Carta de Consejo de Indias al Rey en lo del oficio de cronista y cosmógrafo de las Indias," September 1588, Madrid.

47. "V[uestra] m[erced] me preguntó el otro día, si había de dejar lo que tengo, en caso de que se me diese otra cosa. Y lo que busco y pretendo es tener con que pasar la vida con retención de lo que tengo no sin ello porque lo paso mal con solo lo que tengo. Y así he pedido el sello de ordenes, porque con lo que vale y tengo, quedaría por lo menos sin necesidad de cansar los amigos, o buscando el sustento por medios indebidos, y desocupado con entrambos oficios para servirlo, y servir en lo que mas se me mandare." IVDJ, Envío 25, f. 585, Juan López de Velasco, "Carta de López de Velasco a Mateo Vázquez [?] sobre petición," January 5, 1588, Madrid.

48. AGS, Escribanía Mayor de Rentas, leg. 27, f. 785–853, "Nombramiento de Juan López de Velasco como secretario de S. M.," September 14, 1588.

interest in the Castilian language. He also had a hand in collecting scientific instruments and rare manuscripts for the king's library.[49]

In 1571, through the initiative of Montano, then residing in Antwerp, the Holy Office of the Inquisition began work on a new index that aimed to rescue, with the judicious use of the red pen, a number of worthy literary works that had been placed in the 1559 Index. Velasco's interest in the Castilian language and his close association with Juan de Ovando, who had served in the Council of the Inquisition, positioned him well for the task. At the Inquisition's behest, he expunged three important literary works that had been placed in the Index largely due to anticlerical references: the *Propalladia* by Torres Naharro, the works of Cristóbal de Castillejo, and the first picaresque novel, *Lazarillo de Tormes*.[50] Velasco turned out to be a benevolent censor, given his admiration of the authors and what he categorized as their respect for the purity of the Castilian language. The censor regretted that the works either circulated in the form of careless manuscript copies or were published outside Spain, denying "the persons in this kingdom" access to these important works. His mission was to "reform and cleanse" all that the Inquisition found "inconvenient" in the books while retaining the texts' elegance and grace.[51] Velasco's censorship consisted in removing the instances where the authors touched on matters of dogma, while also attenuating many of the original text's sexual references and vulgar language.[52] The works sometimes bombastic anticlerical references were excised, but

49. In 1577 he noted that a certain Doctor Aguilera wanted to give the library a singularly large quadrant made by a "gifted artificer." IVDJ, Envío 100, f. 297, Juan López de Velasco, "Carta a Mateo Vázquez sobre varias cédulas incluyendo una de Indias y la descripción de España," July 16, 1577, Madrid.

50. *Propalladia de Bartolomé de Torres Naharro y Lazarillo de Tormes, todo corregido y emendado, por mandato del Consejo de la Santa y General Inquisición* (Madrid: Pierres Cosín, 1573), and *Obras de Christóbal de Castillejo corregidas y emendadas por mandato del Consejo de la Sancta y General Inquisición* (Madrid: Pierres Cosín, 1573). In reward, Velasco received the equivalent of an eight-year copyright for the kingdoms of Castile and Aragon.

51. Prologue to the *Propalladia* in Bartolomé de Torres Naharro, *Propalladia and Other Works of Bartolomé de Torres Naharro*, ed. Joseph E. Gillet, 4 vols. (Menasha, Wisc.: George Banta, 1943), 1:59–60.

52. Menéndez Pelayo's assessment as cited in Pérez-Rioja, "Insigne visontino," 36. Gillet considers that in the *Propalladia* Velasco's "suppressions and changes are made in such a manner as to leave the context intelligible and the pattern unbroken." For some interesting illustrative examples, see Torres Naharro, *Propalladia and Other Works*, 69–71. Not all literary critics agree, however. For Gonzalo Santoja, any alteration to the original text is one too many; see his introduction to Anonymous, *Vida del Lazarillo de Tormes castigado*, ed. Gonzalo Santoja (Madrid: Sociedad Estatal España Nuevo Milenio, 2000).

he wielded the censor's pen judiciously. As scholars have pointed out, in the *Lazarillo* Velasco differentiated between the personal failings of churchmen and generalized anticlericalism and antiauthoritarianism, with the latter two disappearing from the text. The colorful churchmen and their scandalous behavior remained unaltered in Velasco's edition.[53]

Velasco's only published work was the Spanish grammar he referred to when he sought to become the prince's tutor. The *Orthographía*, which he dedicated to Philip II, was an effort, in Velasco's words, "[to] improve and ennoble the Castilian language." The book's prologue is an elegant and eloquent exaltation of the importance of writing and pronouncing well any language, because, as he explained, "he who writes poorly pronounces poorly, and he who pronounces poorly speaks poorly, and he who does not speak correctly seems to not understand."[54] Himself blessed with a graceful humanist cursive handwriting (a welcomed relief to the historian from the tortuous scribal hands of the sixteenth century), Velasco insisted on the importance of the "art of forming words and drawing letters." The book is an attempt to standardize the orthography of a number of commonly misspelled or inconsistently spelled Castilian words. His aim was to propose an orthography and grammar that reconciled common usage with a rationalized etymological study but did not restrict the language's flexibility. As Velasco worked his way through the alphabet, for each letter section he introduced a number of problematic words and explained the etymological reasons that words are spelled in a particular manner. He also explained the proper way to partition words and sentences, as well as the use of punctuation. From his etymological explanations, we can surmise that he knew Latin, perhaps also some Greek and Hebrew, and was clearly familiar with spoken Italian. The book ended with a reference list of common words that had problematic spelling.[55]

53. Augustín Redondo, "Censura, literatura y transgresión en época de Felipe II: El 'Lazarillo castigado' de 1573," *Edad de Oro* 18 (1999): 144.

54. "[P]orque quien mal escribe, mal pronuncia: y quien pronuncia mal, mal habla: y quien no habla bien, parece que no entiende." López de Velasco, *Orthographía*, prologue. For an analysis of Velasco's grammatical theory, see José María Pozuelo Yvancos, *López de Velasco en la teoría gramatical del siglo* XVI (Murcia: Universidad de Murcia, 1981). The book was granted a ten-year copyright in Castile (1578), Aragon (1578), and Portugal (1581).

55. A number of Velasco's personal documents, particularly concerning his linguistic interests, are at the library of El Escorial. These include a number of drafts and letters commenting on the *Orthographía*. These papers were among the documents Velasco filed under "Papeles de curiosidad" and instructed that after his death they should be

In the epilogue to the *Orthographía*, Velasco also unveiled a proposal to reform elementary education in Castile.[56] The reform hinged on selecting teachers who not only understood proper orthography but also pronounced words clearly and correctly, which he considered fundamental if children were to learn a language properly. The Council of Castile acknowledged Velasco's pedagogical interests in 1587, when he was invited to issue an opinion on a proposal presented to the Council that would have required professional certification for elementary school teachers. He not only issued an opinion on the proposal but also assessed the sorry state of elementary education in Madrid and proposed a broad project to overhaul elementary education in Castile.[57]

On this occasion, he again insisted on qualified teachers and proposed that they be tested and certified by local authorities—he would have preferred that they come to court to be tested. Among the reforms he proposed were establishing a minimum number of classroom hours, paying teachers a sensible wage in accordance with the hours taught, and limiting the number of students per class. As far as the curriculum was concerned, Velasco proposed amending the text currently in use to include writing instruction since it taught only reading and catechism, with an emphasis on the latter. To standardize spelling, he proposed that all printing in the kingdom conform to the guidelines set down in the new educational text. It is unclear if Velasco's proposal went any further than the discussion in the Council of Castile.

Velasco's involvement with the library at El Escorial also allowed him to participate in the project of collecting, editing, and publishing the works of encyclopedist St. Isidore of Seville (ca. 560–636).[58] In addition to writing

---

presented to the king. Most are in BME, L-I-13 and K-III-8. Miguélez, *Catálogo de códices españoles*, 2:191, 247–49.

56. "Epílogo e instrucción para enseñar bien a leer y escribir," in López de Velasco, *Orthographía*, f. 309.

57. For a detailed study of Velasco's proposal as that of an *arbitrista*, see Augustín Redondo, "Exaltación de España y preocupaciones pedagógicas alrededor de 1580: Las reformas preconizadas por Juan López de Velasco, cronista y cosmógrafo de Felipe II," in *Felipe II, Europa y la monarquía católica*, ed. José Martínez Millán (Madrid: Parteluz, 1998). A draft of Velasco's proposal is in BME, L-I-13, f. 249–267, "Instrucción para examinar a los Maestros de Escuela de la lengua castellana y enseñar a leer y escribir a los niños," July 1588.

58. At the behest of Alvar Gómez de Castro (1515–80), Philip II ordered that the works of St. Isidore of Seville be collected, edited, and published. Gregorio de Andrés, "Viaje del humanista Alvar Gómez de Castro a Plasencia en busca de códices de obras

the most consulted reference work in the Middle Ages, Saint Isidore was a linguist and wrote on natural science, cosmology, and history. It would not be unreasonable to suppose that Velasco found inspiration for the etymological approach for his *Orthografía* in St. Isidore's medieval encyclopedia, *The Etymologies*. Initial contact with private and ecclesiastical libraries throughout Spain was under the direction of Antonio Gracián, with Ambrosio de Morales, a royal chronicler, traveling to verify, copy, and collect the Visigothic documents. After Gracián's death in 1576, Velasco took over the task of locating and collecting the relevant manuscripts from cathedrals and monasteries.[59] In the summer of 1577 he initiated a set of royal orders to Spanish monasteries and libraries, instructing them notify of and surrender any works of St. Isidore they might possess.[60] In addition to collecting the manuscripts, Velasco arranged to have them corrected[61] and later acted as the librarian of the collected material.[62]

It appears that Velasco also edited at least one of St. Isidore's works himself, the *Ecclesiasticis officiis*, an early description of Catholic liturgy, sacramental rites, and feasts. He went about it with his customary orderliness, first collecting reference materials to help him interpret the saint's work. He requested from the Royal Library a codex on canon law, the *Albeldense*, as well as the oldest missals and breviaries available in the library's collection.[63] Velasco's death in 1598 took place as twenty years of labor on

---

de S. Isidoro para Felipe II (1572)," in *Homenaje a Don Agustín Millares Carlo* (Las Palmas: Caja Insular de Ahorros de Gran Canaria, 1975), 609.

59. IVDJ, Envío 100, f. 311, Juan López de Velasco, "Carta a Mateo Vázquez sobre San Isidoro y por gratificación a su libro," October 14, 1576, Madrid. Manuel Miguélez was the first to chronicle López de Velasco's participation in the project. His assessment, however, is tinged by the historian's effort to rehabilitate Velasco as a humanist and exaggerates Velasco's contribution to the project. Manuel Miguélez, "Sobre el verdadero autor del 'Diálogo de las lenguas,'" *La Ciudad de Dios* 117 (1919).

60. IVDJ, Envío 100, f. 297 and f. 222.

61. Velasco suggested that Fray Luis de León should take care of the matter. See IVDJ, Envío 100, f. 245, 246, and 263.

62. IVDJ, Envío 100, f. 18, Juan López de Velasco, "Carta de López de Velasco to Mateo Vázquez sobre que le devuelvan unos libros de la Librería Real," October 23, 1584, Madrid. Velasco was concerned that some of the books of St. Isidore had been removed from the Royal Library in Madrid and sent to San Lorenzo de El Escorial, to Bartolomé de Santoyo. He asked that he king intervene to make sure the books were returned.

63. The *Albeldense* is BME, D-I-2. Velasco describes the others as "un misal y brevario Mozabes [mozárabe], los mas antiguos que en la librería hubiere." AGS, CSR-281,

the project were coming to an end and on the same year that Isidore was canonized. Just a year after his death, Juan Pérez y Grial became the editor of the project and prepared the material for publication. It is unclear, however, whether Velasco's corrected version of the *Ecclesiasticis officiis* was the one published by Grial.

The Reluctant Historian

The inspired Velasco working at the magnificent library of El Escorial and cherishing his association with the leading humanists of Philip II's court seems to be a different man from the dour and pragmatic cosmographer-chronicler of the Council of Indies. His pragmatism is most apparent in the way he carried out his duties as chronicler of the Indies, duties that often entangled him in the risky politics of the time. Early in his career and in accordance with the duties specified in his appointment order and the *Ordinances*, Velasco was asked by the Council of Indies to issue an opinion on whether Diego Fernández de Palencia's recently published *History of Peru* contained untrue, inaccurate, or offensive passages, as some parties alleged. The opinion he issued in the case survives and is telling of Velasco's thoughts on historical censorship and characteristic of the way he approached his duties.[64]

One of the parties named in the book, Fernando Santillán, considered himself misrepresented and dishonored by the way he was portrayed, particularly in regard to his participation in the insurrection led by Gonzalo Pizarro against the Spanish Crown in 1546. Complicating matters was the fact that Palencia had written the book under the sponsorship of Francisco Tello de Sandoval, then president of the Council of Indies, who had given him access to official documents. In his response Velasco chose not to address the alleged misrepresentations, explaining that neither party had

---

f. 354, Juan López de Velasco, "Juan López de Velasco pide libros a la biblioteca de El Escorial para trabajo de San Isidoro," November 12, 1586. Lawson maintains that the original source for the *Ecclesiasticis officiis* in Grial's edition was the Escorial codex D-I-1 (Gothicus Codex). See his study in Isidore of Seville, *De ecclesiasticis officiis*, ed. Christopher M. Lawson (Turnhout: Brepols, 1989), 116–17. Velasco was paid for this work. APR, Registros, Cédulas, Tomo 7, "Carta de pago a Juan López de Velasco por corregir el tratado de San Isidoro 'Officiis Ecclesiasticis,'" 1587.

64. AGI, P-171, N. 1, R. 19, Juan López de Velasco, "Parecer que dio Juan López de Velasco, cronista mayor de Indias, sobre la 'Historia del Perú' que escribió Diego Fernández, vecino de Palencia. Respuesta de Fernández a las objeciones y reparos que puso el licenciado Hernando de Santillana a dicha Historia," May 16, 1572.

sufficient evidence to establish these beyond doubt; he recommended that if the Council wanted to get to the bottom of the matter, it must conduct an investigation that would require consulting records and interviewing persons still living in Peru who might recall the events. Velasco, however, questioned whether this would be a prudent course of action; he reminded the Council that the events had already been addressed by the courts and that revisiting these divisive events might not be prudent. He suggested that while the book was under review all 1,500 copies already printed should be collected in order to prevent a copy from getting overseas. Ultimately, the Council took Velasco's recommendation and recalled the books rather than revisiting the divisive period of colonial history.

The Council's fickle attitude concerning history made Velasco apprehensive about complying with his historical duties. We have a rare glimpse into Velasco's way of thinking in one of his letters to Francisco Cervantes de Salazar. He confided his misgivings about the standards for history at the Council of Indies and questioned whether the king's support for a history of the Indies was genuine. After thanking Salazar for making some historical material available to him, he added,

> I doubt anything will come of the care I am taking [collecting these historical materials], but until it is worth more, I should diligently collect what is necessary for the history [of the Indies]. Although I do not know if His Majesty, due to events, will care to have one written or published. I do not see how. In the [histories] that have been seen up to now in the Council and by ministers there must be the most careful inquiry possible about past and present events, because of the light they shed for suspecting or preventing future ones.[65]

Velasco knew the Council would closely scrutinize any historical work he wrote. Even maintaining the highest standard of documentary evidence would not exempt him from this scrutiny and the political fallout that could result. He had seen the controversies that could brew around histories of

65. "Lo que este cuidado mío prestará no lo sé, pero paréceme que en duda, mientras no valiere para más, debo procurar con diligencia lo necesario para la historia, que aunque no sé si Su Majestad se servirá que se escriba ni publique ninguna cosa por los sucesos, no sé de qué manera, que han tenido las que hasta ahora se han visto para le Consejo y ministros de el, tiénese por conveniente que haya la más cumplida noticia que se pueda de las cosas pasadas en Indias y de las presentes, por la luz que dan para sospechar o prevenir las venideras." Letter from López de Velasco to Cervantes de Salazar, June 10, 1573, Madrid. Published in Millares Carlo, ed., *Cartas de España*, 107–8.

the Indies—Las Casas, López de Gómara, Fernández de Palencia—and he tried to preempt this in two ways. First, he sought to collect as much material written by others as possible and always referred to firsthand, sworn testimonies should he have to offer a historical perspective on a question before the Council of Indies. Second, he simply avoided writing a comprehensive history and limited his historical narratives to discovery accounts, always in the context of a cosmographical work. The latter course was technically in violation of his letter of appointment, which instructed him to "make and compile" a general history of the Indies. It is unclear what arguments Velasco used to evade this task or even whether he was ever pressured to generate such an opus.

Perhaps in response to what he perceived as the treacherous task of writing history, Velasco favored the anonymity permitted by a collective approach. Nowhere is this clearer than in a memorandum addressed to Philip II titled "Your Majesty Should Have His History Written" (*Que su Majestad debe mandar a escribir su historia*), in which he urged the king to have an official biography written.[66] The document survives as an undated memorandum in Velasco's handwriting in a collection of documents once belonging to Mateo Vázquez. In it he stated that recording the history of the monarch's reign was a matter of some urgency and noted that not doing so would do the king a grave injustice, since the king's memory would then be left to the "reckless judgment of the masses always inclined to judge the worst" and to foreigners who relied on rumors and lies.[67] Velasco acknowledged that the king might be reluctant to undertake such a project, his humility disinclining him to indulge in what some might consider an exercise in seeking temporal glory or perhaps out of concern that such a history might offend some. To preempt the king's possible reluctance, he argued that history can serve to educate others by elucidating the truth and the Christian zeal that motivated the king's actions. History, Velasco told the king, would label him "Philip the Prudent." Aware that "the first rule of history is that it not defame," he noted that history could also be

66. Zab, Altamira, 159, D.107, f. 1–4, Juan López de Velasco, "Informe de Juan López de Velasco a Felipe II dando razones de lo conveniente de que el rey mande a escribir su historia," n/d. For a study that places this document in the context of the work of Habsburg chroniclers, see Richard L. Kagan, "Clio and the Crown: Writing History in Habsburg Spain," in *Spain, Europe, and the Atlantic World: Essays in Honor of John H. Elliott*, ed. Richard L. Kagan and Geoffrey Parker (Cambridge: Cambridge University Press, 1995), 78.

67. "[A]l juicio temerario del vulgo inclinado siempre a juzgar lo peor." Zab, Altamira, 159, D. 107, f. 2.

used as propaganda or to plant misinformation, as the situation warranted. "[I]n effect there are things that history, as it should, can and must be silent about, and thus [history] is a suitable means for discrediting and undoing false rumors by manifesting the truth, while also leaving unwritten others that might not be advisable to be known."[68]

Velasco seems to have been aware that some of the monarch's reservations concerning history might be due to what he called the "inopportune and drawn-out nature of historians." He countered these objections by suggesting that the king could order that his history be written quickly and be subject to his personal review. Velasco saw two good reasons for urgency. First, he said, "time is short and the matter long," and many papers had to be collected and testimonies of pertinent persons gathered before they died. Second, once news of the monarch's history reached the public, anyone currently writing a similar history would be dissuaded from doing so "before they published their inevitable errors." Velasco even suggested planting a bit of misinformation. Should the king not wish to have his history written, he could indicate with a public announcement (*con muestras públicas*) that one was being undertaken but instruct via a secret order that it not be carried out. Velasco observed, not without a tinge of regret, that should the king choose this course of action, "time will pass, and the persons and memories will disappear, and once this age is over it will be impossible to return to [write a history] based on the truth and all will be forgotten."[69]

A single person could not undertake the history of the reign of Philip II, in Velasco's estimation, because "[t]here are so many qualities required of a perfect historian that they coincide only rarely in a single man."[70] Instead, a committee should write the history. They would commence the process by collecting all relevant documents, organizing the material for easy access and preparing synopses. Once this was accomplished, a group of trusted ministers composed of statesmen who knew the reasons behind significant events, men of letters who understood the law, and soldiers to elucidate on matters of war would go over the material and determine which events

68. "[Y] en efecto hay cosas que la historia haciendo lo que debe puede y debe callar y por esto es medio convenible para desacreditar y deshacer los rumores falsos manifestando la verdad y para disimular los que no convengan saberse dejándolos de escribir." Zab, Altamira, 159, D.107, f. 2v.

69. "[S]e pasará el tiempo y se acabarán las personas y memorias y acabada esta edad no se podrá después volver a ello con fundamento de verdad y así se olvidará." Zab, Altamira, 159, D.107, f. 2v.

70. "Son tantas las calidades que se piden en un perfecto historiador que de concurrir sino raras veces en un hombre solo." Zab, Altamira, 159, D.107, f. 4.

should be included in the history. These men should be wise in philosophy and humanities so they could choose the most important events and know how to "qualify" them and present them in the best manner. (And by virtue of their stature, they would exempt the historian from any blame.) Once they determined what events would make up the history, they would consult with the king, and only upon his approval would the writing phase start.

As the committee agreed on the subjects to include, the relevant material could be organized and written in Castilian with a somber style (*grave y llano*) and by two writers working separately. Their work would then be reconciled, amended, and judged by the committee as to its arrangement, language, and accuracy. In addition to the original work in Castilian, Velasco suggested that a version be written in Latin by a "most learned man in the country or abroad" to prevent future translations from corrupting the original account. Finally, Velasco insisted that the undertaking could be carried out cheaply. Ministers asked to review material were already on the payroll and would surely view their participation in the project as an additional honor, while the scribes who prepared the synopses could not, in the year or so Velasco estimated the project would take, incur much expense. "At any rate," he added, the whole project "could never cost as much [money] as is given to a chronicler during his whole life who in the end dies without having written a word."[71]

Whether Velasco saw a role for himself in the project is unclear from the memorandum. He gracefully insisted that he had no pretensions other than preserving the "honor and glory of His Majesty" from posterity's whim. Velasco never would have presumed to join the government ministers commissioned with selecting the material to be included in the biography. Writing the synopses or *relaciones* was an intermediate step that he envisioned as best undertaken by secretaries hired for only this purpose—a good opportunity for a young man, as a similar job had been for himself when he did such work for Juan de Ovando. Nor does he appear interested in composing the final Castilian or Latin version meant to be written under such close scrutiny.

Velasco likely had in mind for himself a role not unlike the one he carried out for the edition of St. Isidore's works, that of coordinator. Although he does not expressly refer to it, the way he structured the project required a person who would coordinate the collection of the sensitive primary sources, arrange the ministers' deliberations on them, supervise the work

71. "[N]o será tanto como lo que se da por toda su vida a un cronista que se muere sin dejar escrito letra." Zab, Altamira, 159, D.107, f. 4.

of the secretaries drawing up the *relaciones*, and finally oversee the Castilian and Latin editions of the history of the reign of Philip II. If indeed his proposal dates from the 1580s, by this time in his career Velasco would have proven himself competent to carry out the task. As was apparent from Velasco's work as cosmographer and chronicler of the Council of Indies, most of his functions involved precisely the tasks he perhaps envisioned for himself in writing his monarch's history: collecting, cataloging, and organizing information. The scheme he proposed was perfect for the reluctant historian, who was clearly apprehensive of carrying forth with a firm authorial voice. He would see the project through but without assuming any responsibility—or blame—for the content.

The Cosmographer as Censor

Velasco saw his duties as cosmographer as not unlike those he had imagined for himself when writing Philip II's history. A cosmography had to be composed using facts as described in documents, which he would carefully collect, evaluate, and synthesize. It was understandable then that his most pressing activity during his first years as cosmographer would be collecting and organizing these documents. He resorted to a number of strategies in his attempts to compel informants to supply cosmographical material. Some, as in the case of Cervantes de Salazar, voluntarily gave the new cosmographer major important works simply upon his request. Others had to be compelled legally to give up the material, and many never saw their works again once these fell into the cosmographer's hands. His diligence in collecting and securing within the confines of the Council of Indies information about the history and physical reality of the New World has been interpreted by some historians as characteristic of an agent working for an obscurantist monarch seeking to censor and control his overseas possessions. In all fairness, however, Velasco's actions were a natural consequence of the ambitious cosmographical project Ovando had set forth, where censorship played a secondary role.

Upon assuming the post of cosmographer-chronicler major, Velasco was caught in an awkward situation. His letter of appointment instructed him to compile the cosmography, as well as the natural and moral history of the Indies, but the complementary set of directives delineated in the *Instructions* had still not been approved by the king and would remain unsigned until 1573. As discussed in the previous chapter, these instructions promised the cosmographer-chronicler a steady flow of meticulously compiled geographic, ethnographic, and historical materials from which he was to prepare the official description of the Indies. But, alas, officials in the New World were in the dark.

Velasco spent 1571 and 1572 collecting cosmographical material about the New World. He requested that city officials in Seville collect and send to the Council of Indies the papers of Alonso de Santa Cruz and Christopher Columbus that were in the city's cathedral.[72] He continued to pursue Santa Cruz's important cosmographic collection and gained possession of it in 1572.[73] Velasco also sought the papers left after the death of Francisco López de Gómara, author of one of the most widely circulated cosmographical descriptions of the New World and the source for Gemma Frisius's description of the New World. In a petition to the local authorities in Soria—Velasco's hometown—he asked that they send a "trustworthy person" to "investigate, make an inventory, bring to the Council of Indies any relevant documents in possession of López de Gómara's heirs," and hand them over to the cosmographer.[74]

Despite his efforts at procuring material, Velasco recognized that the material he needed to comply with the exigencies of his new post lay overseas. At his request, the Council of Indies issued in August 1572 a series of royal orders directing the various governing districts in the Indies to send the cosmographer historical material.[75] More specifically, it asked that they identify persons, "whether lay or religious," who might have written or compiled or might have in their possession any histories, commentaries, or accounts of the region's discovery, conquest, and wars from the discovery to the present time. Likewise, local authorities were instructed to seek

72. "Cosmografía—Una cédula para que se hagan recoger los papeles y descripciones de la cosmografía que tenía el cosmógrafo Alonso de Santa Cruz y así mismo todos los que hay en la Iglesia Mayor de aquella ciudad de los que tenía el Almirante Colón." AGI, IG-1505, f. 307, "Libro de despachos de oficio del Consejo de Indias," October 31, 1571, Madrid.

73. AGI, P-171, N. 1, "Inventario de documentos de Santa Cruz que pasaron a manos de J. López de Velasco," October, 1572, Madrid.

74. AGI, IG-427, L. 29, f. 1, "Cédula ordenando se recojan los papeles de López de Gómara," September 16, 1572, Madrid.

75. AGI, IG-427, L.30, f. 233v-234v, "Cédula a Martín Enríquez, virrey, gobernador y Capitán general de Nueva España y presidente de la Audiencia de México, encargándole que se informe sobre las descripciones e historias de los Indios que existen en el distrito de su jurisdicción y los envíe en original o en copia en los primeros navíos que vayan a los reinos, donde se hará una recopilación de las mismas," August 17, 1572, El Escorial. Similar orders went to Charcas, Chile, Guadalajara, Guatemala, Panama, Quito, Santa Fe, and Santo Domingo. Also in BN, MS2932, #188, f. 273–274v and published in DIE, 1:361–62.

accounts describing the natives' religion, forms of government, rites, and customs, as well as written descriptions of the land and its natural history and "qualities" (*tierra, naturales y calidades de las cosas*). The material was to be collected diligently and with care (*diligencia y cuidado*) and sent forthwith to the Council.

By 1573 Velasco was very busy composing what would become his most significant cosmographical work, the *Geografía y descripción universal de las Indias*. The pages of the daily journal kept by royal secretary Antonio Gracián describe a Velasco assiduously collecting cosmographical information about the Indies, and in particular the documents that had been in the possession of Juan Páez de Castro, the king's former chronicler.[76] During this period, Velasco also gained possession of works about the Indies in Castilian and Latin written by Bartolomé de Las Casas from the Colegio de San Gregorio of Valladolid, including the manuscript of the *Historia general de las Indias*. The eager Velasco was reminded years later via a royal order that the cosmographer was responsible for the sensitive document's safekeeping (*tengais muy a recaudo . . . y no la entregueis a persona ninguna*).[77]

By 1577, the cosmographer-chronicler's role as the locus in Madrid for *relaciones* concerning the geography, natural history, and history of the Indies was well acknowledged. His central role was recognized throughout the court and royal councils. When news arrived in the Council of Indies that the fleet from the New World brought a written description and map (*descripción y pintura . . . de aquella tierra*) by Diego de Frías Trejo, Philip II

76. IVDJ, Envío 58, f. 93, Antonio Gracián, "Nota de Gracián al Rey," February 3, 1573. The document is a list of diverse matters Gracián consulted with the king. (It corresponds with the entries in his journal or *diurnal*.) "También me pide Juan López de Velasco cronista de Indias los papeles de la Florida que tengo entre los del doctor Juan Páez que se me entregaron por mandato de V. M. por esa memoria que me ha dado verá V. M. el efecto para que los quiere. Su Majestad es servido darle estos y algunas otras relaciones y copias simples de cosas de Indias que tengo apartados de los otros papeles." The king replied on the margin: "bien se los pueden dar sabiéndolo el presidente para pena(?) lleven recatados." There is a complementary entry in Gracián's journal on February 4, 1573 : Antonio Gracián, "Diurnal de Antonio Gracián, secretario de Felipe II," in *Documentos para la historia del monasterio de San Lorenzo el Real de El Escorial*, ed. Gregorio de Andrés (El Escorial: Imprenta del Real Monasterio, 1962), 5:75.

77. AGI, IG-426, L. 26, f. 178, "Cédula ordenado a Juan López de Velasco para que tenga en su poder las obras del Obispo de Chiapas que se trajeron de Valladolid," July 24, 1579. The documents passed to Antonio de Herrera in 1597, when he became royal chronicler. AGI, IG-427, L. 31, f. 29, "Instrucción del rey a Juan López de Velasco que le entregue los libros y papeles del Obispo de Chiapas a Antonio de Herrera," September 4, 1597.

wrote on the margin of the notice, "[A]s far as the description [give to] Juan de Velasco for the ones he does."[78] Aiding his role was the fact that the Council of Indies had ordered that all works concerning the Indies had to be reviewed by the cosmographer-chronicler major before they could be granted a publication license. Ship captains, official chroniclers, and private parties dutifully submitted their works to the Council of Indies to be censored and with any luck granted a license for publication. Velasco also actively sought this material. In his zeal to collect information, however, he may have overextended his reach and in the process created some enmity within the small community of chroniclers and cosmographers who sought to profit in the marketplace from their works. In some occasions he failed to comply with his side in the economy of favors that permeated exchanges when an author voluntarily supplied material to the cosmographer. Such was the case with the *History of Peru* written by Pedro Cieza de León, who had given the book to Juan Páez de Castro seeking the chronicler's comments. Years after the author's death, Rodrigo de Cieza, Pedro's brother and heir to the documents, accused Velasco of keeping the books "for his personal end" and asked that the police be sent to collect them.[79]

Velasco often used the authority his position granted him to retain materials that he had censored but perhaps considered valuable reference sources. This was apparently the case with Captain Escalante de Mendoza and his *Itinerario de la navegación de los mares y tierras occidentales.*[80] The extraordinary manuscript explicitly described sailing routes to and from the Indies based on the author's many years sailing the *carrera de Indias.* At one point it contained detailed drawings showing panoramic elevations of the principal ports and coasts of America, which would have made the document eminently useful for pilots. In 1579, as Escalante was seeking a license to publish, the Council acting on Velasco's recommendation deemed that the book contained sensitive information and that it would be prejudicial if it

78. "[L]o de la descripción a Juan de Velasco para las que el hace." AGI, IG-739, N. 108, "Consulta del Consejo de Indias informando al rey de la llegada de la flota . . . que trae descripciones de Perú," September 9, 1578.

79. AGI, IG-1086, f. 30, "Petición de Rodrigo de Cieza sobre libros de la historia del Perú en posesión de Juan López de Velasco," January 29, 1578. The whole affair was recounted first by Jiménez de la Espada and more recently in Pedro de Cieza de León, *Obras completas*, ed. Carmelo Sáenz de Santa María, 3 vols. (Madrid: CSIC, 1985), 3:51–53.

80. The manuscript without illustrations is at the BN and published in Juan de Escalante de Mendoza, *Itinerario de navegación de los mares y tierras occidentales, 1575* (Madrid: Museo Naval, 1985).

fell into foreign hands. Escalante responded that foreigners already were very familiar with the routes and that the book promised to help Spanish sailors.[81] He alleged that Velasco had a "personal stake in the matter" and requested that the book be reviewed by experienced persons in Seville.

The Council took the matter out of Velasco's hands and referred it to Juan Bautista Gesio, a cosmographer in the orbit of the Council of Indies who would become Velasco's nemesis. Years later the Council continued to deny Escalante a license to publish. Velasco's actions were redeemed when the king finally issued an opinion on the matter, instructing the Council that it should reward Escalante but that "the book should be taken from him and placed where you recommend so that it may be kept secret and you will send me another one so I can keep it [with me]."[82] The Council eventually granted Escalante a 500-ducat award for the book, but he was never allowed to publish.[83]

In another instance, Velasco acted to secure a valuable source of natural historical material. When it was clear in 1577 that Francisco Hernández would not complete his original plan to write the natural history of the Viceroyalty of Peru, Velasco, interested in preventing the project from languishing, proposed that the *protomédico* of Peru, Antonio Sánchez de Renedo, might be a suitable candidate to continue Hernández's mission. In a petition to the Council of Indies, which Velasco had first vetted with Mateo Vázquez,[84] the cosmographer informed the council that Renedo, "a man of many letters and good standing," had already composed a good

81. AGI, IG-1086, L. 7, 288v, "Petition by Cap. Escalante de Mendoza that his book be reviewed by others besides López de Velasco," December 2, 1579.

82. AGI, IG-740, N. 78, "Consulta sobre la licencia que pide Juan de Escalante para imprimir su obra intitulada *Itinerario de navegación de los mares y tierras occidentales*," July 14, 1582, Madrid. The king replied: "Bien será que se le tome este libro y que se ponga uno donde decís para que se guarde con recato y otro me enviareis para que le mande poner donde me pareciere y miraréis en que recompensa se le podrá dar y a donde y en que y avisármeis de ello." Despite the Council's concerns, the document circulated as a manuscript, to the point that the Council had to issue a royal order in 1593 to halt its circulation. AGI, IG-426, L. 28.

83. AGI, IG-740, N.85. "Consulta del Consejo de Indias sobre si se podría hacer merced al capitán Juan de Escalante de Mendoza por su libro *Itinerario de navegación de los mares y tierras occidentales*." Madrid, August 18, 1582.

84. "Para lo del doctor Antonio Sánchez de Renedo va el memorial que su [majestad] manda y la revisión al que podrá ser si pareciera a V.M. con orden que se informe de mi en aquel negocio." IVDJ, Envío 100, f. 297. Juan López de Velasco, "Carta a Mateo Vázquez sobre varias cédulas incluyendo una de Indias y la descripción de España," July 16, 1577, Madrid.

part of the work at his own expense and as part of his duties. Velasco asked that Renedo be ordered to send what he had already written and to continue the work—two requests, Velasco assured the council, to which Renedo would gladly acquiesce (*holgará de hazer*).[85] The matter apparently did not advance. A year later, Velasco raised the subject again, but this time he did not suggest that Renedo take over Hernández's project. Instead, he mentioned Renedo's book about the "natural history of the herbs, plants, and notable things of those parts" and asked that the viceroy inquire into Renedo's work and which parts of it the doctor would be willing to send the Council. Velasco also suggested that the doctor should receive a suitable reward as encouragement to continue his labor.[86] The original document suggests that the Council agreed with Velasco's suggestion (*hágase así*), although at this time I have no further information on what became of Renedo's natural history.

Perhaps no other of Velasco's actions within the cosmographical and historical information network operating in late sixteenth-century Spain has been judged more harshly or been more misinterpreted by historians than his involvement with the works of Bernardino de Sahagún. The previous chapter mentioned the 1576 letter by Archbishop Pedro Moya de Contreras in which he offered Sahgún's work to the Council of Indies as a way to comply with the demands of the *Books of Descriptions* project.[87] This letter resulted in the often-cited royal mandate of 1577 that ordered all of Sahagún's work collected and sent to Madrid. It has been referred to as "the infamous order" (*infausta cédula*)[88] and has been much derided by historians Joaquín García Icazbalceta, Georges Baudot, and others as evidence of a monarchy's quest to erase Amerindian identities by sequestering and burying in the Council of Indies Sahagún's ethnographic and linguistic study of Nahua culture.[89] Others have argued that the Franciscan was a victim of others in the order who were intent in humbling their

85. AGI, IG-1387, "Petición sobre Dr. Sánchez de Renedo para proseguir el trabajo del Dr. Hernández en el Perú donde reside," July 28, 1577.

86. AGI, IG-1388, López de Velasco, Juan, "Petición sobre historia natural del Perú," May 28, 1578. Also recorded in the Council's petition book, IG-1086, L.6, f. 200v.

87. Letter from the archbishop of Mexico to the king, March 28, 1576. Published in Paso y Troncoso, *Epistolario*, 12:12–13.

88. AGI, P-275, R. 79, "Cédula real sobre recogida de ejemplares: Obra de Fray Bernardino de Sahagún," April 22, 1577.

89. Georges Baudot, *Utopia and History in Mexico: The First Chroniclers of Mexican Civilization (1520–1569)*, trans. Bernard R. Ortíz de Montellano and Thelma Ortíz de Montellano (Niwot: University Press of Colorado, 1995), 493–504.

erudite brother or, more simply, a victim of the "inhuman indifference of a far-reaching bureaucracy."[90] The severe judgment of Velasco's actions predates modern scholarship, as even a contemporary of Sahagún, Fray Gerónimo de Mendieta in his *Historia eclesiástica indiana* (ca. 1596), accused a "certain chronicler" of using lies and artifice to take the documents from its author.[91]

The reasons behind the efforts to collect the works of Sahagún might be explained differently. The very diligent first cosmographer-chronicler, having learned that Sahagún (whom Ovando knew about and admired) had a significant historical work available, set the machinery of state in motion to collect these works so they would inform the works he was responsible to write. A royal order was thus issued on April 22, 1577, with a follow-up reminder sent in 1578—unnecessary, it turns out, as the manuscript we know as the Florentine Codex (now at the Medici-Laurentian Library) was already on its way.[92] Upon its arrival in Madrid, the Spanish portion of the document was laboriously copied and deposited at the Council of Indies (Tolosa Codex, RAH 9/4812). At this point in its history, and as far as our pragmatic Velasco was concerned, Sahagún's manuscript had fulfilled its duty as the bearer of knowledge to the cosmographer's archive. Philip II then sent the lavish original manuscript, with its intriguing pictograms and depictions of Nahua culture, as a wedding gift to Francis I of Florence.[93] When the manuscript made its final trip to Florence, it did so not so much as a bearer of knowledge but rather as a precious, thoughtful gift to another ruler well known for his interest in natural history. It also served as a tangible manifestation of Philip II's knowledge of—and power over—his patrimony and empire.

Velasco was in an enviable position vis-á-vis other early modern cosmographical practitioners, but also in a very difficult one. He had access to a

90. Respectively, Luis Nicolau d'Olwer, *Fray Bernardino de Sahagún, 1499–1590*, trans. Mauricio J. Mixco (Salt Lake City: University of Utah Press, 1987), 72, and Walden Browne, *Sahagún and the Transition to Modernity* (Norman: University of Oklahoma Press, 2000), 34.

91. As quoted in d'Olwer, *Fray Bernardino de Sahagún*, 77.

92. The manuscript's trajectory and supporting documentation have been studied by Giovanni Marchetti, "Hacia la edición crítica de la *Historia* de Sahagún," *Cuadernos Hispanoamericanos* 396 (1983): 24–26, and Jesús Bustamante García, *La obra etnográfica y lingüística de Fray Bernardino de Sahagún* (Madrid: Editorial de la Universidad Complutense de Madrid, 1989), 340–53.

93. Marchetti, "Hacia la edición crítica," 27.

network that kept him supplied with the latest cosmographical information and had the backing and institutional support necessary to procure this information, yet his personal interests and expectations led him in a direction opposite to that demanded by his official post as cosmographer-chronicler of the Council of Indies. He aspired to join the community of humanists that Philip II assembled at El Escorial and to work alongside them on their historical/linguistic projects. He quickly understood that the secrecy required of his official post limited his prospects of gaining the kind of recognition that would grant him this access. All his work for the Council of Indies was destined to the archive of secrets. He sought to counter this by becoming involved in a number of ancillary projects, and in particular those that could earn him recognition by the Council of Castile, the monarchy's most important and wealthiest council. Given that in this system of patronage Velasco's fortunes were directly proportional to his proximity to the king and to the level at which his work was recognized, toiling on secret cosmographies promised little. This did not mean that he neglected his duties at the Council of Indies—he was eminently efficient—but he understandably grew dissatisfied with his work.

Velasco's actions were in some measure the result of the way Juan de Ovando had conceived of cosmographical practice within the Council of Indies. In the *Instructions*, he had put in place a rigid institutional and conceptual structure that, while allowing for the collection and compilation of knowledge about the natural world on an unprecedented scale, constrained the intellectual boundaries of cosmographical practice to the production of facts useful to the administration of the empire. Conceived in this manner, cosmography at the Council of Indies became increasingly divorced from its origins as a humanistic genre and instead became a bureaucratic exercise, with the concomitant lack of recognition such work, then as today, generally garners. As we will see in the following chapter, during his early years as cosmographer-chronicler Velasco tried to bridge the gulf between being a humanist and being a bureaucrat when he sat down to write his first cosmography.

FIVE

The Cosmographer at Work

Juan López de Velasco wrote his first and most extensive work, the *Geografía y descripción universal de Indias*, during his early years as cosmographer-chronicler major of the Council of Indies. Begun in 1571 and finished in 1574, the book was, in Velasco's words, "a complete account, both general and particular of each discovered and populated territory and province in the Indies, with its towns, [as well as] other material necessary for governing."[1] Velasco wrote the *Geografía* using material he had avidly collected, which at that point consisted of depositions and questionnaires directed to church officials generated during Ovando's *visita*, old documents available in the Council's chambers, those created by Santa Cruz, those that resulted from his 1572 requests to the Indies, and a number of privately authored works he had managed to get hold of.[2] The result of Velasco's effort was a 674-folio text (307 pages in the most recent edition) describing the peoples, lands, and seas of the New World.

During López de Velasco's lifetime fewer than a dozen persons, however, would read the substantial and richly detailed geography he poured onto the 674 folios. In keeping with directives Philip II consistently issued during his forty-year reign, the book's geographical and hydrographical descriptions and maps were considered state secrets. Thus, the *Geografía* remained in the confines of the Council of Indies, and only a small number of copies were made, one for the king and perhaps one for each of the eight or nine members of the Council of Indies.

1. Velasco's dedication letter to the king. Published in Juan López de Velasco, *Geografía y descripción universal de las Indias*, ed. Justo Zaragoza (Madrid: Real Academia de Historia, 1894), vii.

2. For examples of these requests, see Solano, *Cuestionarios*, 11–15.

The book followed the traditional organizing principles advocated by Ptolemy in his *Geography* as viewed through the bureaucratic lens of the 1573 *Instructions*. The geographical material was first introduced in general terms, taking the totality of the areas described into account, with regional or particular descriptions relegated to later sections. The book began with a section titled "Universal Description of the Indies and of the Demarcation of the Kings of Castile." Although it might seem a summary of the material discussed in subsequent sections, it is in reality the general cosmographical description of the New World as Velasco was instructed to compose it in the *Instructions* (article 14). It covered the principal points outlined in the *Instructions*: cosmography, history of the discovery, and a general description of the climate and natural history, which included the Indians and their nations and customs. Velasco also added a brief description of the jurisdictions and duties of the administrative, ecclesiastical, and financial bodies governing the Indies, although the *Instructions* did not expressly direct him to write on such subjects.

The general cosmography identified the part of the terrestrial sphere Velasco intended to discuss in his book: "the land and seas encompassed by a hemisphere or half of the world of 180 degrees of latitude ... and beginning 39 or 40 degrees west of Toledo ... and from north/south from 60 degrees north to 52 degrees south."[3] He added, with overstated confidence, that within this area "all has been discovered and sailed." He next addressed the question of how these lands might have come to be inhabited. He rejected theories (*conjeturas flacas*) that suggested that the native inhabitants on the Indies descended from Plato's Atlantis or that Carthaginians reached the New World or that there had been ancient navigations by tribes of Israel. It might be possible that the lands had been populated through contact with Newfoundland (Terra Nova or Bacallaos) over the course of many years (*en los años del mundo*) by people sailing from northern lands, such as Ireland. But his preferred explanation was the theory that there might be or might have existed in the past a land bridge between Asia and America. He followed this discussion with a brief account of the discovery, suggesting that Columbus learned about the New World from two Spaniards whose ship was blown off course.

After specifying the extent of the land and how it was inhabited, Velasco touched upon the controversial topic of the line of demarcation. In this section he presented a learned and objective analysis of the many factors that made placing the line accurately on a map problematic, and he discussed

3. López de Velasco, *Geografía y descripción*, 1.

the problems concerning the Brazilian borders and the location of the Moluccas and Philippines. He repeated the assertion made years earlier by Alonso de Santa Cruz that the Portuguese altered their maps to suit their pretensions "without any authority or basis"; the Spanish, on the other hand, calculated the location of the east side of the line of demarcation by sailing west from Spain and found that the line passed through Bengal. Based on his understanding of Asian geography, this placed the islands of Trapobane (Sri Lanka), Sumatra, and the Moluccas at least 30° safely inside Spain's side of the line of demarcation. Velasco recognized the problematic nature of longitude calculations and addressed these issues in a section titled "Of the Longitude Used in This Book." Acting in compliance with the guidelines specified by the *Instructions*, he relied on lunar eclipse observations whenever these were available.[4] His key observation is of the 1544 eclipse made by Juanoto(e) Durán in Mexico, which placed Mexico City 103° west of Toledo.[5] Velasco admitted, however, that this longitude was somewhat problematic. According to the *padrón real*, Mexico City was 91° west of Toledo, and there were other observations made by individuals (*de particulares*) that placed Mexico City closer to Spain by four or five degrees. (The distance between the two cities is 95° 11′.) Concluding, he remained committed to three fixed points on his maps: a distance of 103° between Toledo and Mexico City, the location of the eastern line of demarcation through the Cape of Humos on the north coast of Brazil (39 or 40° west of Toledo), and the western line through Bengal (49° or 50° east of the Canary Islands). These critical points complied nicely with the directive given in the *Instructions*.

Velasco's general description of the climate, land, rivers, and natural resources of the New World pictured a land of unlimited bounty and fertility, particularly the zone between the tropics. He wrote enthusiastic descriptions of the native plants and of the prospects of profitably growing plants brought from Spain. The native population cultivated only corn and kept no domesticated animals. This last aspect had been remedied with the importation of horses, cattle, sheep, and pigs from Spain, and since there were no significant natural predators for these, they had become so accustomed to the land that they now ran wild. He painted a picture of almost unlimited abundance, always characterizing the quantity and fertility of plants

4. Ibid., 5.

5. Durán carried out a survey of the bishopric of New Spain between the years 1535 and 1544. UTX JGI-XXII-8, "Demarcación de los límites de los Obispados de la Nueva España." Published in Luis García Pimentel, *Descripción del arzobispado de México hecha en 1570* (Mexico: J. J. Terrazas, 1897), 23–24.

and animals as greater than in Spain. But although he at times appears in awe of the resources in the New World, he does not exaggerate or portray anything that might be considered fanciful or fantastic. For example, his assessment of mineral wealth is far from bombastic. Gold, he pointed out, was rarely mined or collected from rivers anymore, because Indians were now forbidden to work in the mines and black slaves were far too scarce and expensive. Silver mining took place only where the veins were known to be rich, and therefore it was limited to New Spain, Charcas, and Potosí.

Velasco did not shy away from recounting the tragedy that the first years of the conquest wreaked on the American Indian. Wars, forced migrations, diseases borne by the Spaniards, and bad treatment and overwork imposed by Spanish colonizers had depopulated the land. Velasco, however, noted that the Indian population had started to recover, should continue to do so—since they had been taught to abandon human sacrifice—and were no longer under tyrannical rule. He insisted that at the present time the Indian population was no longer abused but was protected by laws made to ensure their well-being. For their own benefit the Indians had been removed to villages, where they could live with more "order." In these villages they were taught religion, and the sons of prominent families received schooling in the hope that their examples would inspire others to abandon their native ways. Velasco observed that many already "wear dress and shoes, and some, even hats." Except for some regions of Peru, native leaders were allowed to remain in leadership positions but were not permitted to bear arms or ride horses.

For the most part Velasco did not acknowledge that the Indians had organized governments, save for the kingdoms of Montezuma and the Incas, and these he qualified as ruthless in their cruelty and mistreatment of their subjects. Although he acknowledged that modes of worship varied throughout the continent, in general he surmised that Indians north of the equator worshiped the devil and those in the kingdom of Peru worshiped the sun. He described most of the native population as servile and without any virtues. But Velasco is equally harsh on Spanish colonists, explaining that even the restrictions currently in place had not prevented "men who are enemies of work, spirited and greedy, hoping to become rich quickly rather than to settle in the land,"[6] from moving to the Indies.

The remainder of the general cosmography explained the structure and jurisdictions of the administrative, ecclesiastical, judicial, and financial units governing the New World. This section of the book, at first reading, seems

6. López de Velasco, *Geografía y descripción*, 19–20.

out of place in what until that point had remained faithful to the exigencies of traditional cosmographical description, but it serves to remind us that the *Geografía* was not written solely to serve as a descriptive geography and natural history of the New World. This section of the cosmography bears the imprint of the document's true purpose as an administrative tool intended for the express use of the Council of Indies. It is a concise picture of how the administrative machinery of the empire governed the Indies and of the intricate financial arrangements between the king, the church, and local government. In Velasco's worldview, these structures were an integral part of the world he was describing. Information regarding the lay of the land, its native peoples, and its natural resources was likewise necessary for governing the Indies.

In the book's second chapter, titled "General Hydrography of the Indies and Explanation of the Preceding Sea Chart," Velasco tackled a general hydrographical description of the New World. He followed closely in this section the guidelines established in articles 15, 68 and 69 of the *Instructions*. As the title of the chapter suggests, the hydrography was accompanied by a sea chart, presumably showing the latitudes and longitudes requested in article 15 and no doubt largely based on the *padrón real* of the Casa de la Contratación. This and all the other maps that at one time formed part of the *Geografía y descripción* have since been lost. We know of the contents of the twenty-three maps only from sections of Velasco's text that were meant to serve as map captions, as well as a detailed critique made of the book by Juan Bautista Gesio, discussed later in this chapter.

Since Velasco intended that the hydrography chapter be read in conjunction with the now-lost sea chart, he limited his comments to setting down the names and relative locations of significant bodies of water. There are three separate sections on the tides, winds and storms, and currents characteristic of the seas surrounding the American continents. The section describing the winds, storms, and current, albeit well organized and coherent, is marked by an almost colloquial quality that suggests Velasco relied on firsthand accounts to compose the section. He noted the most propitious months for sailing in the different zones and the winds that were likely to be encountered, and he described hurricanes, when and where they could be expected, and their ferocious intensity. He concluded the general hydrography with a description of the sailing conditions in the Strait of Magellan. The remainder of the chapter discussed rutters of the typical sailing routes to and from the Indies, including what was known of the Pacific crossing from the port of Navidad in Mexico to the Far East, a crossing that was until then not routinely sailed. The rutters are described in somewhat general terms, but Velasco indicated key latitudes for finding suitable winds, as well as critical distances and landmarks. The chapter closes with a concise account of the function and duties of the Casa de la Contratación and its role in coordinating and overseeing all commerce and travel to the Indies.

Having covered the general description of the land and sea, Velasco began a systematic and more detailed treatment of particular regions, or rather, as he had learned from Ptolemy, he proceeded from the general (*geographia*) to the specific (*chorographia*). From the textual descriptions meant to accompany the maps we can surmise that besides the navigation chart, which must have extended at a minimum from the Iberian Peninsula to the Moluccas and served as a universal map, Velasco included two other general maps, one for the northern American continent and one showing South America.[7] Another twenty regional maps showed the spatial relationships between the administrative jurisdictions, provinces, and islands that by the late sixteenth century had come to define the American space.[8]

The content of the regional maps was for the most part dictated by the administrative partitions in place when Velasco prepared the maps in 1574. Thus we find one map each for the nine *audiencias*: Hispañola, New Spain (Mexico), New Galicia (Jalisco), Guatemala, Panama, New Kingdom of Granada (Colombia), Quito, Los Reyes (Lima), and Charcas (Bolivia). Three large but subordinated provinces or *gobernaciones* also received specific maps: Cartagena, Yucatán, and Chile. Four islands or archipelagos are also depicted in specific maps: Hispañola, Cuba, the Solomon Islands, and the Spice Islands. Territories that were not under specific jurisdictions or that were particularly sensitive were also included in specific maps: Terranova, the Strait of Magellan, Brazil, and the coast of China. The now-lost maps of the *Geografía* constituted in effect the first political atlas of the New World.

7. The map of the northern region likely covered from about 8° or 9° north, that is, Panama and the coast of Venezuela, to Quivira, or modern New Mexico. The southern map encompassed from 8° north to the Straits of Magellan. Each of the maps was accompanied by a brief text that discussed the extent of the territory and its administrative and ecclesiastical units. This was followed by a summary of the situation of the Spanish population, a brief discovery account, and a concise natural history—a category that invariably included a description of the native population. For the text describing the northern territories, see ibid., 47–49, 169–70.

8. For a study of the administrative and ecclesiastical jurisdictions that Velasco used, see Ricardo Beltrán y Rózpide, "América en el tiempo de Felipe II según el cosmógrafo cronista Juan López de Velasco," *Publicación de la Real Sociedad Geográfica*, 1927.

General geographical and hydrographical descriptions make up only 15 percent of the book, with the rest of the text dedicated to a systematic description of each of the *audiencias*, principal provinces, and noteworthy territories. The description of each *audiencia* follows a similar format, beginning with a general geography, describing the limits of the administrative unit using latitude and longitude and geographical boundaries. This is typically followed by a brief hydrographical description, an abridged account of the region's discovery, and a summary of its natural history.[9] The text dealing with each subordinate administrative unit is organized similarly, with each subordinate *gobernación* and its principal Spanish towns described in what Velasco titled "Chorography" or *Descripción particular*, with significant port cities described in separate sections titled "Topography." These particular descriptions specify the region's boundaries, give a brief history of the town's founding, lists it current population and buildings, and provide an inventory of the area's natural resources, identifying its potential and principal means of livelihood and trade. For Indian towns within each *gobernación*, Velasco simply listed the name of the town, its number of taxpayers, and the ecclesiastical order and number of priests serving the town.

Velasco's *Geografía* was an exercise in synthesis limited to the compilation of wholly empirical information. When he addressed topics where a natural philosophical explanation would have been appropriate, such as when he discussed weather patterns, he shied away from doing so and restricted his discussion to a description of the observed phenomena. His approach was at once comprehensive in terms of territorial coverage and within the conceptual and thematic boundaries defined by his post and by the *Instructions*. These demanded that the matters discussed in the cosmographer's work remain firmly within the bounds of facts defined by the testimony of eyewitnesses and recorded in documents. How then did Velasco use these documents to compose his cosmography?

Velasco cited his sources in only a few instances, such as when he mentioned the works of Alonso de Santa Cruz and Martín de Rada and the cartography of the cosmographer of the Casa de la Contratación, Chaves (Velasco does not clarify whether this is Alonso or his son, Jerónimo). It is reasonable to attribute the details of the navigations he described for the *carrera de Indias*

9. This organizing scheme is not neatly apparent in the edited version of the *Geografía y descripción*. The editor was at times arbitrary in his selection of chapter titles and section headings.

to either Santa Cruz or the *padrón real.* For some local descriptions and for population numbers, Velasco relied on the responses to Ovando's 1569 request to the archbishops,[10] some of which carry a telltale annotation in Velasco's hand: "papeles del tiempo de la visita."[11] What does this epistemic paper trail tell us about Velasco the cosmographer?

One of the accounts Velasco used was the "Description of the Bishopric of Antequera of New Spain," written circa 1570.[12] The original document has marginal annotations that I believe to be in Velasco's handwriting. A careful correlation between these marginal annotations and the corresponding text supports this and provides a rare glimpse of the cosmographer at work. Most of Velasco's annotations single out geographical information, pointing out sections in the text where the author refers to coordinates, distances or limits. For example, he singled out "Antequera en 18 grados" and later "corre este obispado de mar a mar y Tlascala tambien." The author gave a rather detailed description of the extent of the territory assigned to the bishopric, from which Velasco extracted his information concerning the region's borders. This required that he manipulate the original account to satisfy the format used through out the *Geografía.* Velasco extracted the distances and boundaries of the territory's outlines and assigned latitude and longitude coordinates to the most southwesterly point

10. See Gerhard's meticulous survey of the results of Ovando's 1569–71 requests. Many of the originals are at AGI, IG-1529, and AGI, Mexico 336. Peter Gerhard, *A Guide to the Historical Geography of New Spain,* Cambridge Latin American Studies 14 (Cambridge: Cambridge University Press, 1972), 31. Another group of documents that resulted from this request are at the University of Texas's Icazbalceta Collection. Some appear in García Pimentel, *Descripción del arzobispado de México,* and Luis García Pimentel, ed., *Relación de los obispados de Tlaxcala, Michoacán, Oaxaca y otros lugares en el siglo XVI: Manuscrito de la colección del señor don Joaquín García Icazbalceta,* 2 vols., Documentos históricos de Méjico (Mexico, 1904).

11. See, for example, UTX, JGI-23-1, and JGI-23-7. Historian Fray Miguel Miguélez identified another collection of Velasco's working papers at the library of El Escorial. One of the documents is a summary of Tomas López Medel's *Libro de los tres elementos.* It has written across the top "de las relaciones del tiempo de la visita." Velasco must have collected Medel's work during his time with Ovando. He prepared chapter summaries but copied completely an extensive list of place names. BME, L-1-12, f. 152–156v, Juan López de Velasco, "Sumario del *Epítome breve suma del tratado de los tres elementos* (de Tomas Medel)." Miguélez, *Catálogo de los códices españoles,* 2:73–76.

12. UTX JGI-XXIII-7, "Descripción del arzobispado de Antequera de la Nueva España hecha por el obispo del dicho obispado," Antequera, ca. 1570. The bishop at the time was Bernardo de Albuquerque, O.P. Published in García Pimentel, ed., *Relación de los obispados de Tlaxcala, Michoacán, Oaxaca,* 59–68.

at Ciutla (16°N, 102°W) and the most northeasterly point at Guacaqualco (18°N, 96°W) with Antequera at 18°N, 100°W. He then defined the region's perimeter as both distances and relative to the bishopric's borders.[13] Velasco, however, did not attempt to reconcile these latitude and longitude coordinates with the perimeter values he cited in previous sections and liberally rounded off coordinate values. For example, the bishop's account said the village of Santo Ildefonso was "twenty leagues from here [Antequera] to the northeast"; Velasco turned this into "twenty leagues north of Antequera at 99° and a half of longitude and 18° and a quarter in latitude."[14]

Velasco paraphrased the bishop's topographic description of the region near the town of Antequera (*esta ciudad entre tres valles* . . .) as he did the information concerning the area's natural resources. Velasco's adaptation of the text remains faithful to the tone of the original, making no attempts to convey a rosier or dimmer picture of the region. His approach can be categorized as an exercise in textual extraction, where he intervened very little as an author and in some instances seriously neglected the task of integrating the textual account into the associated cartography.

Velasco copied faithfully the numbers of tributary Indians reported by the bishop, save for what could be a copyist error citing the population

13. The bishop had reported the limits of the territory as follows: "Tiene de largo este Obispado de una mar á otra, por los confines del obispado de Tlaxcala, ciento y veinte leguas, poco mas ó menos, y vase poco á poco estrechándose la tierra, y metiéndose la mar en ella hasta el fin del Obispado, que es el parage de Teguantepec al Sur y Guazacualco al Norte, donde tiene toda la tierra de travesía sesenta lenguas poco mas ó menos, y de allí adelante se torna á agrandar la tierra, metiéndose en la mar, ansi hacia el Norte, por donde corre el Obispado de Yucatán, como tambien hacia el Sur, donde está el de Guatimala, dejando al de Chiapa en medio. Tiene de travesía este Obispado, por medio de tierra, desde los confines del Obispado de Tlaxcala hasta los del Obispado de Chiapa, ochenta leguas, poco más ó menos, y por la costa del Sur cient [sic] leguas, poco más ó menos, y la costa del Norte cincuenta leguas, poco más ó menos." In García Pimentel, ed., *Relación de los obispados de Tlaxcala, Michoacán, Oaxaca*, 60. Velasco took this information and presented it as "[D]e manera, que de la una mar á la otra, por los confines de Tlaxcala tendrá ciento veinte leguas poco más ó menos, y por la costa del norte como cincuenta leguas, y sesenta desde la mar del Norte á la del Sur por los confines de Chiapa, y ciento por la costa del sur desde la provincia de Soconusco hasta Ciutla." López de Velasco, *Geografía y descripción*, 116.

14. García Pimentel, ed., *Relación de los obispados de Tlaxcala, Michoacán, Oaxaca*, 61, and López de Velasco, *Geografía y descripción*, 118. Velasco had previously stated that Antequera was at 18°N, 100°W "poco más ó menos." Using this as the departing point, Velasco should have placed the town at twenty leagues due north or at 19.14°N, 100°W (or taking the bishop's northeast bearing into account, at 18.8°N, 99.19°W) and not at 18.25°N, 99.5°W.

of the Misteca province at 2,000 when the bishop had reported upward of 47,000. In a similar exercise to the one above, but in order to validate some of Velasco's population data, Jean-Pierre Berthe has compared the description of the district of Michoacán in the *Geografía* with two accounts written between 1571 and 1572 that likely served as Velasco's sources.[15] Berthe found several instances where Velasco overlooked information, misread numbers, and made mathematical errors that resulted in a 25 percent undercount of the native population. Rightfully, Berthe called into question the value of the population numbers Velasco presented in his work. Inevitably, the same doubt applies to his geographical work as well.

Another group of documents López de Velasco used when preparing the *Geografía* are at the Academia de Historia in Madrid.[16] One of them is an anonymous description of Cartagena in draft.[17] In the *Geografía*, Velasco repeated complete phrases—the descriptions of the Indians and of native trees are almost word for word—although the material from the draft is organized differently from the book. In this instance, the cosmographer corrected the longitude values stated in the document, which had been calculated using the conversion of 15.5 leagues per degree of longitude (l/°) to be consistent with the longitude values he used in the book. These were all—regardless of latitude—calculated using the ratio of 17.5 leagues per degree. The draft suggest maps accompanied the document at one time. Unfortunately, neither the original maps nor Velasco's interpretations of them survive.

Another document in the collection of Velasco's drafts appears to be an early version of his description of the La Plata region.[18] The draft discussed the limits of the La Plata region, granted to the person chartered by the Crown to settle the area or *adelantado*, Juan Ortiz de Zárate. Apparently,

15. The bishop's report on the number of local priests and parishioners is at AGI, IG-856, "Relación de los clérigos que hay en este obispado." Published in García Pimentel, ed., *Relación de los obispados de Tlaxcala, Michoacán, Oaxaca*, 30–59. According to Berthe the source was Bishop Antonio Ruiz de Morales de Molina. Berthe, "Juan López de Velasco," 161–63.

16. A number of original documents in Velasco's hand bearing the telltale "cespedes" written across the top are bound in volume 71 (9/4661) of the Colección Muñoz at the BAH. As Jiménez de la Espada first noted, the draft documents are written on discarded letter cover sheets, most addressed to López de Velasco.

17. Jiménez de la Espada thinks Velasco used this document (BAH, 9/4661) as a sample draft (or *borrador*) for the way information was to be presented in the *Books of Descriptions*. Published in Jiménez de la Espada, *Relaciones geográficas*, 183:65–69.

18. BAH, 9/4851, f. 158–168v, Juan López de Velasco, "Historia y sucesos del Rio de la Plata," ca. 1574.

the statute defining the region's northern limits was vague, simply stating that the northern border of the new jurisdiction should not infringe on the territories granted to earlier colonizers Pedro de Silva and Diego Hernández de Serpa. In his draft Velasco had calculated the latitudes corresponding to the territory's northern and southern borders, yet when he addressed the subject in the *Geografía*, he did not cite these nor discuss the possible infringement of the new jurisdiction on adjacent territories. Instead he introduced the subject by saying that "the limits of these provinces are not determined as to longitude and latitude, because the lands are not well discovered or traveled." He then explained that judging from explorations headed by governors of adjacent regions, the La Plata region must extend 500 leagues east-west and 500 or 700 north of the La Plata River and 400 south of it. Realizing perhaps that this defined an incongruously large territory, he added the caveat that because of the "difficulties, tortuousness, and obstacles" of the road, these distances are generally stated as greater than they really are and should be shortened by at least a third.[19] Why, if he was instructed to express the limits of each territory in latitude and longitude, did he not do so?

In this case Velasco had only one latitude observation, that of the La Plata river (36°N), and thus could only compute mathematically the coordinates of the northern and southern limits. This could risk delimiting the territory in a way that impinged on a neighboring jurisdiction, a potentially contentious situation. So he resorted to the epistemic guidelines under which he operated and deferred to the distances cited in legal documents that described the region. In my estimation, another of Velasco's principal sources for territorial boundaries was the *capitulaciones* granted by the king to the region's first governor or those that defined the regions assigned to bishoprics. These documents were far from static; they were revised often throughout years of legal disputes and wrangling. Velasco was well aware of these changes and included these amendments in his descriptions, as in the case of the La Plata region discussed above. Velasco's epistemic hierarchy required that he defer to administrative or ecclesiastical jurisdictional edicts to reconcile his geography, placing these even above documents expressly prepared to provide geographical descriptions.

With the sources at his disposal, it would have been very difficult—if not impossible—for López de Velasco to write a cosmography that met the standard of mathematical coherency demanded by Ptolemaic cartography. We do not know if Velasco was aware of the shortcomings that using these

19. López de Velasco, *Geografía y descripción*, 279.

types of documents would pose a cosmographer. As we will see in the next section, others would point this out. In Velasco, however, the Council had found a cosmographer who spoke the language of the *letrados* (men of letters). Velasco understood that for an audience limited to members of the Council of Indies it was more important to locate people and places correctly within delimited political and ecclesiastical jurisdictions than relative to an absolute cosmographical grid of latitude and longitudes.

It its comprehensiveness and organization, the *Geografía* mimicked Renaissance cosmographies, but it also differed from that genre in fundamental ways. Absent from the text are the anecdotes—of real or fantastic events—that gave a cosmography its power to impress and that had been so successfully used by authors to capture and convey the character of a region and its people. Velasco's synoptic approach to exposition instead performed a didactic function, one limited to transmitting discrete nuggets of vetted information. This style was a direct consequence of the legally defined epistemic criteria upon which Velasco had to construct his cosmography.

The Censor Censored: Juan Bautista Gesio

When Velasco completed the book in 1574, the Council of Indies scrutinized it to identify and censor any sensitive information. The Council did raise objections to some of Velasco's frank discussions and requested that a number of paragraphs be deleted, arguing, "There is no reason that one of our books should have anything written in it against our pretensions."[20] What did these changes entail? Luckily, two versions of the *Geografía* remained accounted for when the book was edited in the late nineteenth century, one drafted before the Council's amendments and a "clean" copy destined for the king's chambers with the changes incorporated, and thus the nature of the passages that the Council found objectionable can be easily identified.[21]

The Council did not question Velasco's placement of the line of demarcation but asked him to remove from the text any references to the legality of Spanish territorial claims to the Philippines and Brazil. Velasco

20. "[A]lgunas de substancia que no hay para que esten escritas por nosotros en libro nuestro contra nuestra pretención y otras que tocan a la graduación y demostración." Justo de Zaragoza prints the undated letter from the Council of Indies to (presumably) Mateo Vázquez in his introduction to López de Velasco, *Geografía y descripción*, ed. Zaragoza, vii.

21. Zaragoza's edition indicates where paragraphs of the original version were crossed out. According to Berthe, the two sixteenth-century manuscripts used to prepare the Zaragoza's edition have since disappeared. Berthe, "Juan López de Velasco," 151–53.

had framed the discussion of Spanish settlement in the Philippines by first admitting that the islands fell within the territories leased to Portugal by the *empeño* of 1529, in essence agreeing with the conclusion reached by the cosmographical junta of 1566 and some maps by Martín de Rada.[22] The Council also asked Velasco to remove a sentence that conceded Brazil to the king of Portugal.[23] He also had to remove references to English and French incursions and settlements in Labrador and delete the name of the town of New London—he used the Indian name instead—from the description of Tucumán (in northwest Argentina).[24]

The other group of significant amendments touched on matters of territorial jurisdiction, particularly where these were contentious or vague. For example, the whole section that described the composition and administrative and legal jurisdiction of the Council of Indies was deleted, as well as the section where Velasco explained the particular areas of responsibility of the two viceroys and nine *audiencias*.[25] These administrators, according to Velasco's description, although equal in rank, did not have the same political power. Further, the viceroys' authority varied between jurisdictions, an inequality that could incite jealousies and discontent. The account of special papal indulgences and royal grants to monastic orders in the New World, and Velasco's comment that some orders had taken advantage of these privileges, vanished from the final version, as did his description of the Inquisition in the New World.[26] Some problems identified by the Council had to do with vague borders between jurisdictions assigned to different conquistadors, such as those of Serpa in Nueva Andalusia (Guyana) and Pedro de Silva (El Dorado).[27] An acknowledgment that Jamaica had been granted to the admiral of the Indies (Christopher Columbus) also disappeared.[28] The Council did not necessarily want all the blemishes of the conquest highlighted, so it crossed out references to "the tyrant" Lope de Aguirre, Gonzalo de Sandoval's brutal quelling of the revolt at Panuco, and

22. López de Velasco, *Geografía y descripción*, 5, 289, 292, 295–96. The name of Fray Martín de Rada also fell under the censor's pen. Rada had sent his observations in 1574; see AGI, Mexico 19, N. 128, f. 5, "Carta del Virrey Conde de la Coruña, Martín Enriquez . . . informa sobre mapa de las islas del poniente y papeles de Martín de Rada," March 24, 1574.

23. López de Velasco, *Geografía y descripción*, 286.

24. Ibid., 89, 259.

25. Ibid., 20.

26. Ibid., 25.

27. Ibid., 78.

28. Ibid., 62.

reference to the Zacateca Indians as "bestial, indomitable, and ferocious peoples."[29]

The Council of Indies also prudently chose to delete a complete section titled "Of the Spaniards Born in the Indies." Velasco had explained that after living in the Indies for many years, Spaniards assume some "differences in their color and qualities of their persons."[30] He attributed these to the "changes in the heavens and the temperament of the region." Perhaps more alarming than the author's astrological and environmental determinism was Velasco's assertion that the children of Spanish parents also appeared different. They were larger and of a color "somewhat low tending toward the earth." He said that some believed that even if the Spanish population never mixed with the natives, their corporeal qualities would come nonetheless to eventually resemble the Indian's. Velasco also stated that in the Indies a Spaniard's temperament became "altered." This he attributed to a sympathy between body and temperament, with the temperament following the body's mutation described above. Velasco was echoing a generalized belief of peninsular Spaniards about the deterministic relationship between environment and physiology, rather than a personal bias against settling in the New World.[31] Not only had his two brothers emigrated, but in his will he encouraged his two nephews to seek a better life in the Indies.

Of equal significance to what the Council censored is what it left as written. Velasco made repeated reference to the crisis caused by severe depopulation among the natives as the result of Spanish abuses during the conquest. In the island of Hispañola, as in Cuba, Velasco explained, most of the native population had died in wars, committed suicide, or died of smallpox, and the countryside was overrun with wild cattle, hogs, and dogs. In addition, the city of Santo Domingo had lost half its Spanish population since gold mining had practically ceased due to the scarcity of natives, and ships no longer stopped at the island to trade.[32] He did point out, however, that the trend appeared to be reversing and in many instances native population had stabilized and even increased.

That the Council found so little to condemn is more a testament to Velasco's self-censoring than to the accuracy of his conquest accounts. The

29. Ibid., 74, 102, and 134.

30. Ibid., 19–20.

31. For competing views of environmental determinism in the Hispanic New World, see Jorge Cañizares-Esguerra, "New World, New Stars: Patriotic Astrology and the Invention of Indian and Creole Bodies in Colonial Spanish America, 1600–1650," *American Historical Review* 104, no. 1 (1999).

32. López de Velasco, *Geografía y descripción*, 51–52, 58.

Council refrained from commenting on the accuracy of his geographical, hydrographical, and ethnographical descriptions, a silence that points to members' ignorance of the specifics of the territory they were chartered to oversee. Perhaps aware that they were unqualified to judge the work's cosmographical merits, the Council asked Juan Bautista Gesio, a cosmographer loosely attached to the royal court, to issue an opinion on the scientific merits of the work. Based on Gesio's report, the Council raised objections regarding the "scale and demonstration," or rather the lack thereof, in Velasco's maps.

Sometimes described as a spy, Italian-born Gesio had worked many years in Portugal but had sought to serve the Spanish since 1559.[33] Researchers today regularly stumble upon letters that Gesio sent Philip II, scattered among surviving documents of the empire's bureaucracy.[34] These attest to Gesio's role as adviser to the king on sensitive cosmographical matters and to his self-appointed role of adviser to the king on geopolitical matters as well.

Gesio entered the Spanish stage when in August 1573 Spain's ambassador to Portugal, Juan de Borja, sent him to Madrid with some highly sensitive Portuguese maps. These showed that the Moluccas fell on Spain's side of the line of demarcation and that the Portuguese had altered maps in their favor.[35] Borja had Gesio make two sets of nautical charts, one according to how the Portuguese "under penalty of death" (*so pena de la vida*) had to make their maps and another "according to the truth as they understand they should be made" (*conforme a la verdad con que ellos entienden se deben hacer*). The difference between the two charts was around 14° in the Portuguese's favor. Borja maintained that the correct charts gave the king of Spain the right to claim the rich territories of China and Japan—Gesio's position as well. The king replied that this was very "convenient" and

33. Gregorio de Andrés, "Juan Bautista Gesio, cosmógrafo de Felipe II y portador de documentos geográficos desde Lisboa para la Biblioteca de El Escorial en 1573," *Publicaciones de la Real Sociedad Geográfica*, ser. B, 478 (1967): 1–12.

34. His letters are at the library of El Escorial, AGI, Instituto Valencia de Don Juan, Archivo Zabálburu, Biblioteca Nacional, and the British Library, to name a few. He lacks a comprehensive biographical treatment. For some studies, see Fernández de Navarrete, *Biblioteca marítima española*, 1:234–35; Jiménez de la Espada, *Relaciones geográficas*, 184:137–66; Vicente Maroto and Esteban Piñeiro, *Aspectos de la ciencia*, 71–72, 111–15; and Lamb, "Nautical Scientists and Their Clients in Iberia (1508–1624)," in *Cosmographers and Pilots*, 9:49–61.

35. Juan de Borja, "Carta al rey donde trata de cartas de marear que adquirio en Portugal," 1573, Lisbon. Published in Fernández de Navarrete, *Colección Fernández de Navarrete*, 18:35–36.

welcomed Borja's efforts. The correspondence with the ambassador also suggested that any kind of overt conflict with Portugal about the demarcation was to be averted and thus all these "inquiries" should be done quietly. This reluctance to confront the Portuguese openly on the matter might have had to do with a possible marriage then in the works between the king of Portugal and Philip's daughter, Princess Isabel.[36]

Gesio had been in Borja's employ in Lisbon for over three years studying the matter when he went to Madrid. Upon his arrival, his cartographical material was reviewed at the royal court by Antonio Gracián, Juan de Herrera, and the Count of Chinchón. Gracián's account of his meeting with Gesio suggests that the Italian was kept in Madrid for reasons other than his talent as a cosmographer. Gracián wrote to Juan de Ovando, "[A]s to the person of Juan Bautista, His Majesty says that because he is already aware of this business and because it could be inconvenient if he went elsewhere, it would be good to occupy him here in something. See what you might do with him."[37] Clearly Gesio possessed some sensitive information and the king deemed it best that the Italian-by-way-of-Portugal be kept close at hand. Ovando gave him a modest stipend but no title.[38] López de Velasco went over Gesio's maps and books and prepared an inventory of the material, writing his report in May 1574.[39] He recommended that some of the books and material be copied and placed in the library at the Escorial; but this apparently never happened, and the documents may have remained at the Council of Indies. Velasco also went on record urging that Gesio should make his theories concerning the demarcation clear, so if these turned out to be important the pertinent material could be secured at the Council of Indies and its author given appropriate reward. In September that year,

36. AGS, Estado-387, f. 24, Juan de Borja, "Carta del embajador de Portugal, Juan de Borja, al rey que trata sobre demarcación," August 1, 1570.

37. Gracián recounted his meeting with Gesio in a fascinating letter sent to Juan de Ovando. BL, Eg 2047, f. 325v, Antonio Gracían, "Copia de una carta de Gracián at Presidente de Consejo de Indias," December 23, 1573, San Lorenzo. Published in Andrés, "Juan Bautista Gesio," 9–10.

38. AGI, P-261, R. 2, "Cédula de Felipe II dirigida al Receptor del Consejo de Indias ordenando el pago de un salario de 10 ducados al mes y una ayuda de costa de 30 ducados a Juan Bautista Gesio," January 19, 1574.

39. Velasco's inventory and opinion is in Zab, Altamira, 217, D. 258. Juan López de Velasco, "Informe de Juan López de Velasco de los libros que trajo Juan Bautista Gesio," May 3, 1574. It is published in Andrés, "Juan Bautista Gesio," 10–12. For a study of Gesio's arguments about the line of demarcation, see David C. Goodman, *Power and Penury: Government, Technology, and Science in Philip II's Spain* (Cambridge: Cambridge University Press, 1988), 61–65.

when Velasco completed the *Geografía*, Ovando had the perfect project to keep Gesio busy.

Gesio's comments on the *Geografía y descripción* center on a general critique of Velasco's now-lost universal chart, or rather the sea chart that Velasco included in the hydrography section and that covered 210° of the earth.[40] In a terse remark, Gesio refused to comment on the regional charts because these did not have "divisions or graduations in longitude and latitude, nor scale of leagues and distances." He explained that they violated the general rules of geography and chorography, being "more commentaries and descriptions of provinces than geographical maps."[41] Gesio also observed that Velasco had drawn the maps using equidistant parallels and meridians, with the resulting elongation in territories near the poles. But more important, Gesio found significant deficiencies in Velasco's use of rutters and nautical charts as sources for geographical information. Gesio pointed out that as a result of using nautical distances to determine distances on land, Velasco's maps exaggerated longitudinal values. The erroneous values Velasco copied from nautical charts, Gesio explained, were in turn the result of sailing using compass and distance while relying on altitude observations to correct the rutter. In reporting the distance sailed in a day, pilots traditionally logged longer distances to compensate for the distortion caused on an apparently straight course at sea by the curvature of the earth. Gesio recommended that Velasco correct his regional maps by adding coordinates and correcting distances using "spherical triangles" (spherical trigonometry) at ten-league intervals.

Gesio's critique is at times confused, but in essence, he argued in favor of two cartographic changes that carried significant implications for the location of the line of demarcation in López de Velasco's maps. The geographical reorientation Gesio proposed would have undermined Portuguese claims to Brazil and not only freed Spain from its contractual obligations concerning the Moluccas but also opened up China, Japan, and of course the Philippines to Spanish colonization. These were the same arguments that had brought him to Spain and landed him a modest stipend from the king. He was thus highly vested in making these points.

40. Gesio's ca. 1575 critique of the *Geografía* was published in Jiménez de la Espada, *Relaciones geográficas*, 184:142–62. Jiménez reported transcribing it from the unsigned original at the Zabálburu archive, bearing the initials of councilman Benito Pérez de Gamboa. Recent attempts to locate the original document in that archive were unsuccessful, so this study relies on Jiménez's transcription.

41. Jiménez de la Espada, *Relaciones geográficas*, 184:151.

In broad terms, Gesio increased the width of the Atlantic Ocean and reduced the width of the Pacific Ocean. For the eastern side of the line of demarcation, the critical geographical distance for Gesio was that defined by the distance between the westernmost point of Africa (Cape Verde or Dakar) and the easternmost point of Brazil (Cape San Agustin or Cabo Branco). Gesio proposed a longitudinal distance in the order of 34° rather than Velasco's 20°. (The distance is 17°.) These changes would have placed the eastern line of demarcation clearly off the coast of Brazil, therefore voiding Portugal's claim to that piece of the South American continent. The changes he proposed for the Pacific did not result in moving the line of demarcation but rather in shifting east most of the coast of Asia. He calculated a new distance between Peru and the Solomon Islands that effectively placed China, Japan, and the Philippines beyond the reach of the 1529 lease to Portugal and inside Spain's semisphere.

To support these two changes, Gesio constructed an argument that relied on a combination of evidence gleaned from early discovery accounts and Portuguese charts, while pointing out a significant error on Velasco's part. Apparently Velasco had overlooked the fact that the rutters he relied on to calculate distances in the Pacific Ocean were measured in Castilian leagues (which corresponded to 3,000 paces) rather than Portuguese leagues (4,000 paces), and Velasco had still converted them into degrees using the standard 17.5 leagues per degree (l/°). Gesio argued that the appropriate conversion factor for Castilian leagues should be 23 l/°, resulting in a reduction in the longitudinal width of the Pacific Ocean by between 20° and 24°. He castigated Velasco on this error, scolding, "[A] Castilian who writes chronicles of Castile should use terms, words, and distances as is customary in his land."[42] Gesio also advocated a return to the Ptolemaic reckoning of 15.5 leagues per degree of longitude rather than the value of 17.5 leagues per degree of longitude at the equator that had become accepted since the midcentury and was used by Velasco.[43] (Gesio does not explain in this document the rationale behind this revision.)

His other major objections to Velasco's geography concerned the outline of the coast of Brazil, the longitude of the Atlantic entrance to the Strait of Magellan, and the location, extent, and size of China and Japan. To defend his position concerning Brazil and the strait, he relied on the rutters

42. Ibid., 184:145.

43. 17.5 leagues per degree of longitude yields a diameter of the earth closer to the actual value of 25,000 miles, with one nautical league equaling 3.4 miles.

of Amerigo Vespucci and particularly those of Andrés de San Martín, pilot of the Magellan/Elcano expedition.[44] Gesio had supreme confidence in these accounts and dismissed Velasco's use of the lunar eclipse observations of Juanote Durán (to establish the position of Mexico) and Santa Cruz's description using Ladrillero's account of the Strait of Magellan, although these were more recent than his sources. He criticized Durán's observations because "[w]e do not have the hypothesis, nor do we know how he observed them, nor the propositions he used to demonstrate [his results], nor the time at mid-eclipse, nor its beginning or end, nor its magnitude, nor the *mora in tenebris* [time in the shadows], nor the ecliptic digits, nor the elevation of the star, nor any position which could to used to make a demonstration," therefore they could be used nor believed.[45] However harsh his critique of Durán's observation, whenever a critical longitude remained undefined, Gesio did advocate the use of lunar eclipse observations. His objections stemmed partly from Velasco's approach, which had been to determine the easternmost point of Brazil by using the location of Mexico City as his departure point, rather than using nautical logs of ships sailing west from Portugal to Brazil.

Turning to the lands of the Far East, Gesio again dismissed the celestial observations used by Velasco, this time those of Fray Martín de Rada.[46] He did identify a significant error in Velasco's map and description where Velasco explained that the Canton River on the coast of China had been formerly known as the Ganges River. Gesio pointed out that this observation could cause significant confusion, since the Ganges had from time immemorial been known to be just east of India and thus clearly within Portuguese territory.

Gesio also took exception with some of Velasco's natural philosophy, particularly his description of the tides and currents, dismissing these as too simplistic. Velasco attributed currents to effects of the wind, the inter-

44. It is likely that Gesio had access to the rutter of Andrés de San Martín, which had fallen into possession of the Portuguese when the ill-fated armada landed in Portuguese territories. João de Barros (1496–1570) used this material in his *Décadas*, which Gesio also cites. João de Barros, *Décadas*, ed. António Baião, 4 vols., Colecção de clássicos Sá da Costa (Lisbon: Livraria Sá da Costa, 1945–46), 4.

45. Jiménez de la Espada, *Relaciones geográficas*, 184:147.

46. Gesio nonetheless remained very curious about the Augustinian's work and petitioned the Council to have his papers brought from the Philippines. AGI, IG-740, N. 103 (4), "Orden al gobernador de Filipinas que recoja los papeles de Fray Martín de Rada y envíen al Consejo de Indias," April 24, 1580. Recall that in 1584 Jaime Juan was also instructed to recover these papers during his expedition to the Philippines.

position of landmasses, and the coastal topography. He explained that the two daily tides (*flujo y reflujo*) in the Atlantic Ocean were much smaller that those of the Pacific Ocean; all, however, grew in proportion to the moon's approximation to opposition or conjunction with the sun, given that the sphere of earth and water was one and completely circular.[47]

Gesio preferred instead, so as "to save all appearances,"[48] to assume that the "sphere" of water actually had an oval form. He explained that water had two natural movements, one linear, or "like respiration," and another circular.[49] The circular motion of water was of two types, one caused by the first mobile (*primer móvile*) and another contrary one according to the "order of the signs."[50] His reference to the effect of celestial motions on the tides corresponded to the then prevailing astrological/lunar theory

47. López de Velasco, *Geografía y descripción*, 30.

48. For the significance of this phrase in classical and early modern science, see G. E. R. Lloyd, "Saving the Appearances," *Classical Quarterly* 28, no. 1 (1978).

49. The remarkable passage, titled "Del fluxo y refluxo de la mar," is worth including here in its entirety: "Para salvar todas las apariencias de la mar es menester darle otra forma al agua que no es redonda y que una parte tenga en igual distancia de la otra del centro del mundo; la forma que debe tener es oval o de la cesión cónica o cilíndrica que se dice elipse, y alguna vez forma lenticular y porque el agua tiene dos especies de movimientos uno de por si y propio por línea recta que es de sulevación o tumor y depresión que es como respiración que hace el hombre en alzarse y bajarse a manera de fuelles; este movimiento es de diversas maneras; tiene otros movimientos circulares por accidente según el movimiento del primer móvile y otro al contrario según el orden de los signos; con las dos especies de movimientos susodichas se salvan todas las apariencias del agua, mas es menester *dar movimiento a la Tierra*; puede ser dar razón y causa porque la alta marea es del Nordeste Subdeste y la baja Norueste Sudeste y porque las aguas más altas y más bajas son en el tercero después de la conjunción y oposición del Sol y de la Luna. Este material es de gran placer y contento placer platicarlas y dar las razones las cuales se dejan de decir acá porque hasta el día de hoy no se ha hallado hombre que haya podido dar las razones de todas las apariencias de la mar y la causa porque en unas partes hay flujo y reflujo y en otras no." In Jiménez de la Espada, *Relaciones geográficas*, 184:148. It is unclear if the italics are original or if they were inserted by Jiménez de la Espada.

50. He attributes one of the circular motions to the first mobile or *primum mobile*. He was likely referring to the tenth sphere of the fixed stars, which in medieval interpretations of Aristotelian cosmology accounted for the daily east-west motion of the stars and imparted movement to the rest of the celestial orbs. The second motion Gesio describes is contrary to that of the first mobile, and therefore west-east, which he attributes to the "order of the signs." This was probably a reference to the second principal motion of the heavens corresponding to the movement of the sun on the ecliptic. Edward Grant, *The Foundations of Modern Science in the Middle Ages* (Cambridge: Cambridge University Press, 1996), 108.

of the tides, whereby the moon's occult properties acted upon the water.[51] Gesio stated that these circular motions "saved all the appearances" of the seas' movement but added the caveat that for this to be the case, *the earth had to be in motion*.

Gesio quickly dismissed his speculations about the earth's motion as "pleasurable and happy discourses" in the following clause, explaining that to date no man had been able to explain completely all the "appearances" of water. It remains unclear to me, however, what he meant by the earth's having to be in motion for this theory to account for all the known motions of the water. If indeed in this passage Gesio was conceding the possibility of the earth's motion to explain the tides, his explanation preceded Galileo's by more than forty-two years. One wonders whether as fellow Italians Gesio and Galileo had access to a now-forgotten common source for this theory.[52]

Gesio likewise found fault with Velasco's assertion that there were two different places on earth where the compass needle appeared fixed, one in the Northern Hemisphere along the meridian of the Azores and a different point somewhat east of the meridian of the Azores in the Southern Hemisphere. Gesio maintained that should the compass needle indeed exhibit such a mutation (*dado que haga mutación*), these two points had to be on the same meridian and maintain a relationship of 180° between them.[53] Gesio does not explain any physical reasons for this but argued that the compass had to maintain symmetry about the earth's axis. In Velasco's description of the behavior of the compass needle it sufficed to describe the phenomenon so that it coincided with the narratives written by sailors who had testified to the phenomenon. For Gesio, on the other hand, the natural phenomenon had to be reconciled not necessarily with what sailors witnessed but with natural philosophy. His arguments suggest that he came

51. Rodrigo Zamorano gives a characteristically succinct explanation, in Rodrigo Zamorano, *Compendio del arte de navegar, Colección primeras ediciones* (Sevilla: Alonso de Barrera, 1581; reprint, Madrid: Instituto Bibliográfico Hispánico, 1973), 48v–49.

52. Copernicus never mentions the tides in *De revolutionibus orbium coelestium* (1543), and neither does Rheticus in the *Narratio prima* (1540). The first theory that linked the earth's rotation and the tides is wholly attributed to Galileo. He revealed the theory in the *Discourse on the Tides* in 1616, more than forty years after Gesio's comments. (Galileo had hinted to Kepler about the connection in 1597, however.) For a study of Galileo's theory of the tides, see Ron Naylor, "Galileo's Tidal Theory," *Isis* 98 (2007): 1–22. Galileo Galilei, "Discourse on the Tides (1616)," in *The Galileo Affair: A Documentary History*, ed. Maurice Finocchiaro (Berkeley: University of California Press, 1989).

53. Jiménez de la Espada, *Relaciones geográficas*, 184:149.

from a cosmographical tradition that considered natural philosophical explanations integral to the practice of his craft; Velasco's circumscribed empiricism must have seemed perplexing to him.

Gesio's harsh critique would have not come as a surprise to Velasco, who had reviewed Gesio's documents beforehand and knew the Italian was intent on proposing a new interpretation of the line of demarcation. However, it would be naive to suppose that Gesio's motivation for writing these critiques was solely based on the weight of the geodesic coordinates he had determined. He unsuccessfully sought a royal appointment for more than twenty years and took any opportunity to demonstrate his capabilities vis-à-vis the official cosmographer. His critique, however, did not win him a royal appointment.

In the revised *Geografía*, Velasco ignored Gesio's critique, save for correcting what can be categorized as the copyist's errors Gesio pointed out, as well as the misstated date of the Treaty of Tordesillas and one longitude value (from an obviously incorrect 34° to 54°). For its part, the Council of Indies did not order its cosmographer-chronicler to incorporate Gesio's observations into a "clean" version of the *Geografía* manuscript destined for the king.[54] The historical record is mute about how Velasco responded to Gesio's objections or even whether he was asked to do so. Did he counter Gesio's observations with cosmographical material Gesio did not have access to? Did he point out that Gesio was relying on mostly public information—Vespucci, Cabot, Magellan—that was at least fifty years old? Despite Gesio's critiques and the Council's censorship, two years later the Council qualified the *Geografía y descripción* as a "very good and convenient" work (*muy buena y conveniente*) and granted Velasco a 400-ducat reward.[55]

The *Sumario*

Velasco's second cosmographical composition was a shorter version of the *Geografía y descripción universal* known as the *Sumario* or sometimes referred to as the *Demarcación y divición de las Indias*. It was not an attempt to write a Renaissance cosmography but rather a summary of the *Geografía* prepared to instruct a new cadre of members at the Council of Indies, five of whom had joined the Council since Velasco's earlier work had

54. Ibid., 184:162–63.

55. AGI, IG-738, R. 17, f. 249, "Informe favorable del Consejo de Indias para que se concedan 400 ducados a López de Velasco por el libro de 'La geografía y descripción de las Indias,'" December 7, 1576.

been copied and distributed. It was intended as a general introduction to the geography and the administrative divisions of the New World—essentially a geographical primer for new councilmen. It survives in two slightly differing and undated manuscript versions, one at the Biblioteca Nacional in Madrid and another at the John Carter Brown Library in Providence, Rhode Island.[56] The latter version has fourteen manuscript maps, which may be the only cartographic production attributable to López de Velasco. Although the *Sumario* survives as an undated manuscript, it is presumed to date from 1580. Therefore, it was written after the lunar eclipse observations of 1577 and 1578 but *before* the bulk of the replies from the questionnaire of 1577 had arrived in Spain, two projects initiated by Velasco discussed in the following chapter.[57]

In the *Sumario*, Velasco dispensed with discussions of the native populations, natural history, and particular geographical features. He did modify the text of the original *Geografía* in a few instances, primarily in response to Gesio's critiques made years earlier. This time he used Castilian leagues, carefully specifying that they were of 3,000 paces of five feet each—adding facetiously and clearly for Gesio's benefit—that these correspond to "sixty Italian miles" each. Velasco also used a new value for the ratio of leagues per degrees used to calculate longitude. Without any explanation, he chose 20 l/° rather than the 17.5 l/° that was customary and which he used in the *Geografía*. Velasco, however, did not use the new value consistently. For example, the location of the line of demarcation over Brazil still corresponded to the longitude value calculated using 17.5 l/°—a significant oversight. The only major addition to the text was a brief section describing sailing routes to Honduras and Guatemala.[58] The sections Velasco excerpted from the *Geografía* were also simplified. In many instances he deleted longitude coordinates and expressly avoided discussing problematic issues of territorial boundaries between jurisdictions. For example, instead of the very mediated discussion of the territorial limits of the La Plata River region from the *Geografía*, he simply stated that the region reaches a point on the South American continent that touches the line of demarcation to

56. BN, MS 2825, Juan López de Velasco, "Demarcación." Published under the title *Demarcación y división de las Indias* in DIA, 15:409–572. The version with maps is at the John Carter Brown Library, Cod. Spa. 7, "Demarcación y división de las Yndias."

57. Only the twelve *relaciones* from Venezuela date from 1578–79. Howard F. Cline, "The *relaciones geográficas* of the Spanish Indies, 1577–1586," *Hispanic American Historical Review* 44, no. 3 (1964): 351.

58. Velasco may have used Escalante's manuscript for this section. Escalante de Mendoza, *Itinerario de navegación*, 146–49.

the north and the mouth of the La Plata River to the south. The city of Asunción now appears without a longitude coordinate (the document has a blank space). He did, however, change the longitudes of the cities of Santo Domingo, Panama, and Santiago (Chile).[59]

Perhaps bowing to Gesio's remarks but more likely as a result of the currency taken on by the situation surrounding the Strait of Magellan, Velasco's discussion of the strait concedes that still very little was known about the hydrography of the area. Rather than quoting from Ladrillero's account as he had done in the *Geografía*, Velasco acknowledged that voyages of exploration in the area had failed to conclusively establish whether there was only one canal traversing the strait from the Pacific to the Atlantic. The Strait of Magellan, although long recognized by Spain for its strategic significance as a gateway to the Far East, was now in urgent need of defense from foreign incursions. Just months before the *Sumario* was completed, the Council of Indies learned of Francis Drake's raids on Spanish settlements on the Pacific coast.

Drake had set off from Plymouth with four ships in November 1577, and after pillaging the coasts of Brazil and the La Plata region, he crossed the Strait of Magellan, arriving on the coast of Chile in November 1578. From there Drake continued to attack and sack Spanish ports, the Callao in February 1579 and Guatalco in Mexico in April. Afraid that the Spanish were lying in wait for him in the Strait of Magellan—Sarmiento de Gamboa had indeed been sent to the strait for this purpose—Drake continued sailing along the North American coast in hopes of finding the fabled Northwest Passage. Failing to find it, he set out across the Pacific to the Moluccas and, after a harrowing passage via the Cape of Good Hope, arrived to fame and fortune back in Plymouth three years almost to the day after he had left. Despite Spanish protest, Elizabeth I celebrated Drake's return and knighted him.[60] The news shocked and alarmed the Council of Indies, who had always considered the Pacific coast of the American continent immune to such attacks. Calls went out immediately to fortify the strait, but it became evident that very little was known with any certainty about the geography of the area. Pedro Sarmiento de Gamboa's arrival in Spain in late 1580 remedied the situation. Sarmiento brought with him new maps and rutters

59. Velasco changed the longitudes as follows: Santo Domingo from between 75° and 78° to 70°, Panama from 82° to 89°, and Santiago (Chile) from 77° to 73°. It is unclear whether Velasco made these adjustments as the result of 1577–78 eclipse observations, since no record exists that these areas submitted observations.

60. Henry R. Wagner, *Sir Francis Drake's Voyage around the World* (San Francisco: John Howell, 1926).

of the strait he had prepared while waiting for Drake in the difficult waters at the ends of the earth.[61]

Velasco was well aware that the material contained in the *Sumario* was confidential and sensitive, even more so now that the enemy was actively seeking to engage Spain on its western border. Thus he was alarmed when copies of the *Sumario* began to circulate before the appropriate warning had been written on the manuscript cover. He wrote to Mateo Vázquez asking him to make sure that his copy had the following notice added to the cover: "It is in His Majesty's service that this not be copied or moved without authorization from the Royal Council of Indies," the early modern version of a red "top secret" label.[62] When the king became aware of the book's existence, he went much further in securing the document's confidentiality. He instructed the Council of Indies to collect all the books and to keep them in a locked cabinet in the Council chambers, to be used only as needed by members of the Council.[63] He was concerned that the books might end

61. Fernández de Navarrete, *Biblioteca marítima española*, 2:618. Velasco did not receive, was not ware of, or ignored the eclipse observation Sarmiento de Gamboa made in 1578, which according to Sarmiento's testimony had been "notarized and sent" to Spain. In the *Sumario*, Velasco kept the longitude of Lima at 82°, the same value he used in the *Geografía*.

62. "Alvarado me envió a decir que ha enviado a v.m. el libro de Indias en que quisiera en la primera hoja bajo del título como se pondrá en los que se hacen para los del consejo se pusiera estas palabras 'Conviene al servicio de su majestad no se pierda ni traslade sin licensia del Consejo Real de las Indias.' A v.m. suplico que si ya no le tiene Su Maj. y aunque le tenga a si se puede hacer las mande a escribir de buena letra porque aunque es compendio del libro de otro grande que yo hice, de aquel mismo argumento describiéndose en el tan particularmente todos los pueblos y cosas de Indias que no conviene que se publique en la forma que está y si viniese en manos de extranjeros al punto le imprimieran porque creo que les seria buen acepto y v.m. me mande avisar de lo que le ha parecido de él." IVDJ, Envío 25, n. 561, Juan López de Velasco, "Carta de López de Velasco a Matheo Vázquez sobre libro de Descripción de Yndias," November 21, 1580, Madrid.

63. The king replied: "Y habiendo antes de ahora pensado en estos libros de la descripción de todas las Indias, me ha parecido que por ser de la calidad que son y por el inconveniente que se podría seguir si anduviesen en muchas manos como podría ser faltando alguno de los que los tienen o mudándose de ése consejo pues para solos los del son a propósito sería bien que todos se recogiesen en el Consejo y se pusiesen en algún cajón cerrado a donde cuando se ofreciese necesidad los pudiesen tomar para ser con el espacio que conviniese lo que quisiesen, volviéndolos después a su lugar que sería tenerlos como en sus casas y se remediaría que faltando alguno no se perdiesen o trasladasen y así pareciéndonos que esto está bien como a mi me lo parece lo ordenaréis recogiendo todos los dichos libros en la parte que digo y avisándome como se hiciere." AGI, IG-740,

up "in many hands" and considered that the information in this book was pertinent only to officers while they served at the Council of Indies.

The fourteen maps that accompany the *Sumario* are a subset of the twenty-three that Velasco had prepared for the original *Geografía*.[64] Velasco selected three general maps: the universal map he titled the *Carta de Marear* (plate 5) and the two showing respectively the North American and South American continents (plate 6). The remaining maps were regional maps corresponding to the nine audiencias, plus Chile, and the islands of the Far East (*Islas de Poniente*; plates 7 and 8). Specific maps of the Hispañola, Cuba, Terranova, Yucatán, Cartagena, Strait of Magellan, Brazil, China, and the Solomon Islands were not included in the *Sumario*. The maps, again, were meant to complement Velasco's text by providing a graphical representation of the textual narrative. Gesio's criticism six years earlier had failed to persuade Velasco to abandon the use of schematic maps in favor of mathematical cartography.

Once again, the Council of Indies asked Juan Bautista Gesio to evaluate Velasco's work.[65] Gesio had not been idle during the intervening six years. He had been in frequent communication with the monarch, through Mateo

---

N. 91, "Memorial del Consejo de Indias desaconsejando la concesión de nuevas gratificaciones a López de Velasco por su libro. Contestación al margen de Felipe II sobre que se recojan los libros y se pongan en lugar seguro," September 28, 1582.

64. The attribution of the maps that accompany the JCB manuscript is somewhat problematic. Part of the problem stems from inconsistencies between Gesio's 1580 critique of Velasco's *Sumario* and the maps currently with the JCB manuscript. Gesio describes Velasco's universal chart as having being drawn using a projection where the parallels increased in widths toward the poles "almost double in proportion to longitude." His assessment likely stemmed from an effort to scale the map using the tropic parallels that appear in the manuscripts. However, it is difficult to say with certainty the projection used in the universal chart now with the JCB manuscript, since it lacks meridians and parallels. The faint grid that appears in the manuscript was likely used to help copy the document. Some of the other maps at the JCB, however, contain features that Gesio criticized, particularly the maps titled "Tabla universal de las Indias de Norte," "Tabla de la Audiencia de la Española," and the "Tabla de las Islas de Poniente."

65. The critique of the *Demarcación* or *Sumario* dates from 1580, the year of Gesio's death. A sixteenth-century copy (it is not in Gesio's hand, nor does it have his distinctive signature) is in the AGI, Patronato-259, R.79, "Parecer del matemático Juan Bautista Gesio sobre un libro titulado *Sumario de las Indias* tocantes a las ciencias de geografía," April 11, 1580. It also exists as a eighteenth-century copy in volume 24, f. 8–31v. of the Colección Muñoz at the Academia de Historia de Madrid.

Vázquez, on a wide range of topics.[66] On several occasions he warned the king about foreign events threatening Spanish interests, among them news that the French were amassing an armada of twenty ships with 2,500 men on the coast of Brittany under the direction "of a Breton named La Roche [sic]" with the purpose of sailing through the Strait of Magellan and on to the Spice Islands.[67] Gesio was somewhat misinformed. Indeed, in the spring of 1578 a Breton, Troilus de Mesgouez, Marquis de La Roche, was outfitting a "fleet" under the auspices of the king of France, but its destination was Newfoundland rather than the Far East—and the marquis managed to outfit only one vessel, which was captured by the British as it left the coast of France.

As the Council of Indies debated in 1579 on how best to thwart English advances in the Pacific, Gesio issued an opinion—perhaps at the monarch's request—on how best to defend the Strait of Magellan.[68] He agreed with the consensus opinion that it was necessary to fortify and settle the strait, but he cautioned, based on the accounts at his disposal, that it was unclear whether there was only one channel connecting the Atlantic with the Pacific. He warned that certain accounts stated that the lands at the southern part of the strait were actually an island, in which case a fortress protecting the main channel would be ineffective. If this were the case, the Spanish would have to build a large fortress on the western entrance to the strait and a series of signal towers stretching from that entrance all the way to Mexico to relay news of enemy ships. But first, Gesio advised, the strait needed to be surveyed inch by inch (*palmo a palmo*). He included a detailed topographical description of the strait based on the accounts of Magellan, Andrés de Santa Martín, and Antonio Pigaffeta (who believed Tierra del Fuego was an island). Once properly defended, Gesio predicted, the strait would become the preferred route from Spain to the Pacific and the principal route for goods coming from Peru.

Gesio's political/cosmographical commentary sometimes caused him difficulties with the Council. In one instance the Council ordered him not to write to the king about having concluded, after inspecting the deed to

66. Gesio addressed mostly political topics, as is evident from his correspondence conserved at the British Library (BL Add 28359–60). In these letters he advises the king on matters ranging from astrological forecasts to the Kingdom of Naples, the African League, and of course, Portuguese pretensions in the Pacific.

67. AGI, P-24, R. 43, "Parecer de Juan Bautista Gesio sobre importancia de tener las islas Filipinas," April 14, 1578.

68. AGI, P-33, N. 2, R. 7, "Parecer de Juan Bautista Gesio sobre la navegación al Estrecho de Magallanes (expediente sobre Sarmiento de Gamboa)," August 27, 1579.

the 1529 lease of the Moluccas to Portugal (*escrituras del empeño*), that Spain had legal claims on Japan. For Gesio this implied that Japan and China were open to Spanish trade, evangelization, and even conquest. He wrote despite the Council's strictures.[69] He continued to insist that Spain could construct a strong cosmographical case against Portuguese claims in the Far East,[70] as well as in Brazil.[71] He accompanied his detailed explanations with geographical descriptions and maps, all which have since been lost or are currently not attributed to the Italian cosmographer. But by 1578 he realized that his difficult relations with the Council made his financial prospects dim. He wrote to (whom else!) Vázquez asking to be taken on as *criado del rey* and allowed to work independently. Among his complaints was the Council's reluctance to sanction (*desautorizar*) Velasco.[72] When the Council asked him to review Velasco's *Sumario*, Gesio took the opportunity to further his own claims by systematically undermining Velasco's, and therefore the Council's, official cosmographical view of the world.

In the intervening years, Gesio had gained access to some of the cosmographical material at Velasco's disposal, as we know from a series of requests Gesio made in early 1578 which were registered in the Council's Petition Book.[73] In these petitions Gesio asked the Council to make available to him the rutters for voyages from the Strait of Magellan and Mexico to the Far East, in particular from the expeditions of Villalobos, Juan Gaetan, Loaysa, and Juan de Ladrillero. He also asked that Velasco give him access to some eclipse observations that had recently come from Mexico. Considering that Gesio made his request in February 1578, these must have been observations of the lunar eclipse of September 1577, which

69. BL, Add 28359, f. 239–239v, "anonymous letter in Italian about Japan and China to Mateo Vázquez," undated. Gesio's handwriting is easily recognizable. Gesio discussed the topic with Vázquez in early 1576.

70. Gesio mentions that he prepared a detailed geographical assessment of the situation. Zab, Altamira, 243, GD. 20, f. 170–172, "Informe de Juan Bautista Gesio sobre presencia de portugueses en Oriente," ca. 1579. Published in Fernández de Navarrete, *Colección Fernández de Navarrete*, 18:103–4, 207–10.

71. AGI, P-29, R. 32, "Descripción geográfica del Brasil de Juan Bautista Gesio," November 24, 1579.

72. "[E] abasciaare e disfare Juan de Velasco cosmógrafo che han fatto maggiore de le Indies, et per la amicicia tengono con il detto et per ser io forastero mirano a sustentarlo in honore e credito, et non mirano al servicio di vm." BL, Add 28341, f. 148–49, "Carta de Juan Bautista Gesio pidiendo otro trabajo," August 9, 1578.

73. AGI, IG-1086, L.6, f. 54v–55, 61, "Petición de Juan Bautista Gesio sobre papeles que tiene Juan López de Velasco," February 18 and 25, 1578. Some entries published in Jiménez de la Espada, *Relaciones geográficas*, 183:281–82.

yielded longitudes for Ciudad de los Angeles (Puebla) and San Juan de Ulúa (fort of Veracruz). Apparently Velasco was not forthcoming with the material, for the petitioner had to ask the Council again on February 25 to set a time to meet at the cosmographer's house.

During the same month, Gesio learned that the author of a remarkable geographical description of New Spain had arrived at court to procure a printing license for the book. Gesio never mentions the author's name in the letter, but it is likely that he was referring to the now-lost geographical description written by Francisco Domínguez de Ocampo as part of Francisco Hernández's natural history.[74] The work, according to Gesio, was a "universal and particular geography and description of [New Spain] with maps graduated with scales of latitude and longitude."[75] The maps were drawn based on eclipse observations made by cosmographers in Mexico. Gesio warned that the publication license should be withheld until the geographical material in the book and its maps were reviewed by "a very intelligent person in this profession." This review should confirm that the astronomical observations were "made correctly and demonstrated"; he insisted that not only this work but every "map, history and *relaciones* that are made in the future and deal with matters of distances and longitude" should not be allowed to go to print before being carefully reviewed. Gesio was concerned that publication of a book containing controversial material "concerning the longitudes" written by a Spaniard and authorized by Castile could be used by the Portuguese to undermine Spanish claims—precisely the strategy Gesio had himself used to undermine the Portuguese claims.

74. Hernández arrived in Spain in 1577 and actively sought to publish his work. We have a 1581 statement by Domínguez explaining that Hernández had taken a "draft" of his description and maps when the naturalist returned to Spain. Domínguez said that he had now finished the work but the viceroy wanted to keep it. "Carta del geógrafo Francisco Domínguez sobre S. M. mande al Virrey D. Martín Enríquez remita la descripción de Nueva España que trabajó, Mexico, December 30, 1581," published in DIE, 1:379–85. We also know that Domínguez's geographical work—whether the version Hernández brought to Spain or the later version is unclear—was known in Spain. In his 1629 bibliography, León Pinelo listed the following entry: "Francisco Domínquez. *Description of New Spain.* He [went to New Spain] by order of the Council of Indies the year of 1570 and he made it and sent it, very ample, with that of China and other provinces. Manuscript." Antonio de León Pinelo, *El epítome de Pinelo, primera bibliografía del Nuevo Mundo*, ed. Agustín Millares Carlo, facsimile of 1629 ed. (Wáshington, D.C.: Unión Panamericana, 1958), 176.

75. AGI, P-259, R.72, "Carta de Juan Bautista Gesio sobre que no se imprima tablas y obra de geografía de Nueva España sin ser vista antes," February 18, 1578.

Gesio's concerns were neither novel nor unique; others found the release of geographical information troubling. Months before, the viceroy of Mexico, Martín Enríquez, had recognized the sensitive nature of Domínguez's work, but for different reasons. He wrote to the king expressing concern that the descriptions of the Caribbean coasts and the Sea of Campeche had such "precision and clarity" that it could cause "inconveniences" if published since the coast was frequented by pirates."[76]

Given the new geographical material that had come to the Council of Indies since the 1574 *Geografía*, it is not unreasonable to suppose that Gesio expected López de Velasco's new works to reflect this material. This was not the case. Thus, when in 1580 the Council of Indies asked Gesio to review the *Sumario*, the Italian cosmographer launched into a long series of "castigations" of the cosmographer major's work.[77] He explained that he had been asked to examine a book titled *Sumario de las Indias tocantes a las sciencias de geografía* to see whether it complied with the principles and conventions (*principios y términos*) of geography, as well as whether there was anything prejudicial or beneficial in the work.[78] Gesio never mentioned the author by name, but he made his opinion of the book perfectly clear: "Having seen and studied with much diligence and carefully considered this book according to the principles and to mathematical terms, I have found that it is not written according to the precepts of geography and contains almost nothing of this science. Thus, it is not a geographical book, only an abbreviation of histories and commentaries."[79]

76. AGI, Mexico-69, R.5, N.90, Martín Enríquez, "Carta del Virrey de México," December 6, 1577.

77. Gesio may have been trying to emulate Ermolao Barbaro's (1454–93) critique of Pliny and Mela, which the Italian humanist titled *Castigationes* (1492). Although Barbaro's approach was almost exclusively philological, the work was structured as a rationalized and systematic questioning and resolution of each item of the classical authors' work.

78. Gesio was paid 50 ducats for his work. AGI, P-426, L. 26, f. 211, "Carta de pago a Juan Bautista Gesio por 50 ducados por lo que ha servido y sirve al consejo en cosas de cosmografía," June 27, 1580. Gesio had been receiving meager and sporadic payments from the Council. For example, in 1577 he was paid 150 ducats "for having done cosmographical work assigned by the council." AGI, P-426, L. 26, f. 32v, 63v–64.

79. "Yo habiendo visto y revisto con mucha diligencia, y muy bien considerado este libro segun la ciencia con los principios, y términos matemáticos, he hallado que no está compuesto según los preceptos de la geografía y contiene casi nada de esta ciencia, y por esto no ser libro Geográfico, solo ser una abreviación de historia y comentarios." BAH, Colección Muñoz, v. 24, f. 8.

Gesio was clearly approaching the work from a perspective that considered mathematical cartography *the* essential element of geography. Failing this test, Velasco's work was, in his view, simply a collection of "histories and commentaries." His long list of misgivings about the *Sumario* can be reduced to three fundamental problems: the maps were not drawn according to any established projection or the "precepts of the science," distance values were calculated incorrectly, and the text and maps in many instances contradicted each other. Gesio recommended that the maps be drawn according to one of two projections: one similar to an oval projection[80] or another drawn based on great circles but with the land maintaining the appropriate graduation in terms of latitude and longitude. Instead, he complained, Velasco's maps were drawn without regard to graduation and without scale, rather by "chance and luck" (*a caso y suerte*).

Gesio equated Velasco's maps to a painting of a naval battle by a practical painter (*pintor solo práctico*)—who simply captures the "memory" of the event—as opposed to a painting done by a scientist (*scientífico práctico*) who paints according to the "figure, order, and circumstance that existed." Perhaps wishing to end his scathing critique on a positive note, Gesio suggested that should the king ever order his kingdom portrayed in a "precise and polished Geography," the *Sumario* could be used to establish the true limits of his kingdom. Realizing that his comments would directly affect Velasco's fortunes, Gesio remarked that despite these faults, the Council should reward the author because "he has worked very hard composing the tables, compiling longitudes and distances from *relaciones*, all which he has done in good will, having done what he could and knew how to do, for if he knew more, he could do it better."[81] In addition, concluded Gesio, if the Council indeed rewarded the author, it would thus encourage others to apply themselves in the science. By 1580 Gesio was aware that many would take his criticism as "only his opinion, without reason and motivated by envy"; to assuage these criticisms he demonstrated the book's faults in a series of mathematical proofs.

In addition to the charts not being drawn to scale, Gesio repeated his complaint about the longitudinal values cited in the *Geografía*. Distances were not corrected "according to the customs of geographers" by subtracting a third or fourth of the distance stated in a pilot's logbook to convert

80. "[C]on los paralelos y meridianos por proporción al otro y entre si y restringiendo según la latitud procediendo de austral al septentrión." BAH, Colección Muñoz, v. 24, f. 8.

81. BAH, Colección Muñoz, v. 24, f. 10.

from "estimated and tortuous paths into straight paths." To illustrate the many instances where the text and maps did not correspond, Gesio explained that he drew a scale for each map from coordinates mentioned in the text and tried to correlate distances cited on the text with those shown on the map, all to no avail. In many cases, he said, the longitude values in the book were incorrect simply because "they are estimates, and taken from navigation and apparent angles." The longitude of Mexico, since it was derived from a lunar eclipse observation, "might be given some credit" and used to correct the locations of adjacent provinces.[82]

Gesio also took issue with the method Velasco used to calculate "straight distances" (i.e., distances along a great circle or orthodrome). These, Gesio complained, did not correspond to any of the known and accepted ways of calculating distance along a great circle on a sphere. On this point, however, Gesio's criticism was unfair. Mathematical analysis of the distances computed by Velasco show that he used a method—details of which are unknown—with a precision similar to that afforded by the trigonometric equation used for calculating the distance along the orthodrome. See table 5.1.[83]

Later in the text, Gesio explained three different methods for calculating the distance between two points on a sphere, a discussion he carried out in Latin because "the romance language is poorly suited for this purpose."[84] The first method, the one "commonly used by geographers and introduced by Ptolemy," consisted in reducing the spherical triangle in question into a rectilinear parallelogram symmetrical about the mean latitude, where the diagonal of the parallel represented the distance between the two points. The second method, which Gesio claimed to have invented, reduced the problem to triangles where the sides correspond to the chord defined by the difference in parallels and meridians. The last method solved the problem using spherical triangles.

Before commenting on specific features of Velasco's maps, Gesio inserted two sections questioning the theoretical underpinnings of Velasco's cartography, which he considered to be a fundamental problem of Velasco's work. The first addressed the issue of the appropriate value for the length of the meridian at the equator. In the *Sumario*, Velasco had deviated

82. Ibid., f. 9.

83. All latitudes and longitudes are as they appear in Velasco's *Sumario*. The great circle equation used for the comparison is: cos (dist.) = sin (lat.1) * sin (lat.2) + cos (lat.1) * cos (lat.2) * cos (long. 1 – long. 2).

84. "[M]uy mal se pueden exprimir los términos y vocablos matemáticos en romance." BAH, Colección Muñoz, v. 24, f. 18v.

Table 5.1. Comparison between orthodromic distances in leagues as computed by Velasco in the *Sumario* and as computed using the great circle equation

Toledo 39°N, 0°W	Santo Domingo 19 ½°N, 70°W	Santa Marta 10°N, 74°W	Mexico City 19 ½°N, 103°W	Lima 12 ½°S, 82°W
Sumario	1240	1420	1740	1820
Calculated using orthodrominc (great circle) equation	1251	1426	1748	1835

from the traditional 17.5 l/° used by the Castilians and Portuguese and instead used 20 l/°. Gesio continued to argue, as he had in his critique of the *Geografía*, that Spaniards did not translate leagues to the proper number of paces and thus all distances derived from pilot logbooks were fundamentally incorrect.[85]

The other fundamental problem Gesio found with Velasco's work was his placement of the line of demarcation and its relationship to the Brazilian coast. First, Gesio counseled that in interpreting the Bull of Demarcation and the subsequent agreement between the kings of Spain and Portugal, Portuguese leagues (4,000 paces, rather than 3,000 paces for each Castilian league) should be used, since the king of Portugal was the one who had requested that the original bull be amended. Second, for Gesio it was

85. Gesio thought it necessary to explain the historical origins of this value and the problems that resulted when Castilians used it without understanding how it originated. Gesio begins his explanation by comparing the values used by Erastonese and Theodosious ($29^{1}/_{3}$ l/° or 87,500 paces) with those of Ptolemy and Marinus of Tirus ($20^{5}/_{6}$ l/° or 62,500 paces). Modern geographers, Gesio argued, rounded Ptolemy's value to 60,000 paces and therefore to 20 l/°, largely to make it easier to carry out sexadecimal arithmetic. This, he maintained, was the common practice of "cosmographers in Europe, Italy, Germany and other parts." The Portuguese (like the Castilians, since they tended to copy Columbus, who had in turn learned from the Portuguese) remained committed to the value of 17.5 l/°, not because they had carried out "observations or experiments" to justify the value but as a compromise between the values of Eratosthenes and those of Ptolemy. Gesio concluded that the longitudes shown in Castilian navigation charts based on pilot logbooks were thus all incorrect. He acknowledged that in the past he had advocated different values for the degree of meridian, but he pointed out that he had judged the present work using the value the author chose, 20 l/°.

evident that the science of geography dictated that distances along a parallel should be calculated based on the value assigned to one degree of longitude for that particular parallel. Therefore the degree per meridian had to correspond to Cape Verde at latitude of 14° south, or 19.4 l/°. The third issue was whether the meridian of origin should be measured from Cape Verde in the coast of Africa (the point of origin preferred by Castile) or the Cape Verde *islands* (the point of origin preferred by Portugal). Gesio wisely left this matter for those who could examine the exact wording used in the bull, and he limited himself to calculating distances from both points.

His biggest adjustment comes when he adjudicates the fourth and final question of how many degrees correspond to the two different interpretations of the point of origin of the demarcation, which the Portuguese interpreted as 527 leagues from Cape Verde (470 from San Antón in the Cape Verde Islands) and the Castilians as 470 leagues from Cape Verde on the African coast. Here he castigated Velasco on two points. He argued that Velasco—inexplicably for a Castilian—placed the line of demarcation as per the Portuguese distance of 527 leagues calculated from the Cape Verde Islands. In addition, despite stating that all geographic coordinates in the *Sumario* were based on the Castilian league, Velasco had failed to recompute the location of the line using this value, again against Spain's interests.

Gesio's corrections placed the line of demarcation at either 47.5° or 43° from Toledo, rather than the 39° Velasco used; as Gesio admitted, by itself this would place more Brazilian territory under Portuguese jurisdiction. But he added that according to his calculations Velasco had overextended the eastward reach of the South American continent by almost 13°.[86] To support this assertion, Gesio quoted from his own *Description of Brazil*[87] but made some self-serving modifications. This time he referred to only Juan de Barros, Sebastian Cabot, and Vespucci, all whom he claimed coincided in the opinion that the easternmost point of Brazil was 30° west of Africa (or 26.25° west using 17.5 l/°). Gesio took some significant liberties to make the three accounts agree. He amended Barro's account to say that the distance between the two continents was 450 leagues (rather than 400 leagues, as he had stated in his own *Description of Brazil*). Vespucci's calculations did not escape Gesio's manipulations either. In the 1579 *Description of Brazil*, Gesio

86. BAH, Colección Muñoz, v. 24, f. 16.

87. AGI, P-29, R. 32, "Descripción geográfica del Brasil de Juan Bautista Gesio," November 24, 1579.

had said that the bay at latitude 5° S where Vespucci made landfall was 150 leagues west of the easternmost point on Brazil; now he said that the bay was only 2° (40 leagues) away.[88] With his new computations, only between 1° and 5 1/2° of Brazilian territory fell on the Portuguese side.

Over the years Gesio had gradually shifted from maintaining that Brazil lay beyond the line of demarcation to being willing to grant the Portuguese "no more than 5° or 100 leagues" of Brazilian territories. It is likely that the impending annexation of Portugal that year encouraged Gesio to moderate his position. On this occasion he did not discuss the matter of who had legal possession of the territories in question, and he was willing to at least consider the reasons given by the Portuguese for shifting the line of demarcation westward.

For as much as Gesio was willing to "adjust" discovery accounts to suit his agenda, in his critique of the *Sumario* he willingly bows to any geographical coordinates derived from astronomical observations. Whereas he had criticized Velasco's use of Durán's 1544 eclipse observations in the *Geografía*, by 1580 Gesio had done his homework on the circumstances surrounding this observation. He had carefully examined the corresponding 1544 observations made in Spain by Alonso de Santa Cruz and Jerónimo de Chaves, which referred to Durán's observation. Santa Cruz had noted that Durán's observation took place in Guadalajara, not Mexico City. They understood Guadalajara to be 4.5° farther west than Mexico, thus suggesting a distance from Toledo to Mexico City of 97.5°. (Mexico City is 95° 11′ from Toledo.) Velasco's mistake, Gesio explained, propagated the error through the whole document, since Velasco had used this as the longitude from which to calculate all the distances in the American continent. Panama, for example, was shown at 89°, and the rest of the longitudes for South America were incorrect by default. Clearly exasperated, Gesio refused to comment any further on the maps of the southern continent.[89]

Gesio had also tracked down some Augustinian friars who had been with Martín de Rada when he observed a lunar eclipse in the Philippines.

88. To confuse matters more, Gesio took Vespucci's longitude computations from the bastardized versions of the pilot major's letters published as the *Four Voyages*. Pohl discusses the forgery in Frederick J. Pohl, *Amerigo Vespucci, Pilot Major* (New York: Columbia University Press, 1945), 147–67.

89. Gesio overlooked one glaring mistake in both the *Geografía* or the *Sumario*: the longitude of 79.33° for Coros (Caracas), which was clearly a copyist error for 69.33°. (Velasco clearly states that Santa Marta is *west* of Coros at a longitude of 74° and shows this on his map.)

They claimed the observation yielded a distance of 216° between Toledo and Zebu. Based on this information, Gesio reluctantly concluded that the Moluccas were only 13° inside the Spanish demarcation. He recognized that this contradicted his previous opinion, but he excused his mistake by explaining that in that instance he had stated the distance "according to probability and accounts from pilots and mariners and now I affirm this based on observations."[90]

By comparing the two surviving manuscripts of the *Sumario*, we can see that Velasco was aware of Gesio's critique and as a result corrected several values. One error Gesio pointed out, which was clearly a copyist's error, showed the total north-south expanse of discovered territory as 1240 leagues. It was corrected to read 2240. (The JCB mss has the word "dos" inserted in a superscript.) Velasco also changed the longitude of Panama from 89° back to the 82° that he had used in the *Geografía*.[91] However, neither the distance to Mexico nor the distance to the Moluccas was ever changed to reflect Gesio's critique.

When he completed the *Sumario*, Velasco approached the Council of Indies once again seeking a reward to supplement his salary. This time the Council recognized that there was very little new material in Velasco's latest work and in a note to the king recommended that Velasco not be given a special reward for it. They judged that "the new books . . . did not entail more that copying the first [book] and making smaller maps, and even these are not to scale."[92] The Council's conclusions regarding the *Sumario* were as biting as they were accurate. If Velasco, as of 1580, still harbored illusions about the prospects for rewards and advancement based on writing a cosmography of the New World, the Council's attitude toward his latest work sealed the fate of any future cosmographical production. Although he served as

90. BAH, Colección Muñoz, v. 24, f. 30.

91. Both manuscripts now list the distance from Toledo to Panama as 82°. However, one inconsistency in the JCB manuscript suggests that Velasco had originally indeed used 89° for the longitude of Panama. The cosmographer failed to correct the orthodromic distance that corresponds to 82°, from the original 1670 leagues to 1570 leagues.

92. "[L]os nuevos libros que allí dice no tuvo mas trabajo de trasladar el primero y poner en punto menor las tablas de ellos y aún éstas no están graduadas." AGI, IG-740, N. 91, with a copy at AGS, GA-137, f. 256, "Memorial del Consejo de Indias desaconsejando la concesión de nuevas gratificaciones a López de Velasco por su libro," September, 28, 1582.

cosmographer of the Council of Indies until 1588, he never wrote another work.

The *Geografía* and, to some extent, the *Sumario* are examples of the cosmographical tradition discussed in chapter 1, but adapted to meet the political requirements of the Council of Indies as formulated by the *Ordinances* and *Instructions*. In them we see the intellectual tradition of Renaissance cosmography reformulated into an administrative tool of empire. Its traditional mode of representation with its goal of encompassing all—the land, sea, peoples—and presenting it as a harmonious whole was transformed by the need to put in "good order" and quantify the New World for the sake of bureaucratic expediency. Walter Mignolo has noted the *Geografía*'s departure from classical models and sees it as a "genre modeled by necessity: to collect and organize information of newly conquered lands."[93] Karl Butzer points out the contrast between Velasco's work and that of contemporary cosmographers, such as Sebastian Münster, showing how "the modernity of [Velasco's] secular and empirical synthesis . . . contrasts with the continuing use of an obsolete Ptolemaean framework and theological paradigm to the end of the century for presenting new geographic information in Central Europe."[94]

These assessments, which focus on the "modern" attributes found in Velasco's work, understate how much the *Geografía y descripción* owes to the humanistic cosmographical tradition. As cosmographer-chronicler Velasco was under no obligation to compose such a comprehensive book but rather had to prepare yearly compendia using the material sent from the Indies in response to the *Instructions* of 1573, and these were to eventually form the *Books of Descriptions*. But as a humanist he took the opportunity presented by his first royal appointment to compose a comprehensive work worthy of joining the Renaissance cosmographies that preceded him and that he surely regarded as models. His personal inclination and career, however, made him more comfortable among words than among numbers. In the *Geografía* Velasco displayed his talent for doing archival research, extracting vast amounts of information from myriad legal records, and organizing the material cohesively—this time not in the form of a legal brief but with the textual style and structure of a cosmographical opus. His *tablas* or maps

93. "[M]odelos forjados por las necesidades del caso: recoger y ordenar la información de nuevas tierras conquistadas." Mignolo, "Cartas, crónicas y relaciones," 75.

94. Karl W. Butzer, "From Columbus to Acosta: Science, Geography, and the New World," *Annals of the Association of American Geographers* 82, no. 3 (1992): 554.

were intended as visual aids to orient the readers to the spatial relationships between governing units in the New World, not as documents to be studied with compass and ruler in hand. Thus for a practitioner of the science such as Gesio, who conceived of cosmography as a discipline whose foundation lay in mathematizing space, the *Geografía* was a profoundly flawed cosmography.

Velasco's actions following the *Geografía* perhaps give us some indication of how he decided to address the objections made against his work. He initiated a series of projects to ensure that the cosmographer-chronicler major of the Council of Indies had the most timely and accurate information at his disposal. Rather that continuing to insist that dependencies send him archival information and copies of local descriptions, he set out to collect fresh material, not personally but by using the bureaucratic machinery in place. His efforts yielded two of the most ambitious scientific programs of its time: the questionnaires of the *relaciones geográficas de Indias* and the project to determine the location of Spanish territories by systematically observing lunar eclipses.

SIX

Constructing a Cosmographical Epistemology

In the *Geografía y descripción universal de las Indias*, Juan López de Velasco sought to neatly describe the New World in one comprehensive cosmography and thereby reconcile the intellectual and literary heritage of Renaissance cosmography with the bureaucratic exigencies of his post as cosmographer-chronicler major of the Council of Indies. The task proved frustrating. Not only did Velasco feel that the king and the Council of Indies inadequately rewarded his extraordinary effort, but he must have felt stung by the Council's criticism. The Council's statement regarding Velasco's reliance on material previously collected by Alonso de Santa Cruz carried the implication that the *Geografía* was dated. Juan Bautista Gesio was also critical of Velasco's cosmographical methodology, cartography, and use of available data. Though he was clearly advancing his own agenda, his comments amounted to a veiled critique of Velasco's competency. Given this criticism, and the fact that as of early 1577 the *Books of Descriptions* project proposed by Ovando's *Instructions* had elicited—at best—a lukewarm response, Velasco decided to radically alter the methods employed to collect cosmographical information.

Juan de Ovando died in 1575, and the presidency of the Council remained unoccupied until 1579. Without Ovando's sense of mission and direction, the Council, perhaps at Velasco's insistence, quietly stopped urging colonial administrators to comply with the *Instructions* and send their books of descriptions to the Council. In place of Ovando's project, Velasco set in motion what was perhaps the most ambitious and successful program for collecting geographical, ethnographical, natural, and historical information about the New World during the early modern era. Rather than organizing data collection around the formation of the cumbersome books of descriptions, Velasco resorted to frequent, short, and direct requests for cosmographical information. These requests were structured as questionnaires and as specific instructions for carrying out astronomical observations for determining geographical coordinates. This emphasis on delivering atomized facts—as replies to specific questions or as discrete lunar eclipse observations—signaled a rupture with the epistemological and methodological practices associated with Renaissance cosmography and ushered in a new concept of what constituted a cosmographical fact. The reason for this shift lay in a substantive reassessment of the practices associated with cosmography in late sixteenth-century Spain.

Questionnaires and the *relaciones geográficas de Indias*

Velasco's use of questionnaires to collect cosmographical information was a significant departure from the methods traditionally used to compile cosmographies. For decades, Spanish cosmographers had relied on firsthand, or seemingly firsthand, discovery accounts and *relaciones*, especially those written by expert observers, as their sources. But such sources had their shortcomings. They lacked comprehensiveness, commonality, and a shared standard for exposition. To a humanist cosmographer, these issues had posed some epistemic difficulties that the author could gloss over in the composition, much as Santa Cruz had done in the *Islario general*. But as cosmography became the vehicle for providing useful knowledge to the Council of Indies and the law specified an epistemic criterion with only verifiable eyewitness accounts as suitable sources, Velasco had a serious problem to solve.

The use of questionnaires in Spain was not new; they had been used frequently in the past to gather information about the New World. Santa Cruz's guidelines mimicked the format. Ovando had often structured requests for information during his audit of the Council of Indies as lengthy itemized lists and adopted a similar format when conducting depositions or interviewing persons arriving from the New World. Records of testimonies given in the Council of Indies in early 1571 clearly suggest that the respondents were replying to the same list of two hundred questions.[1] This questionnaire—resembling a list of the questions asked by a notary during interrogations—was solely for the Council's internal use, since it was never promulgated to the Indies.

These precedents for collecting information via questionnaires, however, pale in comparison to the scope of the project and the results

1. The first to study these from a historical perspective was Jiménez de la Espada. Jiménez de la Espada, *Relaciones geográficas*, 183:48–50.

obtained when Velasco adopted this method to collect cosmographical information about the Indies. Early in his career, as Velasco was struggling to extricate cosmographical information from the New World, he noticed the responses to a questionnaire issued to Castilian towns in 1574 for the purpose of collecting geographical and historical information.[2] The originator of the project, known as the *relaciones topográficas*, remains unclear. Some historians consider it the work of royal chronicler Ambrosio de Morales working from the papers of the royal chronicler of Charles V, Juan Páez de Castro, among which was a lengthy questionnaire designed to collect information about Spanish towns for the purpose of composing a geography of Spain.[3] Others see the project as stemming from Juan de Ovando, Antonio Gracián, and Juan López de Velasco.[4] In any case, a hitherto unpublished letter reveals who carried out the project after Gracián's and Ovando's deaths in 1575. As the text of the letter suggests, Velasco was solely responsible for preparing the royal orders that accompanied a revised version of this questionnaire sent to the towns of Castile in 1578.[5] He later took complete stewardship of the project, urging the king in 1583 to send similar printed questionnaires to Aragon and Portugal, so that "[h]aving all the kingdoms of Spain for the first time united under Your Majesty, no more honorable a work in writing nor a more convenient work to guide the government could be made during your time than an illustrated description showing the location of all the towns and giving in writing an account of their memorable things."[6]

2. A brief bibliography of the Spanish *relaciones topográficas* must include Miguélez, *Catálogo de códices españoles*; Juan Ortega Rubio, *Relaciones topográficas de los pueblos de España* (Madrid: Sociedad Española de Artes Gráficas, 1918); Kagan, "Philip II and the Geographers"; and Alfredo Alvar Ezquerra, María Elena García Guerra, and María de los Angeles Vicioso Rodríguez, eds., *Relaciones topográficas de Felipe II: Madrid*, 4 vols. (Madrid: CSIC, 1993).

3. Alvar Ezquerra, García Guerra, and Vicioso Rodríguez, *Relaciones topográficas*, 1:31–32.

4. López de Velasco, *Geografía y descripción*, xxvi–xxviii.

5. IVDJ, Envío 100, f. 297, Juan López de Velasco, "Carta a Mateo Vázquez sobre varias cédulas incluyendo una de Indias y la descripción de España," July 16, 1577, Madrid.

6. "Habiéndose juntado en su majestad el primero todos los Reinos de España no se podría hacer en su tiempo obra más honrada en letras o para todos ellos ni más conveniente para guiar el Gobierno que una buena descripción que por pintura muestre los lugares de los pueblos y por escrito de relación de lo que hay notable en ellos," October 26, 1583, Madrid. BAH, Leg. 4409. Published in Alvar Ezquerra, García Guerra, and Vicioso Rodríguez, *Relaciones topográficas*, 1:38.

Using the Castilian questionnaires as a model, Velasco formulated a similar questionnaire to collect information about the New World. His legal background and approach to historical evidence inclined him toward questionnaires. As discussed in chapter 3, witness testimony in the form of legal depositions was an accepted legal method to establish matters of truth and facts. On repeated occasions, Velasco insisted that only firsthand accounts could be considered reliable historical sources. He applied this same reasoning to cosmographical knowledge. The questionnaires promised such a specific response—a signed eyewitness account—with the added benefit of timeliness. In 1577, and in substantially the same form again in 1584, Velasco issued a printed questionnaire, consisting of fifty queries concerning cosmography, ethnography, and natural history, that was sent to all Spanish dependencies in the New World.[7] The respondents were to collect precisely the information that Velasco as cosmographer of the Council of Indies was instructed to "gather, compile, and put in good order."

The replies to the questionnaires resulted in a corpus known as the *relaciones geográficas de Indias*, a well-studied aspect of Spanish interest in the New World.[8] Almost two hundred replies to the questionnaires still exist;

7. For an English translation of the questionnaire, see Cline, "*Relaciones geográficas*," 364–71. In one of those curious instances when archives yield far more than a historian needs, the payment order survives, calling for 330 ducats to reimburse Velasco for printing the hundreds of questionnaires. AGI, IG-426, L. 26, f. 37v, "Orden de pago de 330 ducados a Juan López de Velasco por impresión que ha hecho hacer por orden de este consejo de ciertas instrucciones para la Descripción de las Indias u otras para averiguar la hora en que habrá en ellas dos eclipses de la luna este presente año y el venidero de setenta y ocho," June 8, 1577.

8. The *relaciones geográficas de Indias* have an extensive bibliography. Since the original studies by Jiménez de la Espada in the nineteenth century, there have been periodic efforts to collect and publish existing *relaciones*. Some of the more comprehensive collections in addition to Jiménez de la Espada's *Relaciones geográficas* include René Acuña, ed., *Relaciones geográficas del siglo XVI*, 10 vols. (Mexico: Universidad Nacional Autónoma de México, 1982); Ponce Leiva, ed., *Relaciones geográficas–Quito*; Mercedes de la Garza, ed., *Relaciones histórico-geográficas de la gobernación de Yucatán* (Mexico: UNAM, 1983); Antonio Arellano Moreno, ed., *Relaciones geográficas de Venezuela: Recopilación, estudio preliminar y notas de Antonio Arellano Moreno* (Caracas: Academia Nacional de la Historia, 1964); Francisco del Paso y Troncoso, ed., *Relaciones geográficas de la Diócesis de Michoacán, 1579–1580*, 2 vols. (Guadalajara: 1958). An inventory of replies was published in Howard F. Cline, "The *relaciones geográficas* of the Spanish Indies, 1577–1648," in *Handbook of Middle American Indians: Guide to Historical Sources* (Austin: University of Texas Press, 1964), and updated in Howard F. Cline, "*Relaciones geográficas*: Revised and Augmented Census of *relaciones geográficas* of New Spain, 1579–1585," in *Handbook of Middle American Indians* (Washington, D.C.: Library of Congress, 1966).

they constitute a treasure trove of ethnographic, economic, and administrative information about territories in the Indies and are in many cases the only surviving ethnographic account of some native peoples.[9] They provide valuable information on the demographic situation of native and Spanish populations and their towns, villages, and natural resources fifty years after colonization began. In many cases the respondents included maps—most drawn with strong Amerindian influence—and descriptions of the natural world and medicinal plants. It is only reluctantly that a historian can turn away from these fascinating testimonies and return to studying the methodological and epistemological aspects of the project.

Under a heading indicating clearly that the directive was issued by the king, the printed broadsheet first explained that the request was intended for the Indies' "good government and ennoblement" (*para el buen gobierno y ennoblecimiento de ellas*; figure 6.1). The questionnaires were addressed to ranking officials in the different governmental jurisdictions in the Indies, who were asked to distribute the questionnaires to all the towns in their district inhabited by Spaniards as well as those "of Indians." We know that just in the case of the viceroyalty of Peru, Velasco sent six hundred of the printed broadsheets.[10] Respondents were asked to read the list of questions carefully and answer only relevant questions. Once the response document was completed, they were to date it, record the name of the person or persons who prepared the answers, and return the responses along with the original questionnaire to the local ranking authority, who in turn would forward the originals to the Council of Indies.

9. Over the years historians have emphasized different aspects of the 1577 and 1584 questionnaires. Jiménez de la Espada, who rescued the project from its long hibernation, saw the questionnaire as another manifestation of a long Spanish tradition of soliciting information from overseas domains. Recently, Álvarez Peláez studied the questionnaire as the culmination of the reforms of the Council of Indies initiated by Juan de Ovando and, more generally, as the result of a climate of scientific inquiry brought to the Council by López de Velasco. In addition to the studies in the previously cited collections, modern scholarship ranges from Mundy's study of the maps that accompanied many of the *relaciones* to Álvarez Peláez's emphasis on medicine and natural history discussed in the documents, while Barrera-Osorio sees the questionnaires as examples of empirical procedures developed by Spain to understand the New World. Barbara M. Mundy, *The Mapping of New Spain: Indigenous Cartography and the Maps of the relaciones geográficas* (Chicago: University of Chicago Press, 1996); Álvarez Peláez, *Conquista de la naturaleza*; and Antonio Barrera-Osorio, "Empire and Knowledge: Reporting from the New World," *Colonial Latin American Review* 15, no. 1 (2006): 50.

10. AGI, Lima-30, cuaderno 3, f. 28–36v, Francisco de Toledo, "Carta del virrey del Perú a Su Majestad," April 18, 1578.

Inſtruction, y memoria, de las relaciones qu[e] ſe han de hazer, para la deſcripcion de las Indias, que ſu Mageſtad manda hazer, para el buen gouierno y ennobleſcimiento dellas.

RIMERAMENTE, los Gouernadores, Corregidores, ò Alcaldes mayores, a quien los Vireyes, o Audiencias, y otras perſonas del gouierno, embiaren eſtas inſtructiones, las diſtribuyran por los pueblos de Eſpañoles, y de Indios de ſu juriſdiction, embiando las a los concejos, ò a los curas ſi los vuiere, y ſino a los religioſos, a cuyo cargo fuere la doctrina, mandandoles de parte de ſu Mageſtad, que dentro de vn breue termino, las reſpondan, como en ellas ſe declara, y les embien las relaciones que hizieren, juntamente con eſtas memorias, para que ellos como fueren recibiendo las relaciones, vayã, embiandolas a las perſonas de gouierno, y las inſtructiones y memorias las bueluan a diſtribuyr ſi fueren meneſter por los otros pueblos adonde no las vuieren embiado.

Las perſonas a quien ſe diere cargo en los pueblos de hazer la relacion, reſponderan a los capitulos de la memoria, que ſe ſigue por la orden, y forma ſiguiente.

Primeramente, en vn papel a parte, pondran por cabeça de la relacion que hizierẽ, el dia, mes y año de la fecha de ella: con el nombre de la perſona, o perſonas, que ſe hallaren a hazerla, y del Gouernador, v otra perſona que les vuiere embiado la dicha inſtruction.

Y leyendo attentamente, cada capitulo de la memoria, eſcreuirã lo que huuiere q̃ dezir a el, en otro capitulo por ſi, reſpodiendo a cada vno por ſus numeros, como van en la memoria vno tras otro y en los que no huuiere que dezir, dexarlos han ſin hazer mencion de ellos, y paſſaran a los ſiguientes, haſta acabarlos de leer todos, y reſponder los que tuuieren que dezir: como queda dicho, breue y claramente, en todo: affirmando por cierto lo que lo fuere, y lo que no, poniendolo por dudoſo: de manera que las relaciones vengan ciertas, conforme a lo contenido en los capitulos ſiguientes.

¶Memoria de las coſas, a que ſe ha de reſponder: y de que ſe han de hazer las relaciones.

1. PRIMERAMENTE, en los pueblos de Eſpañoles ſe diga, el nombre de la comarca, o prouincia en que eſtan, y que quiere dezir el dicho nombre en lengua de Indios, y porque ſe llama aſsi.
2. Quien fue el deſcubridor y conquiſtador de la dicha prouincia, y por cuya orden y mandado, ſe deſcubrio, y el año de ſu deſcubrimiento y conquiſta, lo que de todo buenamente ſe pudiere ſaber.
3. Y generalmente, el temperamento y calidad de la dicha prouincia, o comarca, ſi es muy fria, o caliente, o humeda, o ſeca, de muchas aguas o pocas, y quando ſon mas o menos, y los vientos que corren en ella, que tan violentos, y de que parte ſon, y en que tiempos del año.
4. Si es tierra llana, o aſpera, raſa o montoſa, de muchos o pocos rios o fuentes, y abundoſa o falta de aguas, fertil, o falta de paſtos, abundoſa o eſteril de fructos, y de mantenimientos.
5. De muchos o pocos Indios, y ſi a tenido mas o menos en otro tiempo que ahora, y las cauſas que dello ſe ſupierẽ, y ſi los que ay eſtan poblados en pueblos formados y permanentes, y el talle y ſuerte de ſus entendimientos, inclinaciones, y manera de biuir, y ſi ay differentes lenguas en toda la prouincia, o tienen alguna general en que hablen todos.
6. El altura o eleuaciõ del polo, en que eſtan los dichos pueblos de Eſpañoles, ſi eſtuuiere tomada, y ſi ſe ſupiere, o vuiere quien la ſepa tomar, o en que dias del año el ſol no hecha ſombra ninguna al punto del medio dia.
7. Las leguas que cada ciudad o pueblo de Eſpañoles eſtuuiere de la ciudad donde reſidiere la audiencia, en cuyo diſtricto cayere, o del pueblo donde reſidiere el gouernador a quien eſtuuiere ſugeta: y a que parte de las dichas ciudades o pueblos eſtuuiere.
8. Aſsi miſmo las leguas que diſtare cada ciudad o pueblo de Eſpañoles de los otros cõ quien partiere terminos, declarando a que parte cae dellos, y ſi las leguas ſon grãdes o pequeñas, y por tierra llana o doblada, y ſi por caminos derechos ò torcidos, buenos ò malos de caminar.
9. El nombre y ſobrenombre que tiene, o vuiere tenido cada ciudad ò pueblo, y porque ſe vuiere llamado aſsi, (ſi ſe ſupiere) y quien le puſo el nombre, y fue el fundador della, y por cuya orden y mandado la poblo, y el año de ſu fundacion, y con quantos vezinos ſe començo a poblar, y los que al preſente tiene,

10. El

Figure 6.1. Questionnaire of the Indies, 1584. Spain. Ministry of Culture. General Archive of Indies. Patronato, 18, N. 16, R. 2.

The questionnaire's internal structure at first appears somewhat repetitive but becomes clearer when interpreted in light of the function for which it was designed—to gather first "general" and then "particular" information as outlined in the *Instructions*.[11] The questions address the same subject matters listed in the *Instructions* and follow the thematic structure Velasco used in his *Geografía*. The first ten questions addressed general geographical matters, followed by forty inquiries concerning specific, "particular" matters related to each location described. The general questions ask about the province's or region's name and what the name meant in the Indian language. Who discovered and conquered the land? What is the province's general climate and quality? What is its general landscape? Questions in this first group also inquired about the Indian population, asking specifically whether there had been changes in the population and the cause for

11. Álvarez Peláez has also noted that the questionnaire's structure—with some questions appearing to be repeated—made the document prone to misinterpretation. Álvarez Peláez, "Relaciones de Indias," 300. For a similar interpretation, this time related to the document's language, see Sylvia Vilar, "La trajectoire des curiosités Espagnoles sur les Indes," *Mèlanges du Casa de Velázquez* 6 (1970).

the change, whether the Indians lived in permanent towns, and about their "intelligence, inclinations, and modes of life." These were followed by another five, still general, cosmographical questions. Question 6 requested the latitude of the town in the province. The wording of this section proved to be somewhat problematic, particularly for respondents not trained in cosmography. It asked for

> 6. [t]he height or elevation of the Pole for Spanish towns, if it has been taken, or if it is known or if anyone there can take it, or in which day of the year the Sun does not cast a shadow at all on the Midday point.[12]

The original sentence in Spanish has an awkward construction, consisting of four clauses that depend on the reader's understanding that the "height or elevation of the Pole" (*altura ó elevación del Polo*) refers to the town's latitude. If the respondent did not understand these terms, he would have also been baffled by another technical term: *punto del Medio-dia*. This phrase was in common usage in navigation manuals and meant when the sun was at its zenith. Thus the request was for the date when the sun did not cast a shadow when at its zenith. In regions located between the tropics, there are two days between the March and September equinoxes when the sun is directly overhead and thus does not cast a shadow at noon. Knowing the date allowed someone with access to a table of solar declination to determine the observer's latitude.[13]

Perhaps in tacit acknowledgment that few observers could perform the technical measurement, the next two questions asked for the town's location to be ascertained by its distance in leagues from the seat of government and also relative to nearby towns, with indication whether the roads between the towns were flat or "tortuous." Question 9 asked for a listing of each town in the province and a brief account of its founding and its current population. In the tenth question, the 1577 questionnaire asked for a drawing (*traza*) of the town's location:

> 10. Describe the site and state the situation of said towns, if it lies high or low or in a plain, and give a draft or painting showing the streets, squares, and other significant places and monasteries

12. "6. La altura o elevación del Polo en que están los dichos pueblos de españoles, si estuviese tomada, y si se supiese ó hubiere quien la sepa tomar, o en que día del año el Sol no hecha sombra ninguna al punto del Medio-día." BN MS 3035, f. 42v, "Instrucción y memoria de las relaciones . . . ," 1577.

13. Zamorano, *Compendio del arte de navegar*, 13v–14v, 25.

sketched any which way on paper. It is to be noted which part of the town faces south and which north.[14]

Until this point, the questionnaire was collecting information for a "general" cosmography: geographic coordinates, regional maps, and concise descriptions of each province and its peoples (plates 9 and 10). The next five questions sought general information about Indian towns: names of the towns, their location (distances only; clearly Velasco did not expect Indians to know how to use an astrolabe), and what the names meant in the native language. Interest in cultural matters extended to the natives' former and current modes of government, their language, customs, "good or bad" rites, dress, and whether they had been healthier in the past than at present and what explanation they might have for this.

Questions 16–30 concerned the local, "particular" geography and natural history of each town and were to be answered for both Spanish and Indian towns. Respondents were asked to describe significant geographical features surrounding the town, such as mountains, lakes, volcanoes, and rivers, and specifically if the latter were navigable or good for irrigation. Next, the questions sought information about foodstuffs: native trees and grains and the state of cultivation of trees and plants introduced from Spain (for producing wheat, barley, wine, olive oil—staples of the Spanish diet). The questions on natural history covered medicinal plants, native and domesticated animals, and minerals (gold, silver, and precious stones). These questions echoed the *Instructions*' article 17 and its emphasis on the utilitarian aspects of a region's natural resources. The questionnaire moved on to inquire about the town's buildings, both domestic and defensive, and its commerce. Questions 34–37 concerned the relationship between the town and the Church and asked for a listing of churches, monasteries, and convents.

The next ten questions had to do specifically with coastal towns and the hydrography of the region. Here again the questions solicited information that would complete the hydrographical description of the coast required in the *Instructions*. Was the coastline beach or was it rugged? Were there hidden reefs or other dangers to navigation? What were the times and heights of tides? What were the names and features of significant promontories? Where were the ports and suitable landings, and could the

14. "10. El sitio y asiento donde dichos pueblos estuvieren, si es alto, o en bajo, o llano, con la traza y designo en pintura de las calles, y plazas, y otros lugares señalados de monasterios como quiera que se pueda rasguñar fácilmente en un papel, en que le declare, que parte del pueblo mira al medio día o al norte." BN MS 3035, f. 42v.

respondent describe their size, capacity, orientation, and depth? Finally, the questionnaire requested any other relevant information about unpopulated areas or other "notable natural features."

Velasco used the questionnaire to collect *only* cosmographical, geographical, and ethnographic information, plus information about natural resources. He did not solicit historical accounts other than basic information about the region's discovery or the town's founding. Instead, and in a roughly parallel effort, he solicited historical material directly from the ranking authorities, the communication channel that the *Instructions* specified (article 83). He raised a petition to the Council of Indies, which was followed a month later by a royal order.[15] In so doing, Velasco was following the Council's directive considering secrecy, as the sources used to write the history of the New World were considered politically sensitive and best handled solely by prudent government officials. Local officials were thus instructed to collect from their archives any histories, commentaries, or accounts concerning the region's discovery and governance. If possible, the originals were to be sent to the Council of Indies to the attention of Juan López de Velasco, who—as the royal order indicated—was to use the material to write a *General History of the Indies* "based on the truth" (*con fundamento de verdad*).

The questionnaire project's success hinged on whether the viceroys and the local *audiencias* distributed the questionnaires and urged compliance. Response to the 1577 questionnaire was uneven,[16] for whereas the viceroy of New Spain, Martín Enríquez, complied with the order, the viceroy of Peru, Francisco de Toledo, refused to distribute the six hundred questionnaires Velasco sent him. (It is not a coincidence that most of the twenty-one replies from Peru and Quito were written after Enríquez succeeded Toledo as viceroy of Peru in 1580.)

In a long letter addressed to the king in care of the Council of Indies, Toledo seemed somewhat confused about the new directive. He acknowledged receipt of the broadsheets, the royal order, and a letter from "a certain Juan López de Velasco who says he is chronicler major."[17] He explained that he had been diligently preparing the materials Juan de Ovando had

15. AGI, IG-1388, Juan López de Velasco, "Petición para que se mande a reconocer los papeles en las distintas audiencias pertenecientes a historia," May 23, 1578. AGI, IG-427, L. 30, f. 281v–282, "Real cédula a los oficiales de Indias que envíen historias al Consejo de Indias," June 25, 1578.

16. Cline, "*Relaciones geográficas*," 1964, 193.

17. AGI, Lima-30, book 3, f. 28–36v, Francisco de Toledo, "Carta del virrey del Peru a Su Majestad," April 18, 1578.

requested in years past; for four or five years he had been assembling a four-part description of the viceroyalty of Peru. In fact, he had entrusted the whole project to the very capable Pedro Sarmiento de Gamboa. The first part consisted of a general and particular description of the land both "in writing and in painting" (*en pintura y estampa*), which mapped all the land between the "mar del Sur" and "mar del Norte" (i.e., Pacific and Atlantic) from the Strait of Magellan to Nombre de Dios (Panama) based on the "authority of witnesses who have sailed it and sworn before a judge." Likewise, each province within the territory had been described "in canvas paintings with annotations" by local "caciques y visitadores y corregidores." As of 1578, Toledo claimed all this material was in his possession. Unfortunately, what was surely a stunning geographical description of Peru is now feared lost.

As for the historical aspects of the project, Toledo stated that he had collected all the chronicles, whether in manuscript or in print, and had noted what was "true and what was false in them." This material, according to Toledo's account of the project, was used for the second and third part of the project, a history, including a description of the rites and religion (*idolatría*), of the Incas. This part of the project was finished in 1572; we now know it as Sarmiento de Gamboa's *History of the Incas*. The fourth and final part was to be a history beginning with the arrival of the Spanish. Toledo said that he was having particular difficulty with this final part of the project because of the "many lies" circulating and the fact that many of the early conquistadors who could give true testimony had died.

In conclusion, Viceroy Toledo said that his histories and geographical descriptions, rather than replies to the questionnaires, would be sent to comply with the royal order. Toledo was committed to using the best means possible to include within the documents currently under way "all the curiosities the Your Majesty orders in the instructions sent to describe this land."[18] To excuse his decision not to use the questionnaires, he argued that the new instructions inquired about so many things that if all the appropriate local officials were ordered to do so it would take too much time away from their duties and cost a lot of money. In addition, he said, it was hard to find persons well versed in the "many different professions" needed to understand the instructions. Therefore, the viceroy asserted—leaving little room for negotiation—that the foregoing explained the course of action

18. "[T]odas las curiosidaded que V. M. manda por las instruciones que se envian para la descripcion de la tierra." AGI, Lima-30, cuaderno 3, f. 29.

he intended to take, adding that he would reply to the chronicler under separate cover.[19]

Velasco's strategy did not fare well in Peru. The viceroy's objections reflected a refusal to change course midstream and perhaps a desire to retain control and authorship over the project. Despite Toledo's refusal, Velasco had structured the project to encourage compliance, by leveraging royal authority and by integrating the questionnaires within the bureaucratic machinery of colonial Spain. Perhaps more significant, Velasco made the questionnaire accessible and straightforward, as opposed to the vast and complex series of books of description called for by the *Instructions*. The questions called for knowledge that was within the realm of what a literate and somewhat curious observer could obtain. This accessibility was the key to the project's more general success.

Velasco's training, intellectual disposition, and workload did not incline him toward personally making the cosmographical observations his post required. His position, one defined by legislation and clearly codified, did not require him to venture any farther than his desk when carrying out his duties. The statutes governing the office of the cosmographer-chronicler drew from the Ptolemaic and descriptive cosmographical traditions of Renaissance humanist cosmography, which relied on written sources rather than on personal *experiencia*, which, as we have seen in the cases of Jaime Juan and Francisco Hernández, produced a parallel but distinct style of cosmographical practice.

With the questionnaire Velasco managed to address the epistemological challenge posed by his sources—discovery accounts, miscellaneous histories, and pilot descriptions—which ultimately involved determining in which, in the case of many accounts, and where, in the case of one account, the truth lay. To Velasco's legal sensitivities, the responses to the questionnaire were the ideal source of an unmediated, witnessed, firsthand account. He designed the questionnaires to elicit thematically and structurally organized cosmographical information in a manner consistent with the *Instructions*. In doing so, he sought to overcome the limitations of the sources he had used in the *Geografía y descripción*, which had exposed his work to harsh criticism from a mathematical practitioner.

The problematic question persists whether the information collected by the questionnaires, the corpus of the *relaciones geográficas de Indias*, was ever

19. For a recent study of Sarmiento's contribution, see Brian S. Bauer and Jean-Jacques Decoster's introduction to Pedro Sarmiento de Gamboa, *The History of the Incas* (Austin: University of Texas Press, 2007).

"used," as it does not appear to have been routinely consulted by the Council of Indies. This semblance of neglect should not be surprising, since the questionnaires and the corresponding responses were not at each council member's disposal but were to be consulted solely by the cosmographer-chronicler of the Council of Indies. Velasco certainly handled the materials, as is evident from a memorandum dated 1583 confirming the receipt of the documents.[20] If this was the case, then why did Velasco not write another cosmographical oeuvre along the lines of the *Geografía y descripción universal* based on the replies?

Historians have judged Velasco's sin of omission harshly. Some have concluded that the *relaciones* were quickly filed away and ignored because of little interest in or concern about the peoples under Spanish rule.[21] This apparent neglect has in turn been used to undermine arguments that the project was intended as a mechanism for social or political control and to argue instead that at best it served only as a marginal tool to govern the far-flung empire.[22] Even the Amerindian-style maps that accompanied many *relaciones* have been deemed as having fallen so far short of Velasco's expectations that he was "disappointed" with them and thus never used them.[23]

Given that we have no textual evidence of what Velasco, or any of his contemporaries, thought of the responses to the questionnaire and their maps, any conclusions about the project's "usefulness" need to be handled as speculation. I prefer to posit another theory regarding Velasco's failure to use the replies to compose a cosmography in the traditional style. Perhaps Velasco never intended for the questionnaire replies to serve as source material for a cosmography. Instead, the replies themselves were meant to serve as the cosmographical corpus, standing as individual and unmediated testaments of the reality of the New World as articulated by firsthand observers. The fact that the subjects inquired about in the questionnaire remained essentially unaltered over the span of seven years suggests that the information Velasco received from the 1577 questionnaire did not

20. AGI, P-171, N. 1, R.16, f. 11–14v, Juan López de Velasco, "Relación de las descripciones y pinturas de las provincias del distrito de Nueva España que se han traído al Consejo y se entregó a Juan López de Velasco," November 21, 1583.

21. Poole, *Juan de Ovando*, 202.

22. Pilar Ponce Leiva, "Los cuestionarios oficiales: ¿Un sistema de control de espacio?" in *Cuestionarios para la formación de las relaciones geográficas de Indias, siglos XVI–XIX*, ed. Francisco de Solano (Madrid: CSIC, 1988), xxxiv–xxxv.

23. Mundy, *Mapping of New Spain*, 215, and David Turnbull, *Masons, Tricksters, and Cartographers* (Amsterdam: Harwood Academic, 2000), 110.

disappoint him and that he was more preoccupied with ensuring compliance than with the content of the replies.

Carefully filed and inventoried—as we know they were[24]—the replies to the questionnaire functioned as self-compiled, comprehensive "particular" cosmographical descriptions with unquestionable credentials of having been written by eyewitnesses. By using the questionnaire format Velasco achieved a general commonality of subject matter and structure among the responses, which facilitated access to the information if and when the members of Council requested him to furnish it. Adopting questionnaires into cosmographical practice in the Council of Indies also exempted Velasco from having to compile the material into a traditional cosmography, since the cosmographical information about any particular region of the Indies was now technically "known" by the Council, though it existed in the form of responses to fifty questions rather than within one comprehensive text.

Seen in this light, Velasco's questionnaire project was a solution to the perceived shortcomings of the Renaissance cosmographical genre that we first explored in Santa Cruz's *Islario*. For a cosmographer such as Velasco, with the author's professional fortunes tied directly to the apparent accuracy of the content of his work, the comprehensive cosmography had become a genre increasingly difficult to compose. After experiencing with the *Geografía* the epistemological difficulties inherent in the genre, Velasco resorted to a method of collecting and organizing cosmographical information that addressed the genre's growing limitations: comprehensiveness, timeliness, and veracity. Further, he understood that any cosmographical work he authored would receive only limited exposure, given the monarchy's secrecy policy concerning the geography, natural resources, and history of the Indies. Denied fame, he could be rewarded only by the king and Council. Discouraged by the slight appreciation they had shown for his first cosmographical work, Velasco simply abandoned the writing of Renaissance-style cosmographies.

With the problem of collecting and organizing the "particular" aspects of his cosmographical duties solved by the replies to the questionnaire, Velasco turned to collecting the information necessary for the "general" cosmography that would relate geographically the various dependencies of the New World. To do this, he needed to compose regional and universal

24. Studies of the provenance of the surviving *relaciones geográficas de Indias* suggest that they were in the possession of whoever served as cosmographer or chronicler of the Council of Indies well into the seventeenth century. Cline explains that the *relaciones* were catalogued at the Archive of Simancas sometime between 1659 and 1718 under the classification "descripción y población." Cline, "*Relaciones geográficas*," 1964, 196–97.

maps based on the unquestionable authority of astronomically derived latitude and longitude values. As the *Instructions* stated—and even Gesio conceded—careful, simultaneous observation of lunar eclipses was the only feasible way of determining longitude accurately. Velasco set the bureaucratic gears of the Spanish empire turning to churn out this information.

Eclipses and Longitude

The *Ordinances* and the *Instructions* adopted by the Council of Indies in 1571–73 mandated the observation of lunar eclipses to determine longitude. However, since the year the *Instructions* became law, the heavens had not presented the cosmographer with the opportunity of simultaneous observations of a lunar eclipse in Spain and in the New World. This would change in 1577 and 1578, when the heavens put forth two lunar eclipses. The scope of the eclipse project initiated by Velasco in 1577 was very ambitious. Over the next eleven years, he prepared a series of printed instructions advising all overseas territories about upcoming eclipses and giving detailed instructions for observing and recording them. He intended for observers throughout the world—whom he assumed had little expertise in Western-style astronomical measurements—to measure the same lunar eclipse by carefully following his instructions. The observers were told how to build a simple instrument—the Instrument of the Indies—perform the observation, log the results, notarize them, and send them to the Council of Indies for analysis. Back in Spain, the cosmographer would complete the necessary mathematical computations, determine each respondent's longitude relative to Spain, and correct the appropriate maps. The project has been described as the "first known large-scale, systematic plan of astronomical observation."[25] Nonetheless, historians have generally assumed that observation results were so imprecise that they were meaningless, or that there were so few respondents that they did little to improve Spanish cartography.[26] These opinions unfortunately contribute little to

25. Esteban Piñeiro, "Los cosmógrafos al servicio de Felipe II," 532–33. The project is often cited in bibliographies and survey histories of Spanish science, but besides recognizing the importance of the enterprise there has been little work done on the subject. For a fine English-language study, see Clinton R. Edwards, "Mapping by Questionnaire and Early Spanish Attempts to Determine New World Geographical Positions," *Imago Mundi* 23 (1969): 117–208. Rodríguez-Sala published what were believed to be the sole surviving observations in *Eclipse de Luna*.

26. This is the conclusion Edwards reached in his 1969 article cited and echoed in Mundy, *Mapping of New Spain*, 55–56. Randles, however, conceded that toward the end of

the study of this complex enterprise, whose significance to the history of astronomy and geography, as well as the important aspects it reveals about the practice of science during the early modern era, has been overlooked.

Since the discovery of the New World, determining the extent and location of its domains was a principal preoccupation of the Spanish empire. Navigating to and from the New World was fraught with danger, not the least of which was the real and frequent problem of getting lost. In addition, knowing with certainty the locations of newly discovered territories carried significant geopolitical implications in sixteenth-century Spain. The problem was compounded by the technical difficulties that longitude measurements presented to early modern cosmographers and navigators. Today's methods of determining longitude rely on GPS satellites or on being able to keep accurate time, but timepieces of the early modern era were not accurate enough, even on land, to determine longitude with certainty.[27]

The theory behind the use of lunar eclipses to determine longitude is deceptively simple. The eclipse serves as a global synchronizing event. Observers in one part of the globe record the local time of the eclipse and compare it to the local time it was observed elsewhere. The difference in time can be easily translated into degrees of longitude, since one hour's difference in local times equals 15° of the earth's circumference. This, of course, assumes that the clocks in both places are synchronized and keep time with reasonable accuracy.

Long before Velasco's project, lunar eclipses were observed in Spain and in the New World for the purpose of determining longitude. Mariners routinely used the method when they chanced upon an eclipse while at sea, but some civilian authorities and cosmographers also contributed eclipse observations, as we have seen in the case of Juanote Durán. One of these early observations dates from the eclipse of November 16, 1537, and was recorded by the viceroy of New Spain, Antonio Mendoza. The viceroy

---

the sixteenth century the Spanish had achieved remarkable accuracy in their longitude calculations, but he still asked, "W[ere] such close figure[s] a fluke or the result of a real mastery of the process of measurements?" W. G. L. Randles, "Portuguese and Spanish Attempts to Measure Longitude in the 16th Century," *Vistas in Astronomy* 28 (1985): 238.

27. Few texts explain the "longitude problem" in its historical context better than the classic Taylor, *Haven-Finding Art*. For an explanation of different methods of finding longitude during the early modern era, see Charles H. Cotter, *A History of Nautical Astronomy* (London: Hollis and Carter, 1968), 180–208. For a comprehensive survey of the topic, see William J. H. Andrewes, ed., *The Quest for Longitude: The Proceedings of the Longitude Symposium, Harvard University, Cambridge, Massachusetts, November 4–6, 1993* (Cambridge, Mass.: Harvard University, 1993).

was acquainted with Alonso de Santa Cruz and likely acted at the cosmographer's urging.[28] In a letter to the king via the Council of Indies, the viceroy noted that the lunar eclipse began in Mexico City "half a quarter-hour after sunset." His observation, it turns out, was very accurate. Computations using modern astronomical models show that the beginning of the partial phase of the eclipse (the earliest occultation visible with the naked eye) took place at 17:28, while seven minutes after sunset corresponded to approximately 17:29, a one-minute difference.[29]

The cosmographical conference held in 1566 to discuss whether the Philippine Islands lay on Spain's side of the demarcation line also relied on eclipse observations. The deliberations included the opinion of several cosmographers that the simultaneous observation of lunar eclipses was a reliable method for determining longitudinal distances.[30] No doubt their opinions contributed to formalizing the practice in the *Ordenanzas* of 1571. Missionaries, particularly in the Far East, also contributed to the steadily growing number of reliable eclipse observations. In addition to Martín de Rada, Jesuit missionary Matteo Ricci recorded and communicated to Spain eclipse observations from China.[31]

28. Mendoza wrote: "Si V. M. quisiere mandar averiguar la longitud que hay desde aquí a España por el eclipse que hubo el 16 de Noviembre pasado, sepa que comenzó en esta ciudad medio cuarto de hora después de puesto el sol." AGI, P-184, R. 27, b. 2 (27); copy at BAH, Colección Muñoz vol. 63, f. 40: "Carta de Virrey Antonio de Mendoza a Su Majestad," December 10, 1537, Mexico. For Mendoza's friendship with Alonso de Santa Cruz and his cosmographical interests, see Manuel de la Puente y Olea, *Los trabajos geográficos de la Casa de Contratación* (Sevilla: Librería Salesianas, 1900), 362.

29. The sidereal circumstances surrounding the viceroy's report can be reconstructed using modern astronomical models. All astronomical data used in this chapter is from Fred Espenak at the NASA/Goddard Center. We know that the beginning of the partial phase of the eclipse took place at 01:05 UT on November 17. Since the time distance from Greenwich (UT) to Mexico City is 6:36 hours, the event would have been visible in Mexico City at 17:28 on the evening of November 16. Sunset that night in Mexico City occurred close to 17:22. The viceroy noted the start of the event at "half a quarter-hour after sunset" ($7^1/_2$ minutes after sunset), thus only a minute from the actual start time of the lunar eclipse. The fact that the eclipse occurred so soon after sunset clearly influenced the accuracy of the reported time, since the observers probably started a sand hourglass at sunset.

30. AGI, P-49, R.12, Bloque 2; copy at BAH, Colección Muñoz vol. 33, f. 304–313: Alonso de Santa Cruz, "Parecer sobre la demarcación," 1566. Urdaneta also cited eclipse observations as part of his demarcation arguments.

31. AGI, Filipinas-29, N. 49, f. 215–229v, "Relación sobre el reino de China de Juan Bautista Román," September 28, 1584, Macao. Copy at BAH, Colección Muñoz vol. 33, f. 249–65.

The principles behind using lunar eclipses to determine longitudinal distances was well understood in antiquity and enjoyed renewed popularity after the publication of Ptolemy's *Geography* in the fifteenth century. Within the community of astronomers, interest in determining longitude accurately was largely driven by the need to adjust astronomical motions listed in the Alphonsine Tables—all based on the meridian of Toledo—to a local reference point as determined by longitudinal distance between Toledo and the point of observation. There is little evidence of systematic efforts to carry out simultaneous observations of eclipses expressly for the purpose of determining longitudes.[32] Often when an astronomer mentioned having determined a new value for a city's longitude using a lunar eclipse observation, at least one of the two necessary reference observations was not an eclipse observation at all but a calculation based on predicted times of lunar occultation taken from ephemerides. For places not far from Toledo, astronomers often relied on estimated distances. After the discovery of the New World, however, global distances could no longer be easily determined using estimated terrestrial distances or their counterparts at sea.

With the renewed interest in determining longitude, both on land and at sea, came new methods that promised to solve the "longitude problem." In 1514, Johannes Werner proposed the method of lunar distances.[33] His method used the moon's motion relative to fixed stars (occultation) to determine local time (and thus longitude). It required precise tables listing star positions and knowledge of the moon's position relative to the sun at any given time and at any given latitude. Werner's method, although theoretically correct, required stellar and lunar tables with a predictive accuracy far beyond what was available at the time.[34] Yet his method was not new. As early as 1499, Amerigo Vespucci had used the occultation of the moon with Mars to calculate longitude while on the coast of Brazil.[35] In a previous chapter I discussed how Andrés de San Martín, the pilot of Magellan's expedition, used a derivative of this method to attempt to determine longitude. In 1519 he observed the moon's position relative to Jupiter, only to

32. Richard L. Kremer and Jerzy Dobrzycki, "Alfonsine Meridians: Tradition versus Experience in Astronomical Practice, c. 1500," *Journal for the History of Astronomy* 29 (1998): 188.

33. Bennett, *Divided Circle*, 53–56. For an excellent study and translations of Werner's and Frisius's longitude methods, including their discussions on using eclipses to determine longitude, see Andrewes, ed., *Quest for Longitude*, 376–92.

34. Gallois, *Géographes allemands de la Renaissance*, 123–24.

35. Pohl, *Amerigo Vespucci, Pilot Major*, 65–67.

find the result seriously corrupted by inaccuracies he rightly attributed to the astronomical tables of Zacut and Regiomontanus.[36]

In his popular 1524 *Cosmographia*, Peter Apian explained Werner's method but also advocated the use of lunar eclipses for determining longitude. To facilitate calculations, he included drawings showing the maximum phase of upcoming eclipses (calculated for the position of the city of Leiden) and instructed the reader to compare these images with local observations. Thus correlating the observer's local time with the ones indicated in the book, an observer could calculate the longitudinal distance between the observer and Leiden.[37] Gemma Frisius continued advocating the method in his many editions of Apian. But it was in his *De principiis astronomiae* (1530) that he introduced the method of determining longitude by transporting mechanical timepieces—the method that would decidedly solve the longitude problem almost two hundred years later.[38]

Other sixteenth-century cosmographers also devised complicated schemes to address the problem. In part 3 of his 1544 *Quadratura Circuli*, Oronce Finé discussed what he considered a refinement of the method of using lunar eclipses to determine longitude.[39] It consisted in measuring the angle between the moon when in total eclipse and its position while transiting the meridian, whether the transit happened before or after the eclipse. Pedro Nuñez, the royal cosmographer of the king of Portugal, noted some errors inherent in Finé's method in his *De erratis Oronti Finé* (1546).[40] Nuñez complained that Finé seemed to ignore the fact that the moon's apparent motion is not uniform and thus a given lunar angular distance from the meridian cannot be used to determine how long the moon would take to reach the meridian. Both Werner's and Finé's methods depended on being able to model the moon's motion accurately in order to construct the necessary astronomical tables—one of the most challenging astronomical problems of

36. Fernández de Navarrete, *Disertación sobre la historia de la náutica*, 77:333–34.

37. Peter Apian, *Cosmographicus liber Petri Apiani mathematici studiose collectus* (Landshut [Germany]: P. Apiani, 1524).

38. Gemma Frisius, *De principiis astronomiae et cosmographiae*, facsimile reproduction (Antwerp: Ioannis Steelfii, 1553), 19.64.

39. Oronce Finé, *Quadratura circuli . . . De invenienda longitudinis locorum differentia . . .* (Paris: S. Colinaeum, 1544), 83–91. In his *Protomathesis* Finé also discussed the mathematics behind the method of using lunar eclipses for determining longitude but did not mention any instruments. Oronce Finé, *Protomathesis: Opus varium . . .* (Paris: Gerardi Morrhij et Ioannis Pet, 1532), f. 145v-46.

40. Pedro Nuñez, "De erratis Orontii Finoei liber unus," in *Petri Nonii Salaciensis Opera* (Basel: Sebastianum Henricpetri, 1592), 347–48.

the time and one that would not be better approximated until Isaac Newton put forth his theory of the moon.

The complexities involved in using lunar eclipses to calculate longitudinal distances were not limited to inaccurate clocks and faulty theories of the moon and lunar tables. A number of other problems concerning the observations, such as parallax, were well understood, while others such as the distortion caused by the effect of terrestrial emanations or "vapors" (atmospheric refraction) were beginning to be better quantified. Astronomer Jerónimo Muñoz suggested that the effect of parallax could alter the results obtained from lunar eclipse observations. Muñoz recommended that parallax tables be constructed for different geographical locations for the purpose of correcting longitude values derived from lunar eclipse observations.[41]

In the Iberian Peninsula, Nuñez in his capacity as cosmographer major of Portugal and as early as 1536 ordered the official rutter from Portugal to the Moluccas recalculated using lunar eclipses to corroborate distances.[42] Yet despite the number of early modern cosmographers who discussed using eclipses to determine longitude in print, I have been unable to find antecedents for the specific instrument and method Velasco selected for the observation. The instrument's simplicity and ease of construction made it particularly suitable for the project, as he intended it to be constructed by inexperienced persons to record the start and end of the eclipse. It is reasonable to suppose that Velasco resorted once again to Alonso de Santa Cruz's cosmographical work for guidance on the topic of lunar eclipses and longitude. After all, Velasco had in his possession Santa Cruz's fascinating treatise on the multiple methods of calculating longitude, the *Libro de longitudes*. Santa Cruz wrote the book as part of the evaluation of instruments for measuring longitude made by Peter Apian and presented to the Council in 1554.[43] In the treatise Santa Cruz explained twelve different ways of calculating longitude, among them the use of lunar or solar eclipses.[44] It

41. Historian Navarro Brotóns identified this manuscript currently at the National Library of Naples. Ms. VIII, 33, f. 1r–19v.

42. Armando Cortesão, *Cartografia portuguesa antiga* (Lisbon: Comissão Executuva das Comemorações do Quinto Centernário da Morte do Infante D. Henrique, 1960), 129.

43. Santa Cruz, *Obra cosmográfica*, 1:93. The *Libro de longitudes* is published in this edition in 1:139–202.

44. For a brief study of the subjects Santa Cruz discusses, see Luis de Albuquerque, "Acerca de Alonso de Santa Cruz y de su 'Libro de las longitudes,'" in *América y la España del siglo XVI: Homenaje a Gonzalo Fernández de Oviedo cronista de Indias en el V*

is particularly vexing for the historian, however, that the only surviving manuscript of Santa Cruz's *Libro de longitudes* is missing the one page where Santa Cruz likely discussed how to measure lunar eclipses to determine longitude![45] The surviving section resumes, "I will only say that astrologers and cosmographers, because of the difference in hours between the observations, came to consider that they could determine it [longitude] in any two places, regardless of how far apart they might be in longitude, by taking into consideration the time in which each could see the beginning or end of the eclipse."[46]

Santa Cruz noted that this was one of the easiest and most accurate known methods for finding longitude but that using this method posed some problems: It was difficult for observers to determine the precise start and end of the eclipse, i.e., when the umbral shadow started or finished obscuring the moon. Also, synchronizing clocks to time the event to the meridian of observation was a difficult and imprecise procedure, while the rarity of the event clearly made it impractical for finding longitude at sea. Santa Cruz believed that the solution to the longitude problem while at sea lay in determining accurately the apparent relationship between longitude and the degrees the compass needle deviated from true north relative to the *punto fijo*, or location where the compass needle pointed to the true north. In his treatise, however, he also discussed the method of using a mechanical timepiece to transport the local time of the prime meridian.

A Global Project

In the *Geografía*, lunar eclipses occupied the top of Velasco's epistemic hierarchy. However, he was also under legal obligation to use this method. In 1577 and 1578, the heavens, Velasco surely thought, offered him the precise conditions to settle once and for all the simmering cosmographical disputes concerning the location of the New World. It gave him the opportunity to standardize the observation of lunar eclipses, as well as the

centenario de su nacimento, ed. F. de Solano and F. del Pino (Madrid: CSIC, Instituto "Gonzalo Fernández de Oviedo," 1982).

45. BN, MS 9441, f. 21–22. Alonso de Santa Cruz, "Libro de longitudes."

46. "[M]as solo diré que así los astrólogos como cosmógrafo, por causa de la diversidad de horas del dicho miramiento, vinieron a considerar que según lo tal podían sacarlo de cualesquier dos lugares por muy apartados que pudiesen estar por longitud, haciéndose consideración en ellos del punto de hora que en cada uno se pudo ver el principio o el fin del eclipse." Santa Cruz, *Obra cosmográfica*, 1:149.

format used to record the event throughout the empire. The project was as remarkable in its simplicity as impressive in its scope. Velasco chose the 1577 lunar eclipse—a spectacular total eclipse clearly visible in Spain and in the northern and southern hemispheres of the New World—to inaugurate the project. He sent a printed broadsheet to all administrative units in the New World explaining the project's purpose. Headed by the phrase "By order of His Majesty in order to determine latitude and longitude" and bearing Juan López de Velasco's elegant signature, the instructions included detailed directions on how to build an instrument, how to take the measurements the day of the eclipse, and how to report the findings to the Council of Indies.

A separate document addressed to the local authority and in the form of a royal order stated the purpose of the project, "in order to situate [the towns] in geographical charts, and to correct navigation and distances and routes, and for other effects common to our service." The printed broadsheet began,

> Instruction and announcement for the observation of lunar eclipses and the quantity of the shadows that His Majesty orders done this year of fifteen seventy seven and seventy eight, in the cities and towns of the Indies; to verify the longitude and elevations of them that while astrology and cosmography have for the aforementioned purpose many and different Mathematical methods but being aware that there might not be persons in the Indies that know these other, we have selected these as being the easiest and commonly used.[47]

Although Velasco tried to put the project in place in early 1577, replies to his petition for a royal order were delayed awaiting the king's

47. "Instrucción y advertimiento para la observación de los eclipses de la luna, y cantidad de las sombras que su Majestad manda hacer, este año, de mil y quinientos y setenta y siete y quinientos y setenta y ocho, en las ciudades y pueblos de las Indias: para verificar la longitud, y altura de ellos que aunque pare el efecto sobredicho tienen la astrología y cosmografía propuestos muchos y diferentes medios Matemáticos pero teniendo respecto a la falta que en las Indias ha de haber de personas que sepan usar de otros se han elegido por mas fáciles y usuales: los medios que se siguen." BN, MS 3035, f. 40–41, "Instrucción y advertimiento para observación de los eclipses de la luna, y cantidad de las sombras," 1577–78. This is the earliest surviving version of the eclipse instructions. Two bibliographies report an earlier version dated 1577; it is likely a bibliographical mistake: Pérez Pastor, *BM*, 1:54–55, and José Toribio Medina, *Biblioteca hispano-americana, 1493–1810*, facsimile ed. (Santiago de Chile: Fondo Histórico y Bibliográfico José Toribio Medina, 1958), items 246, 401.

signature.[48] Therefore, the first broadsheet arrived in the colonies barely in time for the eclipse of September 26, 1577. The viceroy of New Spain complained that the instructions arrived only twenty days before the eclipse; they did not arrive in Peru until after the event.[49] Over the next ten years, there were at least four other broadsheet instructions prepared by Velasco and sent to the Indies. In addition to the broadsheet for the eclipses of September 27, 1577, and September 16, 1578, there were instructions sent for eclipses on July 15, 1581, on June 19, 1582, on May 10 and November 17, 1584, and on May 12 and September 4, 1588.[50] After the 1577 announcement issued the same year as the eclipse, all others arrived in the Indies a year in advance.

The instructions first reassure the prospective observer that the procedures to record the eclipse are simple and the instrument is easy to make. The document continues with step-by-step directions for building the instrument, verifying that it is plumb and square, and installing it correctly by making sure it is oriented true to the meridian. Figure 6.2 shows how the instrument Velasco referred to as the "instrument of the Indies" and used to "observe the quantity of the shadows" at the start and end of the eclipse would have looked. It consisted of a wooden board of one *vara* (approximately 84 cm) on the side, from which rose a perpendicular gnomon of one-third of a *vara* in length. With the base of the gnomon as the center, a semicircle one-third of a *vara* in radius was drawn on the face of the

48. IVDJ, Envío 100, f. 297, López de Velasco, Juan, "Carta a Mateo Vázquez sobre varias cédulas incluyendo una de Indias y la descripción de España," July 16, 1577, Madrid.

49. AGI, Mexico 69, R. 5, N. 83, "Carta del virrey Martín Enríquez al rey," October 19, 1577.

50. These are "Instrucción para la observación del eclipse de la luna . . ." dated July 15, 1581, and reproduced in its entirety in Joaquín Francisco Pacheco, Francisco de Cárdenas y Espejo, and Luis Torres de Mendoza, eds., *Colección de documentos inéditos relativos al descubrimiento, conquista y organización de las antiguas posesiones españolas de América y Oceanía, sacados de los archivos del reino, y muy especialmente del de Indias*, 42 vols. (Vaduz, Liechtenstein: Kraus Reprint, 1964), 18:129–36. "Instrucción para la observación del eclipse de luna," referring to the eclipse of June 19, 1582, reproduced in Carlos Sanz, *Relaciones geográficas de España y de Indias: Impresas y publicadas en el siglo XVI* (Madrid: Bibliotheca Americana Vetustissima, 1962), 7–8, and translated in Edwards, "Mapping by Questionnaire." The instruction for the 1584 events is reproduced in *Instrucción para la observación de los eclipses de luna*, 1584, ed. Biblioteca de Historiadores Mexicanos (Mexico: Vargas Rea, 1953) and cited in Medina, *Biblioteca hispanoamericana*, 450, and Pérez Pastor, *Bibliografía madrileña*, 1:100. I have been unable to locate a copy of the 1588 eclipse instructions.

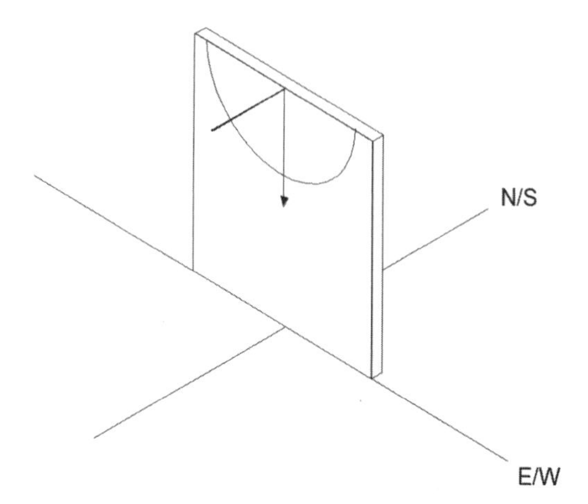

Figure 6.2. Instrument of the Indies per Velasco's 1577 instructions.

board. A plumb line was then hung from the base of the gnomon so that it hung freely and intersected the semicircle, indicating the instrument's vertical reference point.

The directions also instructed the observer to build on a level platform what amounted to a simple vertical sundial (figure 6.3). This setup let observers measure the length of the shadow cast by the sun at noon (relative to a center gnomon of known length). Although the document does not explicitly mention this, the purpose of this measurement was to determine latitude. (Observers could also use this observation to determine local noon and synchronize their clocks.) By following the sun's shadow cast by the center gnomon on the sundial from dawn to dusk and marking where it intercepted the two circles drawn on its surface, the observers could determine the lines representing the four cardinal points. The east-west line was later used to guide the placement of the instrument of the Indies.

The eclipse instructions continued with guidelines for constructing the instrument of the Indies to measure the moon's shadow at the beginning and at the end of an eclipse. Once built, the instrument was to be carefully placed along the E-W line on the sundial, with the gnomon facing in the direction opposite to where the sundial had cast its shadow at noon. This would ensure that the face of the instrument pointed toward the half of the celestial orb where the moon would be later that night. The evening of the eclipse, the instructions advised, a number of people should be invited to witness the eclipse and should all agree on "whether the moon rises perfectly round as it will if it were not eclipsed during

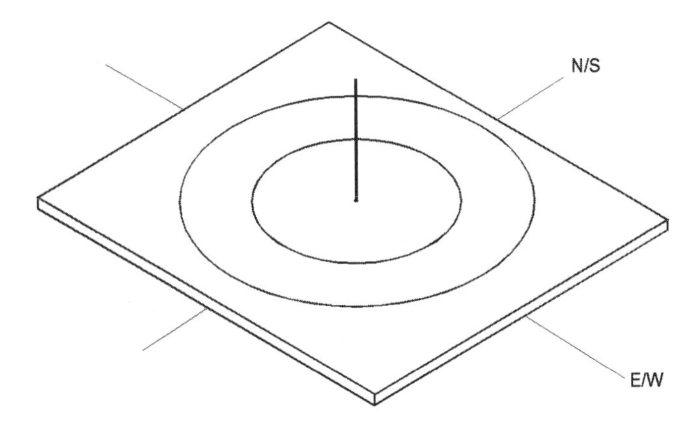

Figure 6.3. Instrument to observe the sun's meridian transit to determine latitude.

rising or whether it appears lopsided on some part of its roundness, or totally obscured."[51] If it rose completely round, the observers were to continue with the measurements. When at some point during the night the moon began to be obscured by the earth's shadow, the observers were to mark the spot on the semicircle drawn on the instrument's face where the shadow of the gnomon cast by the moon fell. They were instructed to do likewise when the eclipse ended and the moon regained its perfectly round shape.

After the eclipse, observers had to copy these marks onto a large sheet of paper made from four sheets (*pliegos*) joined at the edges. They had to submit two sets of measurements, one showing the length of the sun's shadow at noon relative to the length of the center gnomon and another replicating the measurements made of the moon's shadow at the start and end of the eclipse. Observers were further instructed to notarize and make duplicates of the observation, and both copies were to be sent to "His Royal Majesty at the Council of Indies." If because of cloudiness or obstruction the lunar eclipse was not visible, the responsible parties still had to measure the length of the shadow the sun cast at noon and date the observation.

The different versions of the printed instructions have subtle but telling variations. For example, in the 1577/78 version Velasco advised that in order to ensure that the instrument's center gnomon was perfectly vertical the instrument maker should hang a small weight from the top of the gnomon in the manner of a plumb line. In subsequent instructions,

51. Edwards, "Mapping by Questionnaire," 20.

Velasco recommended instead that a compass be used to make sure the top of the stile was equidistant from one of the circles drawn on the plane of the instrument—a far more accurate method for installing the stile perpendicular to the face of the instrument. Perhaps because the eclipse would take place near dusk in some parts of the Indies and the instrument might not cast a shadow, the instructions of 1577/78 also asked observers to have on hand a "geared clock" (*reloj de ruedas*) or, if one was not available, a sand hourglass to time the event. If no clocks were available, Velasco asked that time be estimated "more or less, according to the opinion and judgment of the observers."[52]

The instructions for 1581 described a different instrument of the Indies; this time it was a two-faced instrument, one with gnomons on both faces. Because of the instrument's redesign, Velasco removed from the instruction references to placing the instrument facing away from the direction where the shadow fell at noon. This 1581 instrument, although more complex to build, would be impossible to place facing the "wrong way," that is, away from where the moon could cast a shadow on its face. In all instructions subsequent to 1577, Velasco also advised that the observation reports be written on paper rather than on parchment. (Parchment shrinks if exposed to humidity.) The other instructions were changed to indicate that rather than sending the observations to the "persons in the government who sent you these instructions" (the local *audiencia* or *gobernación*), as the 1577/78 instruction had read, observers were now to send them directly to the king in care of the Council of Indies.

None of the versions of the instructions gave any indication of how the observer could use their observations to calculate their latitude or longitude. Clearly, the intention was for the mathematical computations to be done by cosmographers back in Spain—perhaps an unsatisfying outcome for a diligent observer in the Indies, but also a way to keep the resulting information secret. Sadly, we have no record of how Velasco manipulated these observations to yield latitude and longitude. From the instructions, however, it is clear that the first set of observations that recorded the length of the shadow cast by the sun at noon could be used to determine the latitude of the observer, with the help of solar declination tables. The set of observations taken using the instrument of the Indies recorded the angle made by the moon's shadow at the beginning and end of the eclipse relative to a vertical line that corresponded to the local meridian (figure 6.4).

52. "[A] poco mas o menos, según el parecer, y arbitrio de los que lo miraren." BN, MS 3035, f. 40v.

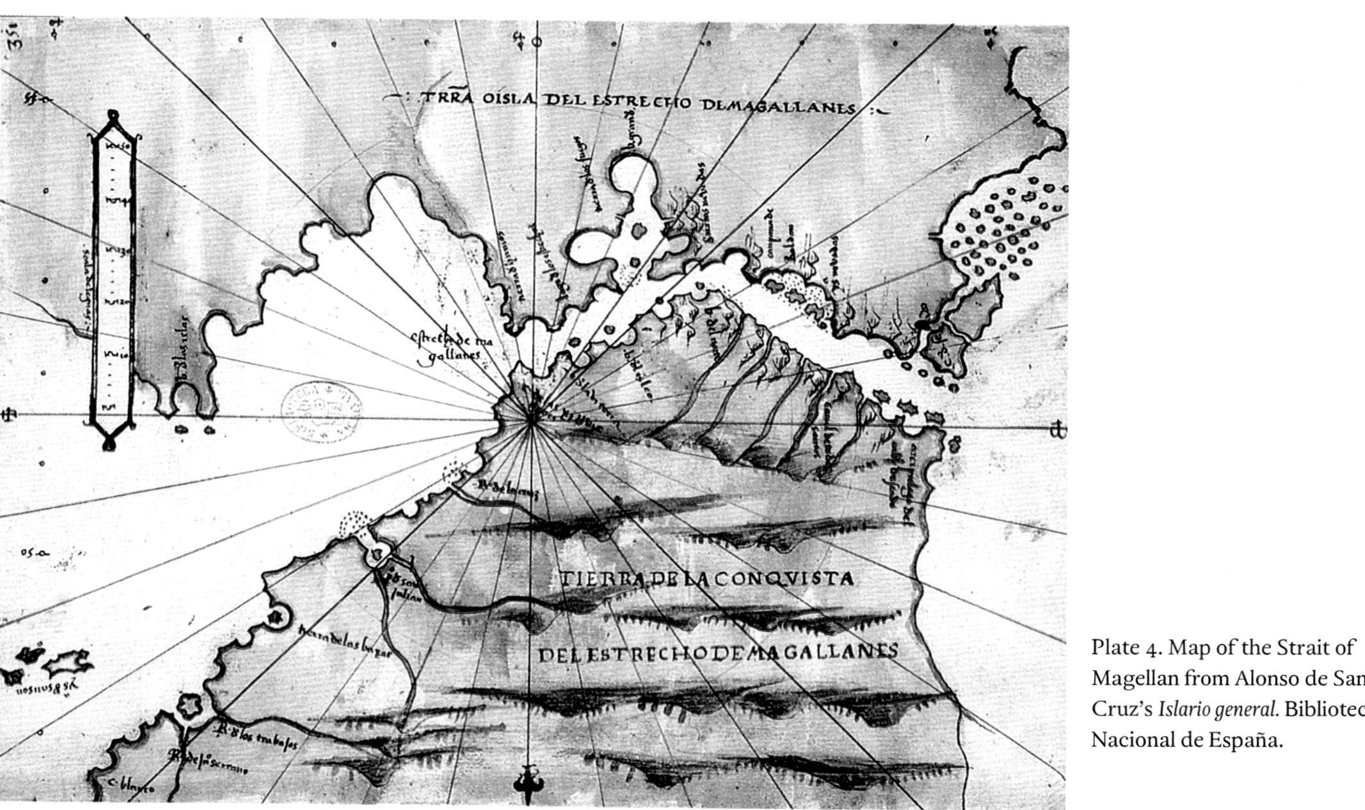

Plate 4. Map of the Strait of Magellan from Alonso de Santa Cruz's *Islario general*. Biblioteca Nacional de España.

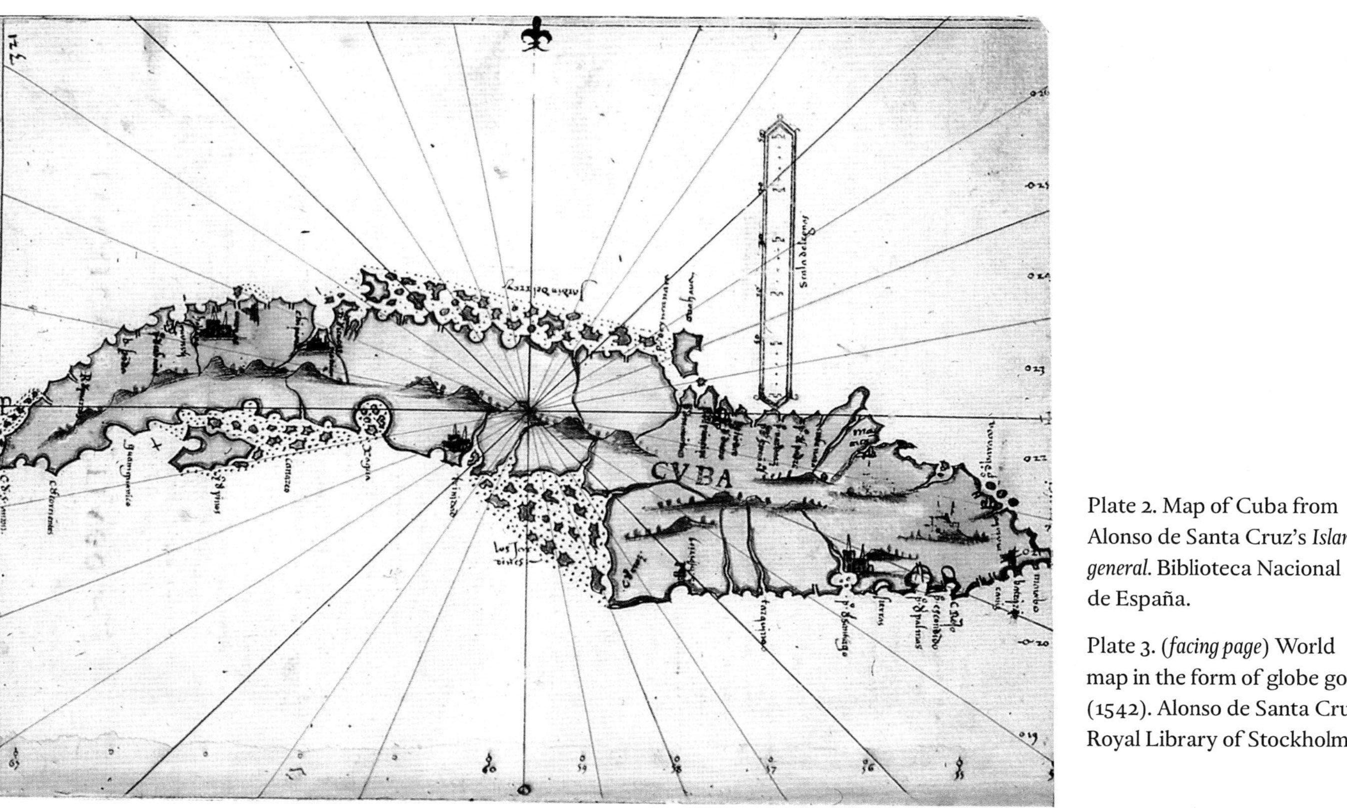

Plate 2. Map of Cuba from Alonso de Santa Cruz's *Islario general*. Biblioteca Nacional de España.

Plate 3. (*facing page*) World map in the form of globe gores (1542). Alonso de Santa Cruz. Royal Library of Stockholm.

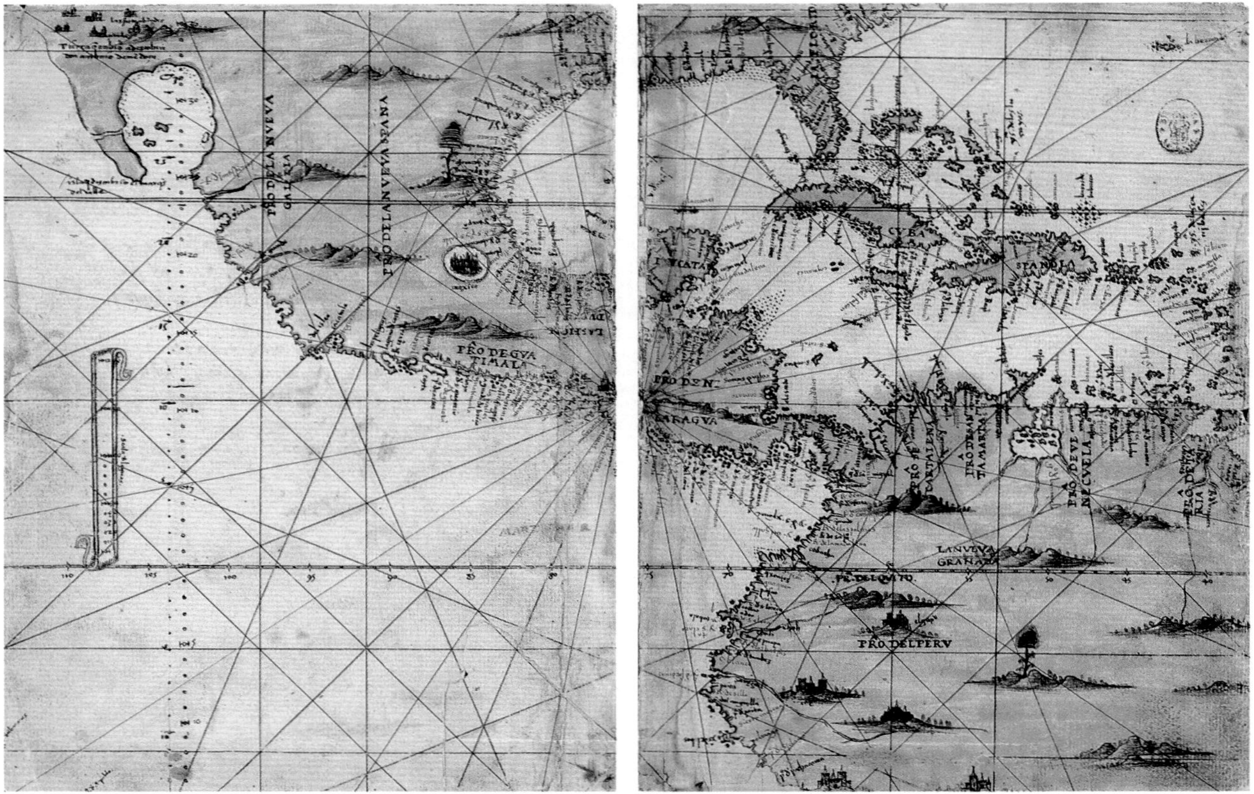

Plate 1. Map of West Indies from Alonso de Santa Cruz's *Islario general*. Biblioteca Nacional de España.

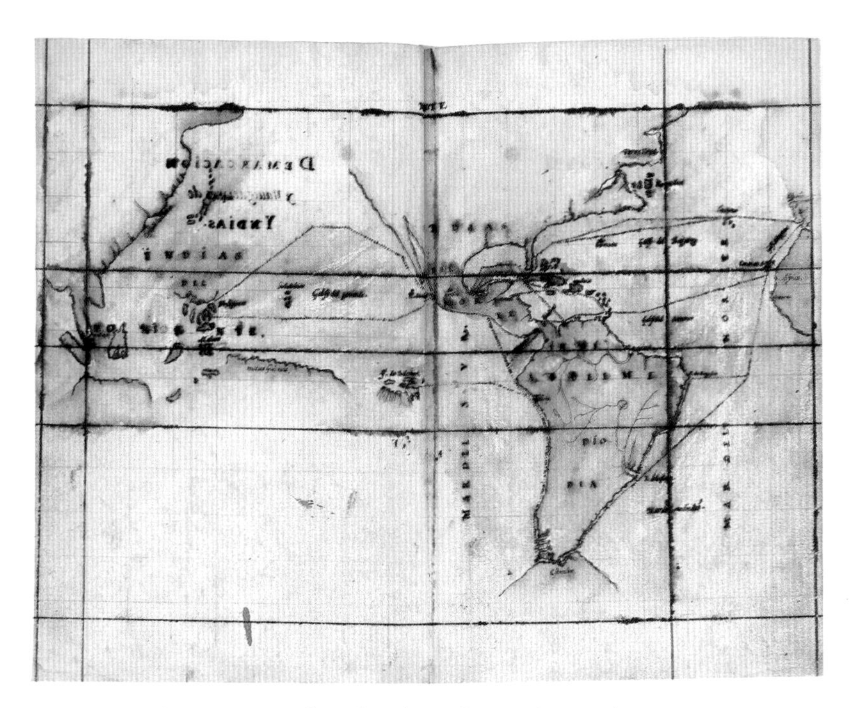

Plate 5. *Carta de marear*. Juan López de Velasco, *Sumario* (ca. 1580). The John Carter Brown Library at Brown University.

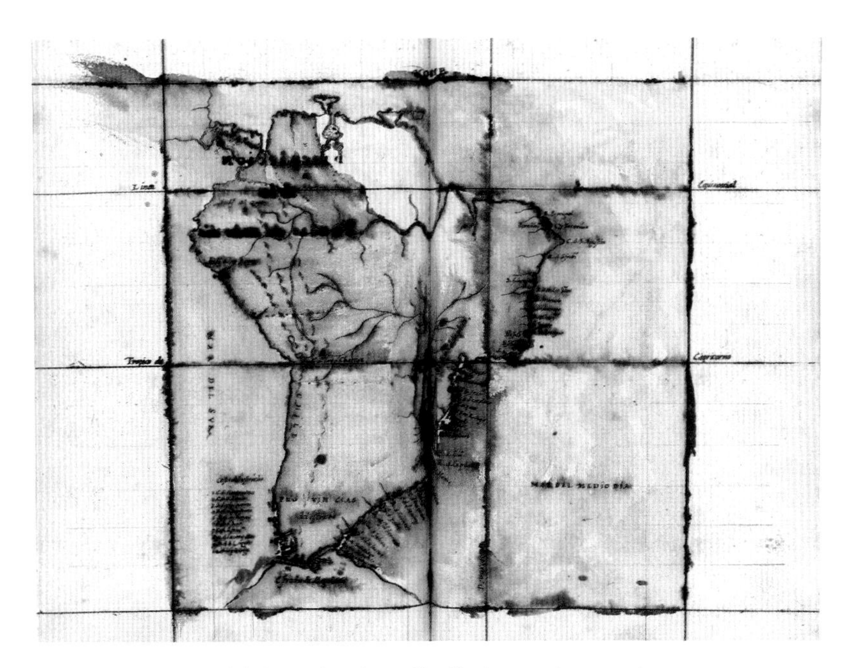

Plate 6. Tabla general de las Indias de medio día. *Sumario* (ca. 1580). The John Carter Brown Library at Brown University.

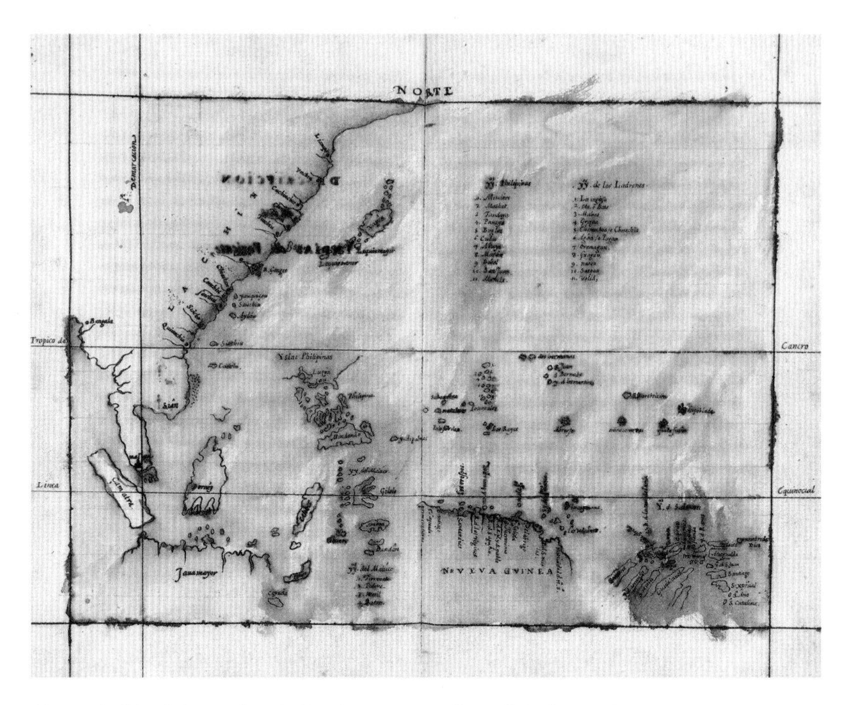

Plate 7. Tabla de las Indias de Poniente. Juan López de Velasco, *Sumario* (ca. 1580). The John Carter Brown Library at Brown University.

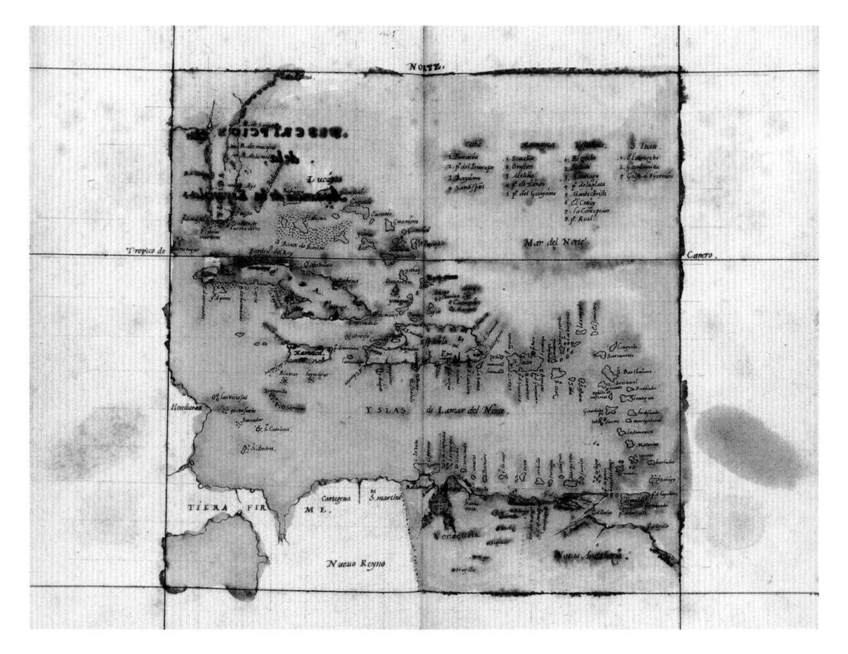

Plate 8. Tabla de la Audiencia de la Española. Juan López de Velasco, *Sumario* (ca. 1580). The John Carter Brown Library at Brown University.

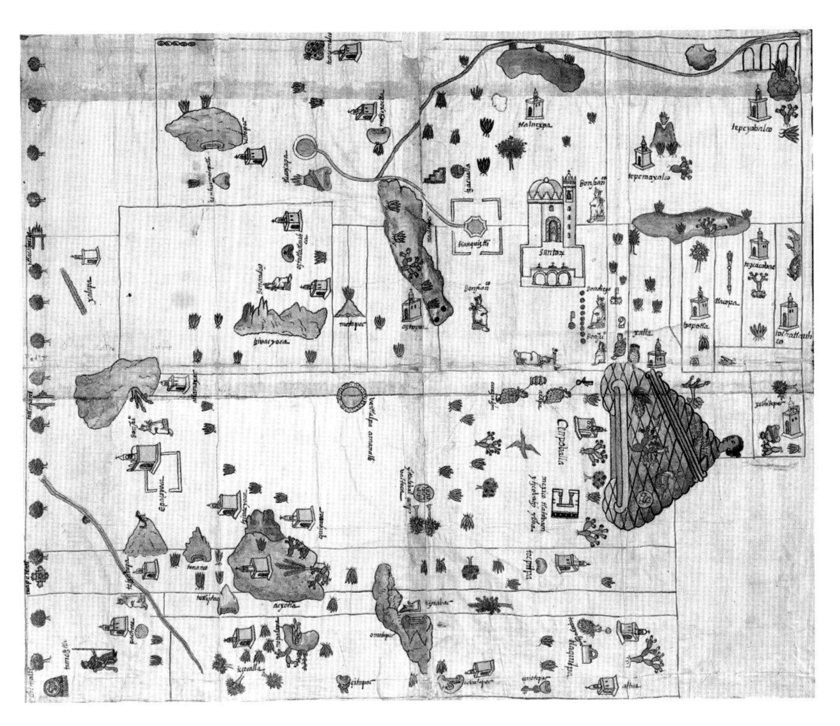

Plate 9. Map of Cempoala. Nettie Lee Benson Latin American Collection, University of Texas Libraries, The University of Texas at Austin.

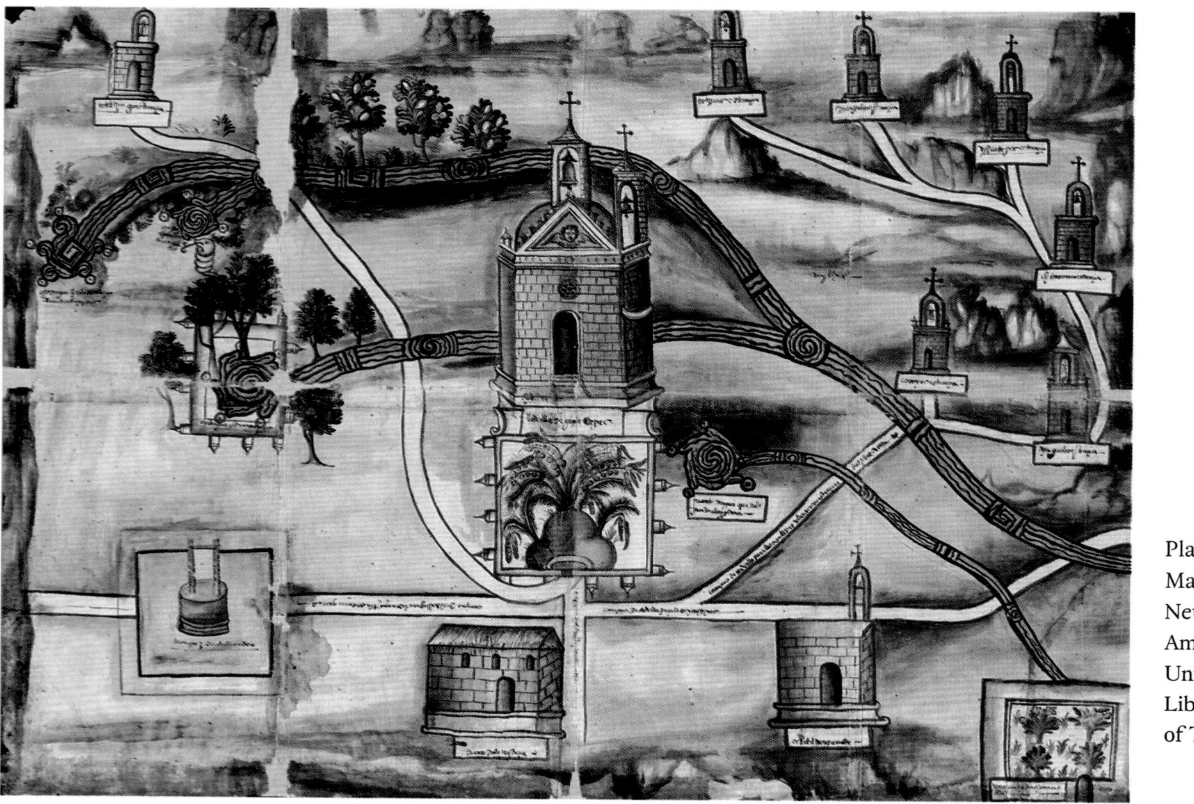

Plate 10.
Map of Guaxtepec.
Nettie Lee Benson Latin American Collection, University of Texas Libraries, The University of Texas at Austin.

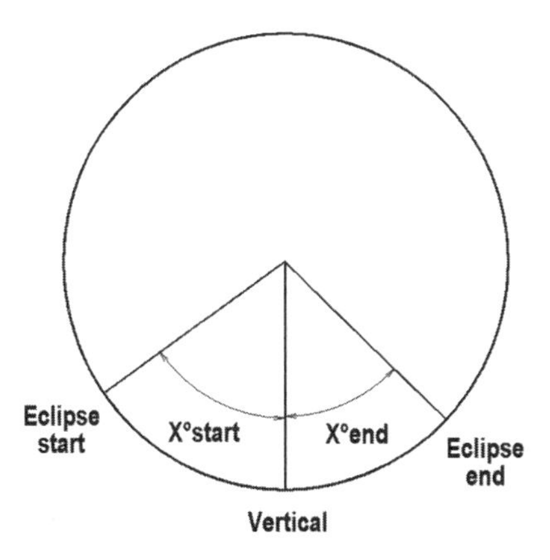

Figure 6.4. Marks on the face of the Instrument of the Indies as determined by the shadow cast by the gnomon at the start and end of the lunar eclipse.

The resulting angle or angles (X°), if the beginning and end of the eclipse were visible, were proportional to the moon's altitude and azimuth.

The instrument's design was based on the premise that because during a lunar eclipse the sun and moon are in opposition, the moon's right ascension has a twelve-hour difference from that of the sun and the moon's declination is the same as that of the ecliptic plus 180°. This sidereal coincidence solved a vexing problem for sixteenth-century astronomers. Both Ptolemaic and Copernican models of lunar motion were recognized as being notoriously unreliable, and thus right ascension and declination values for the moon were equally unreliable. The sun's movement, on the other hand, had been well understood since antiquity. Therefore, a lunar eclipse offered a rare instance when the celestial position of the moon was technically known, as it was 180° away from the sun.

Observers in Spain would know the local time during the event and measure the moon's position using this instrument or perhaps an astrolabe. By comparing the observations taken in Spain with those reported from the Indies and taking into account local latitude—and after some complicated mathematical manipulations—cosmographers could determine the local time of the eclipse in the Indies and thus the longitude.[53]

53. The relationship between the lines drawn on the face of the instrument that correspond to the moon's shadow at the start and end of the eclipse, and the moon's altitude and azimuth relative to the observer's horizon, can be expressed as Tan (X) = Sin (azimuth) / Tan (altitude). I discuss these computations

Although elegant in its simplicity, the method Velasco proposed had some inherent inaccuracies. Determining the precise start and the end of the eclipse is very difficult with the naked eye, since the penumbral shadow that surrounds the total phase can distort the observation. Velasco tried to overcome this by recommending that several people witness the eclipse and concur on when it began and ended. Natural phenomena can affect to varying degrees the accuracy of the observation as well. Distortion caused by atmospheric refraction, however, would have been minimal (~0.1°) unless the eclipse took place when the moon was very near the horizon. Lunar parallax, on the other hand, introduces an error that can be on the order of about 1° on the observed position of the moon, as Jerónimo Muñoz reminded his students. Finally, solar declination tables of the era, although more accurate than positional tables for the moon, were still in need of some refinement.

Not all the lunar eclipses for which instructions were issued turned out to be visible from America. Table 6.1 lists all the lunar eclipses occurring between 1577 to 1588 and shows the moon's elevation or declination relative to the horizon for different cities. Only the ends of the two eclipses of 1577 were visible from Mexico, as was the case again on November 17, 1584. The end of the eclipse was not visible in Spain in the case of the July 16 eclipse in 1581.

On two other occasions the printed instructions indicated incorrect dates for the lunar eclipse. Lunar eclipses were predicted for June 19, 1582, and again for May 10, 1584, but on both occasions, rather than a lunar eclipse there were solar eclipses instead. How did Velasco come to make such glaring, and no doubt embarrassing, mistakes? His papers at El Escorial hold some clues. Among them is a set of astronomical notes on lunar eclipses taken from Cyprianus Leovitius's *Eclipsium omnium ab anno Domini 1554 usque in annum Domini 1606*.[54] The notes are copies of all eclipses Leovitius predicted between 1581 and 1598, although Leovitius's drawings showing the degree of occultation are not included. The copyist (possibly Velasco) noted the

---

and how these observations yield lunar elevations and longitude in a forthcoming article in the *Journal of the History of Astronomy* (2009).

54. BME, K-III-8, f. 311v–318v, "Copias del Eclipsium omnium ab año Domini 1554 usque in annum 1606 de Cypriani Leovitii," ca. 1581. The impressive book Velasco used is still in the library of El Escorial. Cyprianus Leovitius, *Eclipsium omnium ab anno Domini 1554 usque in annum Domini 1606: Accurata descriptio & pictura, ad meridianum Angustanmum ita supputata, ut quibus aliis facillimè accommodari possit, una cum explicatione effectuum tam generalium quàm particularium pro cuiusque genesi* (Augsburg: Philippus Ulhardus, 1556).

Table 6.1. Altitude of the moon during the lunar eclipse for a given terrestrial latitude and longitude.

							Madrid 40° 23′N, 3° 41′W		Seville 37° 22′N, 5° 59′W		San Juan 18° 27′N, 66° 7′W		Panama 8° 48′S, 79° 3′W		Lima 13° 43′′N, 75° 57′′W		Mexico City 19° 24′N, 99° 12′W		
	UT						Lunar	Alt.	Lunar	Alt.	Lunar	Alt.	Lunar	Alt.	Lunar	Alt.	Lunar	Alt.	
Date			Type	Tot.*	RA*	Dec*	Start	End	Start	End	Start	End	Start	End	Start	End	Start	End	
1577	Apr	2	T+	20:05	13.4	-9.00	(4.90)	30.60	(4.75)	32.63	(60.54)	(10.57)	(74.34)	(22.09)	(61.58)	(15.41)	(78.80)	(41.82)	
1577	Sep	27	T-	0:13	0.82	5.50	47.67	45.25	49.70	47.46	2.53	56.96	(11.04)	46.14	(9.94)	45.53	(27.98)	26.09	1
1578	Sep	16	P	0:19	0.12	1.70	50.37	48.36	53.21	51.17	17.68	42.45	5.35	31.52	7.69	33.39	(13.54)	11.41	2
1580	Jan	31	T	21:41	9.6	14.80	25.29	60.57	25.08	62.85	(31.89)	14.71	(47.67)	0.28	(52.48)	(2.12)	(52.63)	(15.05)	
1580	Jul	26	T	11:07	21	-17.30	(45.79)	(66.32)	(46.41)	(69.30)	9.23	(36.61)	24.55	(22.32)	28.32	(18.30)	35.80	(6.21)	
1581	Jan	19	T+	21:22	8.83	17.60	22.47	62.36	21.94	64.43	(34.87)	14.12	(50.68)	(0.64)	(58.10)	(4.06)	(51.88)	(15.08)	
1581	Jul	16	T	4:11	20.3	-19.20	23.24	(6.39)	25.89	(5.11)	42.16	47.79	39.21	61.28	51.07	78.46	17.04	49.14	3
1582	Jan	8	P	22:12	8.03	19.70	48.06	61.83	48.58	63.48	(7.42)	10.80	(22.88)	(4.24)	(27.56)	(8.41)	(33.51)	(17.67)	
1582	Jul	5	N	18:05	19.6	-20.50	N/D	N/D	N/D	N/D	N/D	N/D	N/D	N/D	N/D	N/D	N/D	N/D	4
1583	Jun	5	P	11:10	16.8	-23.20	(56.48)	(72.53)	(57.17)	(75.54)	(1.30)	(35.27)	14.19	(20.26)	19.85	(14.16)	25.09	(5.95)	
1584	May	24	T+	11:43	16.1	-20.90	(52.98)	(63.07)	(53.69)	(65.12)	1.97	(51.82)	17.38	(37.14)	22.28	(30.39)	28.59	(21.50)	5
1584	Nov	18	T+	0:02	3.57	19.10	58.90	58.57	60.32	60.42	6.92	56.52	(8.15)	42.12	(12.13)	35.48	(21.32)	25.81	6
1588	Mar	13	T-	2:21	11.6	3.20	52.74	27.72	55.77	29.03	29.64	73.37	17.58	69.63	19.16	67.38	(1.62)	47.86	
1588	Sep	5	T+	4:13	22.9	-7.00	33.54	(2.91)	35.97	(2.29)	51.66	54.26	45.26	69.88	51.32	73.75	23.38	62.55	

( ) indicates the moon was below the horizon. Generally an altitude of 10° or better is needed to observe an eclipse.

1. Only the end of the eclipse was visible in Mexico.
2. Only the end of the eclipse was visible in Mexico.
3. The end of the eclipse was not visible in Spain.
4. Instruction says June 19, but there was a solar eclipse on that day instead.
5. Instruction says May 10. There was a solar eclipse on that day.
6. Eclipse observed by Juan and Domínguez in Mexico.

*Data from Fred Espenak, NASA/Goddard Center

information for eclipses on 1581 (July 15), 1582 (June 19), and two eclipses for 1584 (April 29 and November 7),[55] as well as for two lunar eclipses in 1588 (March 2 and August 25), 1591 (December 19), and 1598 (February 10). On the page of Velasco's notes with details regarding a supposed lunar eclipse on May 10, 1584, the text is crossed out and a marginal annotation was made that reads, "this one of the Sun was sent by mistake."[56] Consulting a copy of Leovitius's book makes it evident that a careless copyist included two instances of solar eclipses in what was supposed to be a compilation of lunar eclipses and Velasco issued instructions based on the incorrect copy. This error does not reflect well on Velasco, since once the mistake of 1582 (June 19) was detected, the one of 1584 could have been avoided.

The printed instructions were sent to the Indies accompanied by royal orders, which not only urged officials to make the eclipse observations but also reminded them to comply with the request for geographical descriptions and replies to the questionnaires. The reminder to Panama for the first eclipse observations expressively conveyed the king's support for the project: "I order that you have the observations made in that city according to the notes and instructions that for that purpose we have sent, signed by Juan López de Velasco, our cosmographer and chronicler major of the Indies, or by other exact and precise means if there is a person in that city who knows them."[57]

Similar royal orders accompanied each new edition of the instructions. Again, there are subtle differences between the different editions of the royal orders that give some insight into the project's evolution. The 1577 royal order stated the objective of the observations: "so we can determine the distance between these our kingdoms and those parts, for the purpose of accurate navigation and as well as for other purposes concerning and necessary for the good government of [the Indies]."[58]

55. Since it was published in 1554, dates subsequent to 1583 appear in Leovitius's book according to the Julian calendar. The date of the eclipse in Leovitius shows "the night of," while my table 6.1 is based on Universal Time.

56. "Este del Sol se envió por yerro." BME, K-III-8, f. 313.

57. "Yo os mando que en esa ciudad hagáis hacer las observaciones, conforme a los apuntamientos e instrucciones que para ello mandamos enviar, firmado por Juan López de Velasco, Nuestro cosmógrafo y cronista mayor de las dichas Indias, y por otros medios exactos e precisos si hubiere algunas personas en esa ciudad que sepan de ellos." In DIA, 17:506: "Sobre el eclipse," May 25, 1577. Other reminders survive; see DIA, 18:127–28, "Recordatorio a La Plata sobre el eclipse en julio 1581," June 3, 1580, and for Yucatán in DIU, 11:3–4, May 20, 1580.

58. AGI, IG-427, L. 30, f. 278–79, "Real cédula a Martín Enríquez, virrey de Nueva España y presidente de la audiencia de México, mandándole que de acuerdo con las

By 1581 and 1582 the wording about "good government" of the Indies was dropped and replaced by a much more specific objective, "to know the true latitude of all Spanish towns in the *audiencia* and to determine with precision the longitude and distance between these kingdoms and said towns, which up to this time has not been made as it should, and to situate them properly in geographical charts, and to correct navigations and distances and itineraries, and for other purposes to our service."[59] By then Ovando had been dead for over five years, and memory of his comprehensive and ambitious project for *buen gobierno* had faded.

The royal order accompanying the 1584 instructions acknowledged receipt of prior eclipse observations and explained that based on these "it is very important for the description of the Indies" that the next set of eclipses be diligently observed. This time, however, the royal orders that went along with the printed instructions sent to all viceregal and provincial authorities were not identical.[60] The one addressed to the new viceroy of Peru, Martín Enríquez, the cooperative former viceroy of New Spain, acknowledged the care he had taken to ensure that the dependencies under his administration complied with the questionnaire—a little praise might encourage him to collect observations and questionnaire replies from Peru just as diligently![61]

Juan López de Velasco coordinated the last eclipse observations in 1588. The royal order issued on this occasion explained that previous eclipse observations had shown how important the reports were for the

---

instruciones enviadas observe el eclipse de luna que ocurrirá el 26 de septiembre de 1577 y el 15 de septiembre de el 1578," May 25, 1577. Similar orders were sent to Peru, Charcas, Quito, Nueva Granada, Tierra Firme, Santa Marta, Veragua, Tucumán, Gualsongo y Pacamoros, Islas Occidentales, and Cartagena de Indias.

59. AGI, IG-427, L. 30, f. 313v–314v, "Cédula a Martín Enríquez virrey del Peru para que envíen a todos los pueblos de dicho distrito unas instrucciones impresas para que anoten las observaciones astrológicas sobre un eclipse de luna que se producirá en julio de 1581," June 3, 1580.

60. AGI, IG-427, L. 30, f. 357v–358, "Real Cédula al presidente y oidores de la Audiencia de Panamá, mandándoles que distribuyan en el territorio de su jurisdicción unas instrucciones para que se realicen las observaciones de dos eclipses que tendrán lugar el 10 de mayo y el 17 de noviembre de 1584, y se encarguen luego de recogerlos y enviarlos junto con todos los papeles referentes a la historia de dicha provincia," August 16, 1583.

61. AGI, IG-427, L. 30, f. 357–357v, "Real Cédula a Martín Enríquez, virrey del Perú, mandándole que distribuya en las provincias de su jurisdicción unas instrucciones para que se realicen las observaciones de dos eclipses," August 2, 1583, Madrid.

"precise verification" of longitudes. It also noted that "this is the last task to be done for now to accomplish the intent, and without it the work done in the past would be lost and without benefit."[62] The outgoing cosmographer-chronicler of the Council of Indies hoped that one last set of observations would definitively establish the longitude of several dependencies that had either failed to report or been unable to do so. For as we will see, although the observations had yielded a significant number of replies and longitude coordinates, they did not turn out to be as definitive and unquestionable as Velasco had hoped.

The Lunar Eclipse Observations

The success of Velasco's project depended on the number and quality of replies he managed to collect. Were observations made according to Velasco's instructions? And if so, did Velasco receive them, and most important, were they used to correct longitudes in maps and descriptions of the Indies? In fact, the outcome of the project was not publicly known until 1606, when Velasco's successor as cosmographer major of the Council of Indies, Andrés García de Céspedes, discussed some of the observations in his *Regimiento de navegación e hydrografía.* Céspedes had a favorable opinion of the project, stating that the observations were diligently carried out as per the instructions and were undertaken by the most skilled men in each location.[63] Taking Céspedes's assessment of the project and collecting other observations scattered in a variety of sources and archives indeed reveals that the eclipse project succeeded far beyond what its historiography suggests.

Unfortunately, only a sketch of one of Velasco's astronomical observations survives. He never, to our knowledge, compiled or drew a map based on longitudes derived from the eclipse project. (The maps of the *Sumario,* the only cartography attributed to Velasco, do not reflect the results of lunar eclipse observations collected by the project.) Given the secrecy policy under which Velasco operated this is not surprising, nor would one expect to find a discussion of the project in material published during the reign of Philip II. Most of the information we have about the project's outcome

62. AGI, IG-427, L. 30, f. 374v–375v, "Observación de eclipse: Instruciones al Conde de Villar, virrey del Perú," June 1, 1587.

63. "[L]o cual todo se cumplió como iba la instrucción, y se hicieron las observaciones por los mas diestros que en cada lugar había y se enviaron al Consejo de Indias." Andrés García de Céspedes, *Regimiento de navegación e hydrografía* (Madrid: Casa de Juan de la Cuesta, 1606), 140v.

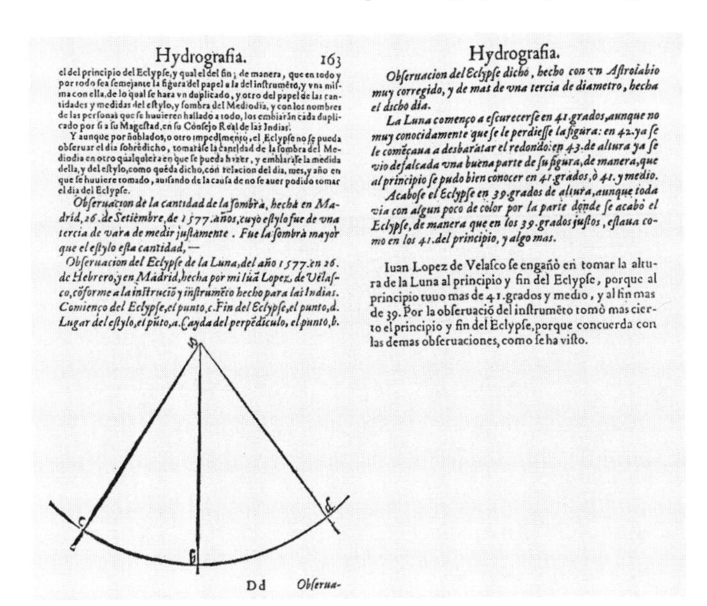

Hydrografia. 163

el del principio del Eclypse, y qual el del fin; de manera, que en todo y por todo sea semejante la figura del papel a la del instrumẽto, y vna misma con ella, de lo qual se hara vn duplicado, y otro del papel de las cantidades y medidas del estylo, y sombra del Mediodia, y con los nombres de las personas que se huuieren hallado a todo, los embiarán cada duplicado por si a su Magestad, en su Consejo Real de las Indias.

Y aunque por ñublados, o otro impedimento, el Eclypse no se pueda obseruar el dia sobredicho, tomarase la cantidad de la sombra del Mediodia en otro qualquiera en que se pueda hazer, y embiarase la medida della, y del estylo, como queda dicho, con relacion del dia, mes, y año en que se huuiere tomado, auisando de la causa de no se auer podido tomar el dia del Eclypse.

Obseruacion de la cantidad de la sombra, hecha en Madrid, 26. de Setiembre, de 1577. años, cuyo estylo fue de vna tercia de vara de medir justamente. Fue la sombra mayor que el estylo esta cantidad,—

Obseruacion del Eclypse de la Luna, del año 1577. en 26. de Hebrero, y en Madrid, hecha por mi Iuã Lopez de Velasco, cõforme a la instruciõ y instrumẽto hecho para las Indias. Comienço del Eclypse, el punto, c. Fin del Eclypse, el punto, d. Lugar del estylo, el pũto, a. Cayda del perpẽdiculo, el punto, b.

Dd *Obserua-*

Hydrografia.

Obseruacion del Eclypse dicho, hecho con vn Astrolabio muy corregido, y de mas de vna tercia de diametro, hecha el dicho dia.

La Luna començo a escurecerse en 41. grados, aunque no muy conocidamente que se le perdiesse la figura: en 42. ya se le començaua a desbaratar el redondo: en 43. de altura ya se vio desfalcada vna buena parte de su figura, de manera, que al principio se pudo bien conocer en 41. grados, ò 41. y medio.

Acabose el Eclypse en 39. grados de altura, aunque toda via con algun poco de color por la parte donde se acabò el Eclypse, de manera que en los 39. grados justos, estaua como en los 41. del principio, y algo mas.

Iuan Lopez de Velasco se engañò en tomar la altura de la Luna al principio y fin del Eclypse, porque al principio tuuo mas de 41. grados y medio, y al fin mas de 39. Por la obseruaciõ del instrumẽto tomò mas cierto el principio y fin del Eclypse, porque concuerda con las demas obseruaciones, como se ha visto.

Figure 6.5. Velasco's eclipse observation of 1577, as published in Céspedes's *Hydrografía*.

comes from Céspedes, who practiced cosmography at the Council during a time that witnessed an erosion of the secrecy policies that had regulated the dissemination of geographical information during the latter half of the sixteenth century. But sadly, Céspedes discussed only a selected number of eclipse observations in his book, after stating tantalizingly that he had a greater number of observations at his disposal.

According to Céspedes's account, it appears Velasco personally observed the eclipses in Spain, using both the instrument of the Indies and an astrolabe. In his *Hydrografía*, Céspedes faithfully transcribed Velasco's now-lost personal observations of the eclipse of 1577 and reproduced in the book the tracings taken from the instrument of the Indies indicating the beginning and end of the eclipse (figure 6.5). The angles from the vertical for the line indicating the start (32°) and end of the eclipse (37.5°) correspond reasonably well with the predicted values of 31.3° and 36° respectively, given that the angles are measured from the published observation. Céspedes found these values acceptable. He noted, however, that Velasco also used an astrolabe to measure the altitude of the moon at the start (41°) and end of the eclipse (39°) and wrote that Velasco "fooled himself" by trusting these values. (This was indeed so, since the moon's altitude at the start was 47½° and at the end 45¼°.) Céspedes reported that Velasco determined the end of the eclipse took place at 2:16, twenty-three minutes too late.

To corroborate eclipse observations taken overseas with those taken in Spain, Velasco recruited a number of other cosmographers throughout the Iberian Peninsula to contribute observations, although it appears they did not use the instrument of the Indies. Céspedes published their observations as well as those made in Mexico for the 1577 eclipse and one for the 1578 eclipse.

Overseas, the viceroy of New Spain, Martín Enríquez, described the circumstances surrounding the two 1577 observations from Mexico:

> I took care of what had to be done right away and I sent [the instructions] to Ciudad de los Angeles, Guadalajara, Michoacán, and Veracruz, because there was no time to send word to Guajaca. And because this city [Mexico] is surrounded by mountains, I sent observers to a hill ten leagues away for there was no other convenient place, [and] just as I feared would happen, there was a great cloud cover and darkness and [the eclipse] could not be seen. What was done in Ciudad de los Angeles is being sent.[64]

Table 6.2 shows the times of the end of the eclipse as reported from Mexico and Spain. When compared with the end time of the eclipse predicted by mathematical models using historical eclipse data, the times reported by Mexican observers are remarkably close. Such was not the case, however, with end times reported by expert observers in Spain. We may presuppose that the observers, both in Mexico and in Spain, used some type of timekeeping device to determine the eclipse end times. The reason for the accuracy of the Mexican times could be that the 1577 and 1578 lunar eclipses ended in Mexico soon after sunset—the point when a clock would have likely been started and thus would have accrued less error. In Spain, the eclipse ended much later in the evening, and thus the clocks would have accrued more error. In this case, the error introduced in the resulting longitude computation resulted from the observations in Spain rather than those in Mexico.

The 1577 observations yielded a distance from Toledo to Veracruz of 95.5° and from Toledo to Puebla of 99°, or within 3–4° of the modern-day

64. "Lo que había de hacer aquí previne luego y despaché a la Ciudad de los Angeles y a Guadalajara y a Michoacán y a la Veracruz, porque para Guajaca ya no había tiempo. Y por estar esta ciudad metida entre tierras envié diez leguas para que los que habían de hacer lo que se les demanda estuvieren sobre una sierra por no haber otro lugar conveniente y sucedió lo que yo me temía que fue muy gran nublado y oscuridad de manera que no se pudo ver. Lo que se hizo en la Ciudad de los Angeles envío hay." AGI, Mexico-69, R. 5, N. 83, "Carta del virrey Martín Enríquez al rey," October 19, 1577.

Table 6.2. Lunar eclipse observations 1577–78: eclipse end times as noted in Andrés García de Céspedes's *Hydrografía* versus theoretical end times

City (modern longitude) Observer	Sept. 27, 1577 Observed end time of eclipse	*(Predicted local time at the end of the eclipse: UT 2:09)**	Error	Sept. 15, 1578 Observed end time of eclipse	*(Predicted local time at the end of the eclipse: UT 01:12)**	Error
Madrid (40° 23′N, 3° 41′W) Juan López de Velasco	02:16	(01:53)	23 min.			
Toledo (39° 52′N, 4° 1′W) Juanelo Alcántara	02:12	(01:52)	20 min.	01:20 (Obs. by Velasco)	(00:56)	24 min.
Valladolid (41° 38′N, 4° 43′W) Doctor Sobrino	02:08	(01:49)	19 min.			
Seville (37° 22′N, 5° 59′W) Rodrigo Zamorano	02:04	(01:45)	19 min.	01:00 (per Gamboa letter)	(00:48)	12 min.
Ciudad de Los Angeles Puebla (19° 24′N, 98° 10′W)	19:36	(19:36)	0 min.	18:46	(18:39)	7 min.
San Juan de Ulua (19° 11′N, 96° 7′W)	19:50	(19:51)	1 min.			

*Historical eclipse data and model from Fred Espenak, NASA/Goddard

values. Adjusting these values using the distances that lay between Puebla, Veracruz, and Mexico City according to Velasco would have yielded a longitude between Mexico City and Toledo in the range of 98.5° to 100°.[65] Yet

65. "Puebla de los Angeles . . . is at 18½° latitude, twenty-two leagues east, somewhat south, of Mexico and forty-four from Veracruz." Later in the *Geografía*, Velasco adds that Veracruz was located at "18½° altitude, 60 or 70 leagues from Mexico." Converting

as of 1580 per the *Sumario*, Velasco continued to maintain that the distance between Mexico City and Toledo was 103°. One must wonder why neither Velasco nor Gesio in his castigation of the *Sumario* referred to the 1577–78 lunar eclipse observations for the longitude of Mexico City.

An answer may lie in a now-lost letter by Francisco Domínguez, the cosmographer assigned to the Francisco Hernández expedition, in which he declared that the lunar eclipse observations taken in 1577 and 1578 were useless and inaccurate. In a later letter he noted, "[I] also took care of Spain's business, which was verifying the lunar eclipses of the years 77 and 78, by order of Your Royal Highness, and to make all the models and their duplicates and make them such that they could be understood over there."[66] He added that he did this for Mexico City and that he "verified" observations made in other provinces. However, he had some misgivings as to their usefulness. He mentioned that along with the observations he sent the Council a letter noting the "variation" among the multiple observations and expressed in his opinion that the findings were uncertain and a waste of time. We know Gesio considered Domínguez a competent cosmographer. Perhaps Domínguez's negative assessment of the eclipse observations weighed heavily on Velasco's and Gesio's confidence in these results.

Rodrigo de Zamorano, the cosmographer at the Casa de la Contratación, also observed and recorded the 1578 lunar eclipse at the request of the Council of Indies.[67] We know from his collaboration with Pedro Sarmiento de Gamboa that Zamorano witnessed the end of the eclipse in Seville at 1:00 while Sarmiento saw the eclipse end in Lima at 20:04 ("ocho horasy un dieciseisavos de hora").[68] This difference in time would suggest a

---

these distances into degrees (at 20 l/°) would place Puebla 1° and Veracruz 3° to 3½° east of Mexico City. López de Velasco, *Geografía y descripción*, 107–9.

66. "[A]sí los negocios de España, que fue verificar los eclipses lunares del año 77 y 78, por mandato de V.R.M., y hacer todos los modelos y duplicados de ellos y ponerlos de suerte que allá se pudiesen entender, así los de esta ciudad como todos los demás que en otras provincias se verificaron, escribiendo y avisando a vuestro Real Consejo, allende de los trasuntos, la variedad y poca certidumbre que tenía semejante regulación, y que era tiempo perdido, y lo regulado sería incierto." AGI, P-261, R. 9, "Carta de servicio y relación de meritos de Francisco Domínguez," 1584.

67. AGI, P-262, R. 11, f. 1–3, "Rodrigo Zamorano cosmógrafo y catedrático de la Casa de la Contratación de Sevilla sobre se le acreciente el salario que tiene," May 14, 1582. Published in Esteban Piñeiro, "Cosmógrafos al servicio de Felipe II," 529–30.

68. AGI, P-33, N. 3, R. 27, Pedro Sarmiento de Gamboa, "Relación de lo sucedido a la Armada Real de S.M. en el viaje al Estrecho de Magallanes . . . el año 1581 hasta 1583." Published in Fernández de Navarrete, *Colección Fernández de Navarrete*, 20.1:212v–214v.

longitudinal distance of 4:56 or 74°, as they calculated. Modern astronomical models, however, show the end of the eclipse occurred at 20:08 in Lima and at 00:48 in Seville. Each observer thus erred only by a matter of minutes, but these compounded for a total error of sixteen minutes, which constitute 4° longitude. Given the difficulty of determining with the naked eye when the moon leaves the umbral shadow, their observations should be considered quite good.

Sarmiento and Zamorano's exercise ran into the problem that plagued longitudinal computations that relied on published eclipse predictions computed using either the Alphonsine tables (Ptolemaic) or newer Prutenic tables (Copernican). In their case, they noticed that Jerónimo de Chaves in his popular *Chronographía* predicted an end time for the lunar eclipse in Seville of 1:24. Had they used Chaves's value they would have situated Lima 10° farther west. Sarmiento articulated the very purpose and design of the eclipse project when he explained why they chose Zamorano's observation instead: "Although Chaves in his calendar has put down the end of the eclipse at one hour and twenty-four minutes of an hour, since science and experience together when they agree are two irrefutable witnesses, we will go with Zamorano, who is learned and experienced it, while Chaves did not observe but rather calculated it."[69] *Witnesses, irrefutable, science, experience*—these are words that would have rung sweetly in Velasco's ears.

Another observer recording the lunar eclipse of 1577 was the renowned professor of astronomy from the Universities of Valencia and Salamanca, Jerónimo Muñoz. That same year he published a small book titled *Summa del prognostico del cometa de la ecclipse de la Luna que fue a los 26 de septiembre del año 1577 a las 12 horas 11 minutos qual cometa ha sido causado por la dicha ecclipse.*[70] It is principally an astrological prediction associated with the comet of 1577, which Muñoz thought had been caused by the eclipse. The book began with a stark remark: "At the request of Juan López de Velasco, chronicler major of the Indies of the Majesty King Philip our lord, having made the observation of the time of the eclipse I also made the following prognostication of it according to Ptolemy's doctrine."[71] He indicated the time

69. "Aunque Chaves en su repertorio pone acabar a la una hora y veinte y cuatro minutos de hora; mas como la ciencia y la experiencia juntos cuando concuerdan son dos testigos irrefutables habemos de ir con Zamorano que es docto y lo experimento y Chaves no lo observo aunque lo calculó." AGI, P-33, N. 3, R. 27, f. 5r–6v.

70. Reproduced in facsimile in the appendix to Jerónimo Muñoz, *Libro del nuevo cometa*, ed. Víctor Navarro Brotóns (Valencia: Valencia Cultural, 1981).

71. "A instancia del señor Juan López de Velasco cronista mayor de las Indias de la Majestad del Rey Philippe nuestro Señor hecha la observación de la hora de la

of the eclipse only in the title and gave no positional data for the moon, nor did he clarify whether this time was for the beginning, total, or ending phase of the lunar eclipse—making the data useless for correlating to the results from other observations. He referred to Velasco as chronicler but not as cosmographer, perhaps a reflection of his poor opinion of Velasco as an astronomer.[72] If we assume the time Muñoz stated in the title is for totality and that he observed the eclipse from Valencia (39°28′N, 0°23′W), his reported end time of 00:11 compares well with 23:50, the time predicted when the lunar eclipse reached totality in Valencia. His twenty-one-minute error is in line with reports from other Iberian observers but, like theirs, resulted in a longitudinal error of between 5° and 6°.

Not all local officials were as diligent as Zamorano and the Mexican viceroy in procuring observations, nor as recalcitrant as Muñoz. In Nueva Salamanca de la Ramada (Colombia), the city councilmen explained, "We are not sending the observation, because when the eclipse took place there was nobody here to tend to it. [W]e are keeping the model here so that if there is another eclipse and someone to observe it, it will be done and sent."[73]

In contrast, there is an interesting record of an eclipse observation made in Panama City in 1581 by a meticulous cosmographer, Alonso Palomares de Vargas, who observed the event with the president of the Audiencia de Tierra Firme, Licenciado Juan López de Cepeda.[74] They sent a detailed memorandum describing the instruments and method that must have at one time also included now-lost drawings showing the moon's shadow at the start and end of the event. Their careful description of how they built

Eclipse hice el siguiente prognóstico de ella según la doctrina de Ptolomeo" (ibid., appendix 2).

72. Navarro Brotóns has studied Muñoz's work and his rocky relations with the court. The paucity of detail in Muñoz's account was likely his response to negative comments following the publication of his *Libro del nuevo cometa* about the nova of 1572. Muñoz complained that following his experience with the book about the "comet" of 1572, "in exchange for my realizations not only have they not thanked me, but I have been showered with lies by many theologians, philosophers, and courtiers of King Philip" (ibid., 108).

73. "[N]o va la observación porque no se pudo hacer porque al tiempo del eclipse no hubo aquí hombre que lo tendiese queda acá el molde para que si hubiera otro eclipse quien lo haga se hará y se enviará." AGI, IG-1528, N. 11, f. 1–18, "Descripción de Salamanca de la Ramada," April 24, 1578, Tenerife.

74. AGI, P-260, N. 1, R.3, f. 1–4, Alonso Palomares, "Descripción de un eclipse de luna, hecha en Panamá, por el cosmógrafo Alonso Palomares de Vargas," July 15, 1581. Published in Fernández de Navarrete, *Colección Fernández de Navarrete*, 27.1:89–94.

the necessary setups and instrument show they understood the purpose and methods discussed in Velasco's instructions very well. The observers explain, for example, that due to cloudiness in the days preceding the event, the latitude could not be taken using the sundial-like instrument as delineated in the instructions. Instead the observers used an astronomer's astrolabe to determine correctly the latitude of Panama City at 9°N (actual 8°58′N, 79°33′W). They noted that because the region had been plagued by bad weather, they decided to time the lunar eclipse using a half-hour hourglass that was started and carefully monitored beginning at noon the day the eclipse was to take place.

As they had feared, the evening of the eclipse was overcast, but at nine in the evening they caught a glimpse of the moon and it was still perfectly round. A little after ten they were able to see the moon and determined that it was somewhat eclipsed, but because of the cloudiness it was too faint to cast a shadow on the instrument. They recorded that the eclipse reached totality at 22:30 and the moon stayed eclipsed a few minutes more than an hour. The observers also noted that their hourglass was in effect running slow when they checked its time when the moon was directly overhead, as indicated by the position of the moon's shadow on the instrument of the Indies. (During a lunar eclipse, the moon's meridian transit would have cast a vertical shadow on the instrument of the Indies, since the sun and moon are in opposition when the moon transits at midnight.) Finally, they noted that the end of the eclipse occurred one hour and seven minutes after totality. All along they had been marking, whenever possible, the location of the moon's shadow on the face of the instrument.

When compared with predicted values, the observers' results determined using the hourglass were indeed quite good. They reported a local time at totality of 22:30 "or somewhat earlier." This compares favorably with a predicted time at totality of 22:48 (18 min. error). They also reported that the eclipse ended at 00:37 but noted, "The clock was running slow." Unfortunately, since the end of the eclipse was not visible from Spain, their efforts could not yield a definitive longitudinal distance based on witnessed observations. Had Velasco consulted Chaves's *Chronographía*, he would have found an end time for Seville of 05:36, ten minutes later than what historical data predicted.[75] He thus would have calculated a distance

75. Jerónimo de Chaves, *Chronographía o Repertorio de los tiempos, el más copioso y preciso que hasta ahora ha salido à luz* (Sevilla: En casa de Fernando Diaz en la calle de la Sierpe, 1581), f. 204v. Based on historical data, the predicted local time in Panama City for the end of the eclipse was 00:31 (6 min. error versus Palomares), and for Seville it was 5:26 (10 min. error versus Chaves).

between the two cities of 5:00 or 75°, when the actual distance is 73½°. In this case, as in the case of the Sarmiento-Zamorano observation, using published values based on faulty ephemeredes inserted significant error into the computation.

There are only two extant sets of drawings of lunar eclipse observations taken using the instrument of the Indies at the Archive of the Indies. One set is from Mexico for the event of November 1584 and has original observations from four witnesses, including Jaime Juan. The other set is a previously unpublished observation taken in Puerto Rico.[76] These are undated drawings made according to Velasco's instructions but without any documentation explaining the circumstances surrounding the observations. Archival records suggest the observation is from Puerto Rico, for either 1581 or 1588. We know both eclipses were observed with their reports sent to Spain. One clue is in the response to the 1577 questionnaire from Puerto Rico dated January 1, 1582. The answer to the sixth question reads, "The height and elevation of the town where the City of Puerto Rico [San Juan] will be known by the eclipse that I, Juan Ponce de León, by order of Captain Juan de Céspedes, Governor that he was of this island, took on the fifteenth day of July of last year, and which is being sent on this same ship to your Majesty."[77] Thus it is clear that the eclipse of 1581 was indeed observed and a report sent to Spain. Additional letters support this.[78] However, another letter, this time from 1588, suggests that the lunar eclipse of that year was also observed and recorded and the "best possible" results were sent

76. AGI, Mapas y Planos, Teóricos 1 and 2. "Observación astronómica de eclipse de luna hecha en Puerto Rico demostrada en círculos, c. 1600." The drawings were taken from AGI, P-175, R. 40.

77. "La altura y elevación del pueblo en que está la Cuidad de Puerto Rico se verá por el eclipse que yo Juan Ponce de León, por mandato del Capitán Juan de Céspedes, Gobernador que fue de esta isla, tomé a los quince de Julio del año pasado, el cual se envía en este propio navío a Su Majestad." AHN, Diversos-25, Doc. 53, "Descripción de la isla de P.R. hecha por el presbítero Juan Ponce de León y bachiller Antonio de Santa Clara, abogado por encargo del gobernadores Juan Melgarejo en cumplimiento de orden real," 1582. Published in Pacheco, Cárdenas y Espejo, and Torres de Mendoza, eds., DIA, 21:255. This Juan Ponce de León should not be confused with the conquistador and former governor of Puerto Rico who died many years earlier. Puerto Rico was the original name of the city of San Juan, Puerto Rico.

78. AGI, Santo Domingo, 155, R. 9, N. 65, Juan de Céspedes, "Carta del gobernador de Puerto Rico al rey," September 20, 1580. Also, AGI, Santo Domingo, 155, R. 10, N. 66, Juan Malgarejo, "Carta del gobernador de Puerto Rico al rey," February 3, 1582.

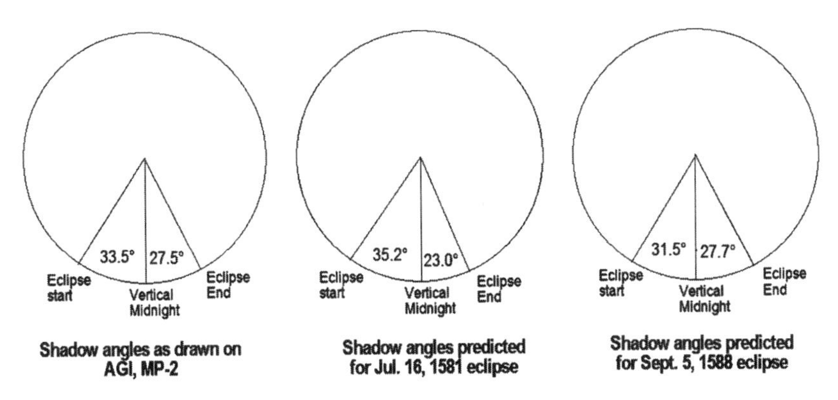

Figure 6.6. Comparison between recorded and predicted shadow angles, as noted on the face of the Instrument of the Indies for lunar eclipses observed from Puerto Rico.

to Spain.[79] The questions remain, do the drawings at the Archive of Indies correspond to the 1582 eclipse or the 1588 eclipse, and do they correspond to the observations made in Puerto Rico?

From the four sheets of paper carefully pasted together showing the tracing of the sundial-like instrument, we can surmise that the observation was taken at a place near the equator (the sun's shadow is short), but because the length of the gnomon was not noted, we have to assume the observers used, as per the directions, a gnomon $^1/_3$ of a *vara* (27.3cm) in length. The drawing suggests they did so and luckily includes a small notation that the sun's shadow was recorded on September 2. (This suggests the 1588 lunar eclipse rather than earlier ones.) Given a solar declination of 7° 40′ for that date, the drawing suggests the observers were near latitude 19°N. The latitude of San Juan, Puerto Rico, is 18°27′N.

From the drawing showing the shadows cast by the moon at the beginning and the end of the eclipse, we can surmise that the eclipse straddled the meridian in the location where the observations were recorded and was visible both at the start and at the end of the eclipse. Only the eclipses of July 16, 1581, and September 5, 1588, meet these conditions for Puerto Rico. Note in table 6.1 the similarity in the moon's elevation for the two events as seen from Puerto Rico. Based on the comparison shown in figure 6.6, I believe the drawings are indeed from Puerto Rico and for the lunar eclipse of September 5, 1588.

79. AGI, Santo Domingo, 155, R. 11, N. 118, Diego Menéndez, "Carta del gobernador de Puerto Rico al rey," October 7, 1588.

Observations also reached Spain for the eclipses of 1584, as a letter from Quito attests.[80] But other than the set of drawings from Puerto Rico, the only other surviving eclipse observations done according to the method outlined by Velasco correspond to the eclipse of November 18, 1584, in Mexico City. One of the observers was Jaime Juan, who, as discussed in chapter 2, was on an ambitious voyage of scientific exploration. Velasco, who had only a marginal role in planning the expedition, had asked Juan to measure lunar eclipses during the voyage. The evening of the eclipse a group of observers, in the presence of the viceroy of New Spain, Archbishop Pedro Moya de Contreras, and including Jaime Juan, gathered on the roof of the *casas reales* in the city of Mexico to observe the eclipse. Along with Juan's extensive report, we have recorded observations from cosmographer Francisco Domínguez, royal armorer Cristóbal Gudiel, and doctor Pedro Farfán, a local official.[81]

Domínguez remained in Mexico preparing a geographical description of the land and making a living selling instruments and local maps for the viceroy after Francisco Hernández returned to Spain. We know of Domínguez's activities after Hernández's departure from an interesting letter he sent the king in 1581 describing his service as cosmographer in New Spain, where he expressed skepticism about the eclipse project.[82] Despite his misgivings, Domínguez participated in the 1584 observation, and his results are among the four sets of drawings made according to Velasco's instructions (see figure 6.7). That year only the end of the lunar eclipse was visible from Mexico—the moon was only 25° 48′ above the horizon when the eclipse ended—and the observations clearly reflect this fact. The set of observations made by Juan is somewhat different from the others. Juan's drawings explain the instrument setup in far more detail, includes latitude calculations, and show he carefully measured the angle cast by the moon's shadow at the end of the eclipse using the graduated edge of an astrolabe

80. "[E]n cumplimiento de una cédula de V. Mg. que vino en pliego del año pasado de 84 se hicieron las diligencias para tomar las alturas y sombras del sol y luna al tiempo del eclipse de la luna que hubo el mes de noviembre. La partes donde se pudieron hacer y se envían los recaudos de ellos en pliego." AGI, Quito-8, R. 19, N. 50, f. 1–4, Pedro Venegas de Cañaveral, "Carta sobre diversos asuntos . . . se envía los datos tomados en el eclipse de luna del mes de noviembre," March 26, 1585, Quito.

81. AGI, IG-740, N. 103. Drawings were with AGI, P-183, N. 1, R. 13, and currently in Mapas y Planos Mexico-34. They were recently published with a comprehensive introductory study in Rodríguez-Sala, *Eclipse de luna*, 67–83.

82. Letter by Francisco Domínguez to Philip II, 1581. Printed in DIE, 1:379–84. For more on Domínguez, see Rodríguez-Sala, *Eclipse de luna*, 67–83.

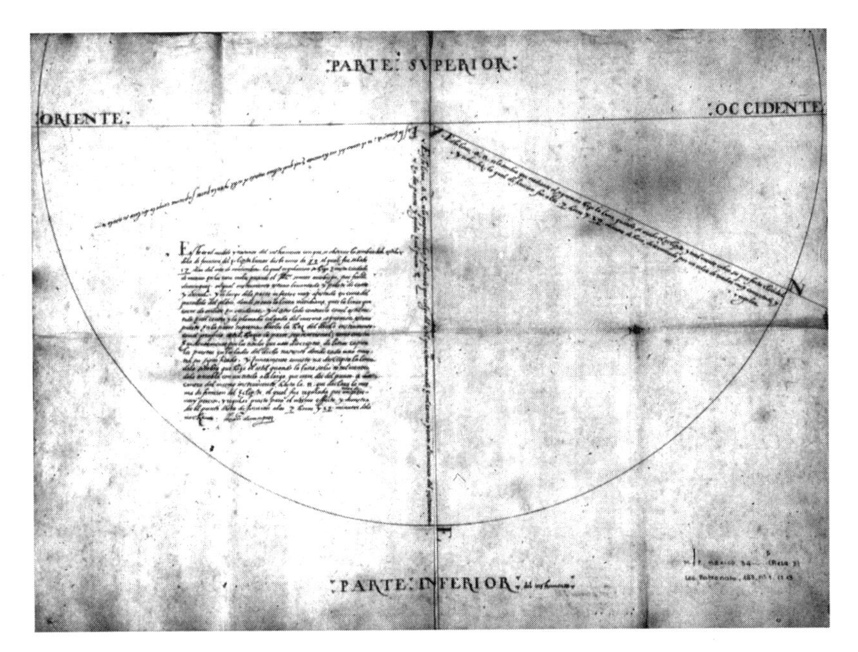

Figure 6.7. Francisco Domínguez's 1584 eclipse observation. Spain. Ministry of Culture. General Archive of Indies. MP-Mexico, 34.

as one would a protractor.[83] He understood that the angle of the shadow cast on the instrument at the start and end of the eclipse was meant to function as a time-recording device using the same principles as those governing a vertical south-facing sundial. Juan recorded on his drawing that at the end of the eclipse the shadow formed an angle with the vertical of X°= 64.25° (figure 6.8). A model of the moon's position computed using historical eclipse data indicate an angle from the vertical of 63.65°. Juan erred by only 0.6°; his was a diligent observation indeed, given that the eclipse took place so close to the horizon, and was in line with the accuracy of observations recorded using the instrument of the Indies.

In addition to dutifully recording the event according to Velasco's instructions, Juan discussed—this time in Latin—two other methods for determining the time of the eclipse's end. He observed the position of the moon at the end of the eclipse using an astrolabe, noting it was 27°44′ in Taurus. Using this information, he determined the local time of the event by observing the angle between the moon's position at the end of the eclipse

83. He placed Mexico City at latitude 19°13′N. The actual location is 19°24′N. He noted that the shadow formed a 25°45′ angle (25.75°) with the horizontal line, the line representing 18:00.

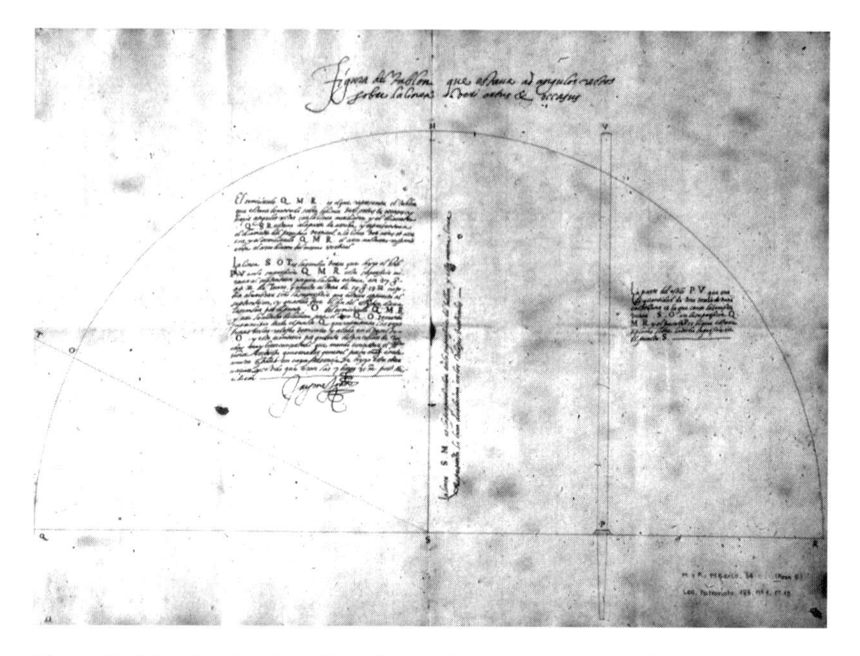

Figure 6.8. Jaime Juan's 1584 eclipse observation. Spain. Ministry of Culture. General Archive of Indies. MP-Mexico, 34.

and its position when it crossed the meridian.[84] This computation suggested that the end of the eclipse took place at 19:20, but Juan was concerned that variations in the moon's motion might have introduced some errors in his calculations.[85] As a third way of determining the end time of the eclipse, he calculated the moon's position using a fixed star as the sidereal point of reference—a method that he considered to rest on "very firm demonstrations discovered by very learned men" (perhaps a reference to Johannes Werner). He suggested using Betelgeuse, the "right shoulder of Orion," as his reference point. This method would correct for the effect of parallax on direct observations of the moon, particularly when the moon was near the horizon, as was the case in the 1584 eclipse. His calculations suggested an end time of the eclipse of 19:22. Both computations are within ten minutes of the end time of 19:12 calculated using historical models. The other observers also used a "geared clock, very accurate," provided

84. His second proposition suggests that he was familiar with Oronce Finés's method. Rodríguez-Sala, *Eclipse de Luna*, 165.

85. "Magnam enim habet propter varium lunae motum varietatem et non levem errandi suspicionem" (ibid., 166).

by the archbishop and reported observing the end of the eclipse between 19:27 and 19:31.

Juan then proceeded to use this value to determine the longitude between Mexico City and Seville. To carry out that computation, he had to know the end time of the event as seen from Seville, which, he explained, was taken from an ephemerides—probably Chaves's *Chronographía*[86]—although he does not mention which. Juan was aware that the end time from the ephemerides could be incorrect, since they in turn depended on the accuracy of the model of the moon used in the ephemerides. His fears proved correct. Until that point Juan's computations had been quite accurate, yet the ephemerides listed an end time of 02:30, more than an hour too late, which threw off his longitude computation by over 16°.

Unaware, although suspicious, that his longitude computations were incorrect, Juan concluded his report by calculating the distance along a great circle between Seville and Mexico, this time using spherical trigonometry. Weighing in on a debate that as of the close of the sixteenth century showed no sign of abating, Juan explained that for these types of calculations he preferred to use the value of 18 leagues per degree rather than the customary 17.5 l/° or even Apian's 20 l/°. This was because his teacher, Jerónimo Muñoz, had confirmed "after many experiments" that one degree equaled 18 leagues.[87] Just as his preceptor had done for the eclipse of 1577, along with his observations Juan included an astrological prognostication of the eclipse.

Eclipse observations continued per Velasco's instructions until 1588. The Philippines reported on a lunar eclipse of 1587,[88] and Panama and Quito sent observations for the 1588 eclipses.[89] Céspedes discussed in his *Regimiento de navegación* additional observations made from Puerto Viejo (Ecuador), Ciudad de los Reyes (Lima), and Arequipa (Peru) from the event of September 4, 1588.[90] The observations from Puerto Viejo were done

86. Chaves, *Chronographía*, f. 204v.

87. The original reads: "In hac computatione unicuique gradui circuli maiori 18 leucas tribuimus, quia experientia comprobavit caepissime institutor noster Geronymus Munnoz" (Rodríguez-Sala, *Eclipse de luna*, 161).

88. This eclipse was not visible from Spain. AGI, Filipinas-18A, R. 5, N. 31, f. 1–8, "Carta de Santiago de Vera sobre la situación en Filipinas . . . anuncia muerte de Jaime Juan," June 26, 1587.

89. "[A]sí mismo haber enviando las observaciones de los eclipses que allá se hicieron." AGI, Quito-209, L.1, f. 87v–91, "Respuesta de la audiencia de Quito," February 27, 1591.

90. García de Céspedes, *Regimiento de navegación*, 153v–154.

according to Velasco's instructions, and the event was observed by members of the judiciary and the military and "many other persons." Céspedes noted that he personally observed the eclipse while in Lisbon and corroborated his findings with those taken in the same city by Doctor Sobrino (Philip II's chaplain). These and the previously discussed eclipse observations became the basis for Céspedes's subsequent work revising the official rutter, the *padrón real*, and the nautical routes used by the ships sailing under the Spanish flag and chartered by the Casa de la Contratación. Chapter 7 considers the implications of this important project for cosmographical practice at the Council of Indies.

In his study of the 1584 eclipse observations, Jaime Juan left his candid impressions about the changes that were overtaking Renaissance cosmography and that by the end of the century would have significant effects on the practice of cosmography at the Council of Indies. He argued against using geographical descriptions of remote regions transmitted by "ignorant sailors." Sailors, Juan complained, based their descriptions on observations taken with defective instruments, and they had scant and poorly assimilated training in the cosmographical arts. Nonetheless, these sailors prided themselves in describing vast and distant territories. One could only hope these accounts depicted the region accurately, Juan continued. He regretted that since few learned persons had traveled to those remote regions there was little choice but to give credit to sailors' accounts.

Given this situation, even "very learned men" could do little else but base their calculations on longitude values derived from these "hopefully truthful" accounts. For Juan the consequences were clearly evident and troubling. Since no proper proofs were given to support geographical coordinates used in descriptions based on these pilot accounts, distant provinces could not be mapped, and where maps had been drawn these could not be considered credible. But for Juan the situation was not without hope. In a statement that can only be a reference to the goals of the expedition Juan de Herrera had sent him on, Juan noted that in order to remedy this situation—and because lunar eclipses were rare—royal astronomers had found other methods, "based on the most firm demonstrations," to determine longitude. These methods, he hoped, would be confirmed presently through experience (*las experiencias*). These experiences, he was confident, promised to yield "no small measure of benefit and utility" whether they were carried out by expert astronomers or by curious sailors.[91]

Jaime Juan's opinion was an indictment of the very methodology used by Velasco at the Council of Indies. The law required that the geographical descriptions Velasco composed be based on eyewitness testimonies, whether by sailors or other nonexperts. Of all the aspects of cosmographical practice defined by Ovando's statutes, only the observation of lunar eclipses yielded what Juan would have considered results based on "firm demonstrations," by which he meant mathematical proofs (such as the ones he had included with the eclipse observations). If we consider that Jaime Juan was articulating the opinion of his mentor Juan de Herrera, his comments significantly undermined the epistemological foundations of the two cosmographical projects currently under way at the Council of Indies: the questionnaires and the eclipse project.

When Juan's reports and comments reached Velasco, they must have been unsettling to the cosmographer, although perhaps not novel. Gesio had argued this point for years. But in this case the remarks carried extra weight: here was a surrogate of the king's right-hand man for scientific and technical projects advocating a purely mathematical approach to cosmography based on experiments conducted by experts using sophisticated instruments, an approach that invalidated twelve years of Velasco's work to describe the Indies.

The questionnaire and eclipse projects marked a midpoint between the Renaissance cosmographical opus and the mathematized cosmography that came to define how the discipline was practiced at the Council of Indies by the end of the sixteenth century. Velasco's preference for atomized facts was a methodological remedy to the problems of representation that had made the cosmographical opus obsolete. Atomization—then as now—was the most efficient way to process a vast amount of information. Nevertheless, in Velasco's eyes, the two projects had similar epistemic criteria. Whether a geographical description in the form of replies to specific questions or an astronomical observation, they constituted cosmographical facts because they were the notarized testimonies of eyewitness. Gesio and Jaime Juan, in contrast, advocated the

91. My analysis is based on a Spanish translation of the Latin original in Rodríguez-Sala, *Eclipse de luna*, 173.

mathematical method whose form of representation was mathematical cartography.

Velasco put projects in place that allowed him to conscientiously pursue the geographical, ethnographical, natural, and historical information he had the responsibility to "gather, compile, and put in good order." But instead of interpreting this directive—as he had in earlier days—as a request to compose a Renaissance cosmography, he came to interpret the duties of his post differently: "to gather" now meant "to ask," "to compile" now meant "to collect," and "to put in good order" meant "to file." The reasons that Velasco reinterpreted his cosmographical duties in this manner go far beyond issues of personal motivation and careerism. Ovando's ordinances, the legal instrument used to provide the Council of Indies with the information it needed to govern effectively, sowed the seeds for the bureaucratic approach Velasco would adopt. Its comprehensive yet strict definitions of what constituted cosmographical facts and its intrusion into the discipline's methodology shaped Velasco's understanding of what constituted cosmographical practice and would lead him away from the humanistic approaches characteristic of the cosmographical genre.

Additional pressure came in the form of the gradual but persistent divergence of the cosmographical discipline into two separate areas of specialization, one mathematical, the other descriptive. After 1580, the mathematical camp unofficially came under the direction of Juan de Herrera when he returned from Portugal convinced that the mathematical aspects of any science, and particularly cosmography, were of vital importance to the empire. His influential position at court likely contributed to a decision that severed the last ties that had bound cosmography to the Renaissance. Upon Velasco's departure from the Council of Indies, the position of cosmographer-chronicler was split into its two constituent parts.

SEVEN

Cosmography Dissolves

With Juan López de Velasco's departure in 1588, cosmographical practice at the Council of Indies came under scrutiny once again. This time, however, the resulting changes did not come about through modifications to the laws regulating the post, as they had during Juan de Ovando's tenure. The duties of the cosmographer and chronicler remained consistent with those outlined in the pertinent articles in the 1571 *Ordinances*. But certain changes took place indirectly. First, the position was split into its constituent posts of cosmographer and chronicler. The next change was subtle but more definitive. Only practitioners who emphasized certain aspects of the discipline, particularly those that relied on mathematics, were named cosmographer major of the Council of Indies. As a result, the epistemological basis on which cosmographical practice at the Council of Indies had relied—unmediated eyewitness accounts—eventually yielded to an empirical and mathematical approach, best exemplified—and examined in this chapter—by the cosmographical practices used by Andrés García de Céspedes, who served from 1596 to 1611. The title of cosmographer of the Council of Indies would now be associated only with practitioners who emphasized the mathematical aspects of the discipline: mathematical cartography, astronomical navigation, hydrography, and geodesy. The descriptive aspects of cosmography and in particular those that used a textual form of representation—descriptive geography, ethnography, natural history—would now come under the purview of the chronicler of the Council of Indies.

With the end of the century, and concurrent with the divergence in cosmographical practice, came a gradual relaxation of the secrecy regulations that had governed cosmographical knowledge during the reign of Philip II. New imperatives, both political and social, contributed to this change during the transition between the monarchies of Philip II and Philip

III. Cosmography was no longer secret. López de Velasco's once closely guarded *Sumario* became the preface to the Council of Indies' first published official history of the New World, something unthinkable three decades earlier. Likewise, the cosmographer major of the Council of Indies would publish the results of the eclipse project and make public, in the form of navigation charts and rutters, the geography of the New World that resulted from the project.

A New Patronage Equation

The patronage structure and courtly context of cosmography in the Council of Indies changed significantly after the death of Philip II. In contrast to the penury that had plagued Philip II's court, it now became possible to earn very generous *mercedes* from the new monarch—a reward reserved, however, for favor seekers enjoying the king's good graces or those of his right-hand man or *privado*. The monarchy of Philip III has become synonymous with the rise of the *privado*, yet the king's reliance on one man taken into confidence to assist in governmental affairs was not new. Philip II had relied on a select number of advisers to handle affairs of state, but he had been careful to appear to rule alone—taking advice, yes, but ultimately ruling as the sole and absolute monarch and thus as the source of all royal patronage. After Mateo Vázquez's death in 1591, Cristóbal de Moura became the aging monarch's conduit for petitions seeking royal patronage, and in the final years of his life he was "a key element in governing the monarchy."[1] But whereas Philip II had relied on close advisers, during the reign of his son, Philip III, Spaniards learned how much power and influence a *privado* could wield. Early in Philip III's reign, the Duke of Lerma, Francisco Gómez de Sandoval y Rojas (1553–1625), parlayed a carefully cultivated friendship with the heir into the most powerful position in the monarchy. Lerma used his influence with the king to control access to the monarch. This allowed him to channel royal patronage to individuals and factions that supported his *privanza*.[2]

At Lerma's insistence, the new monarch appointed members of Lerma's family and associate factions to prominent positions throughout the kingdom. This included Lerma's nephew and son-in-law Pedro Fernández de Castro y Osorio (1576–1622), Count of Lemos and Andrade, who served

1. Feros, "Viejo monarca," 30.

2. Antonio Feros, *El Duque de Lerma: Realeza y privanza en la España de Felipe III* (Madrid: Marcial Pons, 2002), 186–88, 240.

as president of the Council of Indies between 1603 and 1609.[3] The new president shared with other courtiers, particularly those within the circle of royal favorites who gravitated around the Duke of Lerma, an interest in sponsoring artists and writers in what heralded the start of Spain's Golden Age in the arts and literature. The Count of Lemos featured prominently in this new patronage stage, as the many dedications written to him by Miguel de Cervantes, Lope de Vega, and Francisco de Quevedo attest.[4] The Council of Indies also added at least two more chroniclers of the Indies to its payroll. Antonio de Herrera, Gil González Dávila, and Pedro de Valencia, in addition to their salaries from the Council of Indies, drew generous salaries from the Council of Castile.

During Philip III's reign, the cosmographer and chronicler operated under new rules of noble patronage that now rewarded *public* production, since it was through the public's recognition of an artist's talent that his patron gained prestige. Thus, in order to drink from the "magnificent fountain" of patronage, the author had to practice in the public sphere.[5] This situation posed a potentially significant problem to the lone cosmographer of the Council of Indies, since long-standing secrecy policies prohibited royal cosmographers from publishing works on precisely the topic of their expertise, the geography of the Indies. They were thus effectively barred from this potentially profitable and prestigious economy of patronage.

Under the young Lemos, restrictions surrounding geographical information about the New World eased. While serving as the president of the Council of Indies, Lemos authorized the publication of a series of works that would have been inconceivable just years before. One example is the commission Lemos gave chronicler Bartolomé de Argensola (1561–1631), a favorite of the Lerma faction.[6] In 1606 the Spanish had recaptured Ternate, the principal city in the Moluccas, from the Dutch after the Portuguese had lost the citadel in 1601. Lemos asked Argensola to write a history of Spain's involvement in the Moluccas that would feature an account of its recent "conquest" of the islands, a move Lemos had championed.

3. Schäfer, *Consejo Real y Supremo de las Indias*, 1:187.

4. Eduardo José Pardo de Guevara, María del Pilar Rodríguez Suárez, and Dolores Barral, *Don Pedro Fernández de Castro, VII Conde de Lemos* (1576–1622), 2 vols. (Santiago de Compostela: Xunta de Galicia, 1997), 1:245–53.

5. Harry Sieber, "The Magnificent Fountain: Literary Patronage in the Court of Philip III," *Cervantes: Bulletin of the Cervantes Society of America* 18, no. 2 (1998): 94.

6. Pardo de Guevara, Rodríguez Suárez, and Barral, *Don Pedro Fernández de Castro*, 1:138–39.

To this end, Lemos made the Council of Indies' secret archives available to Argensola. Using these sources, Argensola wrote a history celebratory of Lemos's stewardship of the "conquest" of the archipelago, while claiming, by repeated reference to the Council's papers, that he was writing a factual account of the events.

The *Conquista de las Malucas* (Madrid, 1609) not only discusses in detail the geography of the contested archipelago but also includes a synopsis of Pedro Sarmiento de Gamboa's account of his 1579–80 exploration of the Strait of Magellan in the wake of Drake's incursion in the Pacific Ocean. In the not too distant past, knowledge of this geography had been considered of vital military value and kept under lock and key to safeguard Spanish possessions in the Pacific Ocean. In Argensolas's exposition, Sarmiento's account contributed a significant portion of the book's geographical contents, which the author used to portray the territories along the route from Spain to the Moluccas as well-mapped and well-explored parts of the Spanish empire. The preface to the reader states that although the narrative of the events of the recent conquest would have occupied only a few pages, it was necessary to explain things from the beginning (*dar razón de todas las cosas desde su principio*) so that the events of the present would be seen as fully justified and necessary.[7] The detailed geographical narrative enhanced the book's central theme—Spain's mastery over the land and sea, a mastery displayed as much through military might as through intimate knowledge of the territories under its domain.

The shift in secrecy policy signaled a radical reconceptualization of the strategic value of geographical knowledge. Under Philip II, the value of geographical knowledge lay in how it could provide an intelligence advantage in the case of war or defense, or a way of protecting economic assets by keeping then "hidden" from covetous enemies. For Philip III, geographical knowledge was most valuable if it could be deployed, albeit properly contextualized, to create a public image of Spanish domination and prestige. Demonstrating this knowledge to the public worked to create the perception that Spain, because it knew its territories so well, controlled them effectively.[8]

7. In the preface (*Al lector*) written by the author's brother, Lupercio Leonardo de Argensola. Bartolomé Leonardo Argensola, *Conquista de las Islas Maucas* (Madrid: Alonso Martín, 1609).

8. For a study of the strategic use of history for political purposes during the reign of Philip II and III, see Richard L. Kagan, *El rey recatado: Felipe II, la historia y los cronistas del rey*, Colección Síntesis (Valladolid: Universidad de Valladolid, 2004).

The changes in secrecy policy that took place during Lemos's stewardship of the Council were not sudden but were rather the culmination of an erosion of secrecy concerning geographical information that had, despite Velasco's best efforts, begun in the late 1570s. It had proved increasingly difficult for Spain to keep information secret about territories that technically, if only nominally, lay on the empire's western frontier, in particular China and Japan. This pressure came from groups outside the monarchy and principally from religious orders, which could skirt the secrecy regulations of the Council of Indies by invoking papal authority.

Since midcentury, Jesuit missionary activity in the Far East had operated well outside the Spanish Crown's sphere of influence. In addition, an integral part of the Jesuit program was precisely to inform all Christian countries of their missionary work abroad. The Jesuits' annual letters served this public relations purpose; the *Letters from the Indies* (1571) included particularly detailed geographic descriptions that satisfied Europeans' curiosity about the Far East.[9] The society's tactics put it at odds with secrecy policies at the Council of Indies. In fact, the groups were in a collision course when it came to geographical knowledge. For whereas the Council under Philip II was intent on safeguarding geographical information, this same body of knowledge was vital to promoting the Jesuit program—not only to support the logistical aspects of Jesuit missionary activity but as a essential means of gaining patronage and recruits.[10] In 1576, Juan Bautista Gesio argued against the publication of the book of Jesuit letters, which he believed contained geographical information that could undermine Spanish claims that Japan fell on its side of the line of demarcation.[11] (Recall that Gesio had advocated a Spanish conquest of Japan, or at least that its conversion be carried out exclusively by Spanish priests.) His complaints against publication were to no avail; the Spanish version of the book had already appeared in 1575 and was never recalled.[12]

9. Emanuel Acosta, *Rerum a Societate Jesu in Oriente Gestarum ad annum usque à Deipara Virgine* (Dilingen, Germany: Sebald Mayer, 1571). For a study of the role of Jesuit annual letters as vehicles for the dissemination of geographical and natural historical knowledge, see Steven J. Harris, "Confession-Building, Long-Distance Networks, and the Organization of Jesuit Science," *Early Science and Medicine* 1, no. 3 (1996): 306.

10. Dainville, *Géographie des humanistes*, 123–26.

11. IVDJ, Envío 25, f. 22, Juan Bautista Gesio, "Advierte de un libro de cartas de los Jesuitas escritas desde el Japón y ser de nuestro inconveniente para la demarcación," February 6, 1576.

12. *Cartas que los padres y hermanos de la Compañia de Iesus que andan en los Reynos de Iapon escriuieron alos dela misma Compañia, desde el año de mil y quinientos y quareta y nueue,*

The Augustinians followed with Juan González de Mendoza's immensely popular *Historia de las cosas mas notables, ritos y costumbres, del gran reino de la China* (Rome, 1585). Philip II had sent González de Mendoza on a failed embassy to the kingdom of China in 1581. Upon his return, Pope Gregory XIII instructed Mendoza to publish a description of that kingdom "to awaken the desire [in Spaniards] to save so many souls."[13] When Mendoza approached the Council of Indies in 1585 to procure a license to publish in Spain, he argued that he was doing so under a papal directive. His description of China, including the informative itinerary of Ignatius of Loyola, was published in Madrid in 1586. By the 1590s, efforts to keep secret the geographical picture of the empire's western boundary were all but abandoned, as evidenced by the actions of Luis Teixeira (1564–1604), a Jesuit priest and cartographer practicing at the court of Philip II, who working from Japanese and Jesuit sources created a map of Japan and sent it to Ortelius in 1592.[14]

In 1590 another Jesuit, José de Acosta, published a work that mimicked the symbiotic structure of the natural history and moral history that Juan de Ovando had introduced in the 1573 *Instructions*.[15] The *Historia natural y moral de las Indias* (Seville, 1590) is a natural philosophical inquiry into the nature and people of the New World from a firmly Aristotelian and Christian perspective. His often-quoted prologue explains his reason for writing: "Up to now I have not seen an author who attempts to explain the causes and reason of such novelties and peculiarities of nature, nor one who makes a discussion or inquiry in this regard; nor have I come across a book whose subject is the deeds and history of the Indians themselves, the ancient and

---

hasta el de mil y quinientos y fetenta y vno . . . (Alcala: En casa de Juan Iñiguez de Lequerica, 1575).

13. Juan González de Mendoza, *Historia de las cosas mas notables, ritos y costumbres del gran reino de la China* (Madrid: M. Aguilar, 1944), 10.

14. Beginning with the 1584 edition of his atlas, Ortelius had included a map of China based on one by the Portuguese Luis Jorge de Barbuda (Ludovicus Georgius), a Gesio protégé. Benito Arias Montano sent it to Ortelius in 1580. M. P. R. van den Broecke, *Ortelius Atlas Maps: An Illustrated Guide* (Goy-Houten, Netherlands: HES Publishers, 1996). Reference based on Ortelius correspondence in Jan Hendrik Hessels, *Abrahami Ortelii et virorum eruditorum ad eundem at ad Iacobum Colium Ortelianum epistolae*, 4 vols., Ecclesiae Londino-Batavae Archivum (Cambridge: Academiae sumptibus Ecclesiae Londino-Batavae, 1887–97), nos. 62, 99, 210.

15. Fermín Del Pino Díaz, *La historia natural y moral de las Indias como género: Orden y génesis literaria de la obra de Acosta* (1999), http://www.fas.harvard.edu/~icop/fermindelpino.html (accessed April 5, 2007).

indigenous inhabitants of the New World."[16] Acosta wrote the book as part of a corpus of works whose purpose was to educate missionaries. He hoped that by illustrating the "natural works that the wise Author in all nature has made," his reader—presumably European—would give glory to God, and that after learning of the religion and culture of the native people, the reader would also help them to "seek and remain" in the Christian faith.[17] Perhaps Acosta's dual missionary purpose—presenting a religious rationalization that would enhance the faith of Europeans by revealing the wonders of the New World and a manual instructing missionaries about the unconverted—opened doors that other works touching on the New World found closed.

The publication fortunes of Acosta's books suggest that secrecy stipulations had relaxed by the 1590s. The first two books of the *Historia natural y moral* had been written in Latin as *De natura novi orbis* and were intended to serve as an introduction to Acosta's missionary manual, *De procuranda indorum salute* (Salamanca, 1588). Acosta had written *De natura* by 1583, but apparently the publication was delayed due to the "censoring of the small new work and certain difficulties with the printing."[18] The censorship of *De natura* poses an interesting historical question, as it took place while López de Velasco was very much involved in censorship activities concerning books about the New World. Unfortunately, we do not have additional information as to the nature of the censorship the *De natura* underwent or even whether the book was presented to the Council of Indies for review. What does seem clear, however, is that when Acosta sought a publication license for his expanded *Historia natural y moral* (including *De natura*) in 1589, the Council of Castile granted the license without much delay. Tellingly, the book does not have a license from the Council of Indies.

Although Acosta's work is not a systematic or detailed geographical description, it does touch upon several topics that the Council of Indies had censored a few years prior. For example, books 3 and 4 contain an account of Francis Drake's navigation, a description of the Strait of Magellan that speculates whether the land to the south of the strait was an island, and a detailed description of the mining processes used at Potosí. Books

16. José de Acosta, *Historia natural y moral de las Indias, en que se tratan de las cosas notables del cielo, y elementos, metales, plantas y animales dellas: y los ritos, y ceremonias, leyes y gobierno, y guerras de los Indios* (Sevilla: Juan de León, 1590), facsimile ed., ed. Barbara Beddall (Valencia: Hispaniae Scientia, 1977), 29.

17. Acosta, preface to *Historia natural y moral*, 11–12.

18. The publication license was granted on June 22, 1586. Introduction by Mateos to José de Acosta, *Obras del P. José de Acosta*, ed. Francisco Mateos, Biblioteca de autores españoles (Madrid: Atlas, 1954), xxxvii.

5–7—the moral history—contain a careful ethnographic description of the native peoples of Peru and Mexico, similar to accounts by Sahágun, Las Casas, and Cervantes de Salazar that Velasco kept under lock and key in the Council's secret archive.

The 1590s were very different from the 1570s in other ways as well. Spanish territories had been objects of repeated incursions by the English and French, a pattern that would soon be followed by the Dutch. Toward the end of the sixteenth century, keeping geographical information secret was becoming an increasingly futile exercise. In 1579, while pillaging the western coast of South America, Drake captured a Spanish ship bound to the Philippines and, along with its two Spanish pilots, a treasure trove of maps and rutters describing sailing routes in the Pacific. Thomas Cavendish repeated the feat in 1588, this time gaining new geographical information about the northern Pacific that the Spanish had carefully guarded.[19]

Geographical knowledge was central to the budding Elizabethan program of territorial expansion.[20] Projectors sorely needed geographical information, particularly concerning the North American coast and the Northwest Passage. Richard Hakluyt spent the years between the *Diverse Voyages* (1582) and his *Principal Navigations, Voyages, and Discoveries of the English Nation* (1589) in France—as historian George Bruner Parks explains, "he went to France to discover America."[21] Unable to get close to Spanish centers of geographical information—Seville, Madrid, and now Lisbon—he found in France a number of exiled Portuguese who opposed Philip II's claim to the crown of Portugal, including the pretender to the throne Dom Antonio, who were glad to share geographical information with Spain's rival. He also interviewed French sailors familiar with the eastern North American shore and gleaned geographical information and natural history from Acosta's book.[22] For the curious, eager, and motivated Hakluyt, the ramparts that guarded geographical secrets in Madrid and Seville, while still unassailable, were capable of being circumvented.

19. Wagner, *Sir Francis Drake's Voyage around the World*, 35–37, and Henry R. Wagner, *Spanish Voyages to the Northwest Coast of America in the Sixteenth Century* (Amsterdam: N. Israel, 1966), 345.

20. For a study of how Elizabethans valued geographical information, see Cormack, *Charting an Empire*, 200–220.

21. George B. Parks, *Richard Hakluyt and the English Voyages* (New York: American Geographical Society, 1930), 99.

22. Ibid., 99–115, 137–38.

While by the 1590s the Council of Indies was abandoning efforts to keep geographical information secret, it still tried to enforce its right to censor any work intended for publication that dealt with the subject of the Indies. In 1597, the Council wrote to the king asking that he please remind the Council of Castile that the Council of Indies had to approve all printing licenses for books "dealing with matters concerning the Indies."[23] It is unclear what motivated the Council of Indies to issue the petition. It might have been simply a rather lukewarm reminder to the Council of Castile of the Council of Indies' right to opine on such matters, particularly on books that touched upon volatile political or religious issues.

Under Philip III the Council of Indies continued to be aware that news concerning the political situation of the New World was sensitive. Such was the case with a printed account Pedro Fernández de Quirós circulated in Madrid around 1610, after his return from his voyage exploring the South Pacific, in which he claimed to have found the Terra Australis Incognita (1605–6). On that occasion, the Council asked the king to order that the documents be collected, concerned that Quirós had included some discussions about the governance of the Indies that it would be best to keep out of foreign hands:

> [Quirós] has printed in this court several accounts and lately a very long one in which he discusses his voyage and treats indiscriminately many other matters concerning the government of the Indies and other sensitive matters. [He] has given and distributed these accounts to different persons from this nation and foreign, something that is considered very inconvenient, both for the information that foreigners can get from it, coming in the wake of his news concerning those lands and navigations, and because most of the things he discusses have no foundation.[24]

23. AGI, IG-744, N. 214, "Sobre la censura que han de tener los libros que se imprimen sobre cosas de Indias," July 16, 1597. Published in Antonia Heredia Herrera, ed., *Catálogo de las consultas del Consejo de Indias, 1529–1591*, 2 vols. (Madrid: Dirección General de Archivos y Bibliotecas, 1972), 2:374.

24. "[Quirós] ha impreso en esta corte diversos memoriales, y últimamente uno muy largo, en que hace un discurso de aquella jornada y viaje que hizo, y trata indistintamente de otras muchas cosas del gobierno de las Indias y material bien excusadas, y ha dado y distribuido estos memoriales a diferentes personas nacionales y extranjeras; cosa que se tiene por de muy grande inconveniente, así por la noticia que por él pueden sacar los extranjeros, viniendo de mano en mano a las suyas noticias de aquellas tierras y navegación, como por ser cosas las mas de las que trata en el dicho memorial sin fundamento." The king replied: "Dígasele al mismo Quirós que él recoja estos papeles

Philip III replied on the margin of the memorandum that Quirós should be asked to collect these papers himself and give them "with secrecy" to the Council of Indies, so that "these things do not cross many hands." It is worth noting that the Council's main objection was to Quirós's political discussion and not the news of his recent discoveries. The historian Toribio Medina observes that it was not until 1641 that the Council of Indies reasserted again its censorship rights via a royal order.[25]

Mathematical Practitioners Take Over

In 1591, after years of letting the office of cosmographer of the Council of Indies languish in the disenchanted hands of López de Velasco, cosmography both at court and at the Council underwent a significant reorganization. The Royal Mathematics Academy's financial underwriting passed from the coffers of the royal palace to those of the Council of Indies.[26] The Academy, however, continued under the direction of Juan de Herrera until his death in 1597. As a result of a general shuffling of cosmographical appointments at court, Juan Bautista Labaña was sent back to Portugal as principal cosmographer of the Council of Portugal.[27] Juan López de Velasco, after twenty years as cosmographer-chronicler major of the Council of Indies, officially ceased to hold the appointment, and his former post was split in two.

A number of factors went into the decision to split the post into its two constituent offices, cosmographer and chronicler. Toward the end of Velasco's tenure, the Council had expressed on several occasions a general dissatisfaction with the way he carried out the duties associated with his post. As early as 1588, the Council petitioned the king for a new chronicler, alleging that Velasco was too busy with other offices to fulfill his duties at

y los de con secreto á los del Consejo de Indias porque no anden por muchas manos esas cosas." Quoted in Justo Zaragoza, *Historia del descubrimiento de las regiones austriales: Hecho por el general Pedro Fernández de Quirós, el Pacífico hispano y la búsqueda de la "Terra Australis"* (Madrid: Impr. de Manuel G. Hernández, 1876–82; reprint, Madrid: Dove, 2000), 389.

25. Medina, *Historia de la imprenta*, 9.

26. The reorganization is studied in Vicente Maroto and Esteban Piñeiro, *Aspectos de la ciencia*, 98–102.

27. He returned to Spain in 1600 and continued working as a cosmographer and chronicler for Philip III. Among his works in this period is a commission by the Council of Aragon to map its territory, which he did using the method of triangulation. Agustín Hernando, *La imagen de un país: Juan Bautista Labaña y su mapa de Aragón (1610–1620)* (Zaragoza: Institución Fernando el Católico, 1996).

the Council.[28] The perception within the Council of Indies seems to have been that the mathematical aspects of cosmographical practice required the services of a specialist and had to take precedence over the historical/descriptive aspects. Although we do not have documentary evidence supporting Juan de Herrera's direct involvement in the reorganization, his prominent position at court and commitment to mathematical cosmography suggest that he was likely the architect behind the changes.[29]

Herrera's Portuguese sojourn had convinced him that applied mathematics was fundamental to the well-being of the state. As far as cosmographical practices were concerned, to him this meant an emphasis on cartography, instrumentation, navigation, and geodesy. Witness the projects he set in motion while still in Portugal and over subsequent years: a review of Portuguese navigation practices and cartography, the establishment of the Royal Mathematics Academy, and Jaime Juan's voyage. Herrera recruited cosmographers who practiced the mathematical aspects of the discipline—Juan Bautista Labaña, Pedro Ambrosio de Ondériz, and Luis Jorge—to work on his projects at court. By the 1590s, the palace/monastery of El Escorial was finished—to the king's satisfaction—and with it came honor and prestige to its designer, marking the zenith of the architect's influence with the monarch.

When the reorganization took place, the post of chronicler major of the Indies was given to Juan Arias de Loyola, while Pedro Ambrosio de Ondériz became cosmographer major.[30] The new officials were instructed to take over the pertinent functions and to carry out their duties as Velasco "had, could, and should have." To this end they were given all the "histories, *relaciones*, reports, memoranda, letters and other books and papers presently in possession of Juan López de Velasco" concerning their respective offices.[31]

As far as their duties at the Council of Indies were concerned, their functions were divided as follows. The chronicler major was responsible for "compiling and writing the general, moral, and particular history of the

28. IVDJ, Envío 23, caja 1, leg. 144, Hernando de la Vega y Fonseca, "Carta del Consejo de Indias al Rey en lo del oficio de cronista y cosmógrafo de las Indias," September 10, 1588, Madrid.

29. Vicente Maroto and Esteban Piñeiro, *Aspectos de la ciencia*, 390.

30. AGI, IG-874, "Carta de Felipe II nombrando al Lcdo. Arias de Loyola Cronista Mayor de las Indias." October 19, 1591. Also AGI, IG-874, "Provisión de Felipe II con el nombramiento y título de Cosmógrafo Mayor del Consejo de Indias a favor de Pedro Ambrosio Ondériz," September 9, 1591. Both published in Vicente Maroto and Esteban Piñeiro, *Aspectos de la ciencia*, 124–28.

31. Vicente Maroto and Esteban Piñeiro, *Aspectos de la ciencia*, 125.

events and memorable cases that have taken and are taking place in those parts, as well as of the natural things in them that are worthy of knowing."[32] He also had to "see and examine" the "histories of the Indies," a reference to the royal chronicler's duty as censor. In addition, and in a clause reminiscent of a regulation in the *Instructions* of 1573—which had been ignored in Velasco's case—the chronicler had to submit to the Council every year "whatever he is writing or has written concerning said history" before he could receive his salary.[33]

In the case of the cosmographer major, his duties were to "organize, decide upon, and execute cosmographical matters and descriptions of the Indies."[34] In an effort to ensure that Velasco's neglect of navigation matters did not repeat itself, the appointment order expressly instructed the new cosmographer to reform and correct the charts, instruments, and nautical practices currently in use at the Casa de la Contratación for the "universal benefit of those who sail." The appointment order stated that the Council (or possibly Herrera) was aware that there were "notable errors" in the charts and instruments currently in use and that new refinements (*primores*) in the art were being discovered daily which would be of great benefit to navigation. The order also commanded all the cosmographers in the kingdom to supply the documents necessary to complete any investigations that would be helpful to the cosmographer of the Council in carrying out the reforms.

32. "Y que como tal mi Cronista Mayor de las dichas Indias entendáis en recopilar y hacer historia general, moral y particular de los hechos y casos memorables que aquellas partes han acaecido y acaecieren, y de las cosas naturales dignas de saberse, que en ellas hay y hubiere, y veáis y examineis las historias de las dichas Indias." AGI, IG-874, "Carta de Felipe II nombrando al Lcdo. Arias de Loyola Cronista Mayor de las Indias. Toma de posesión de su oficio," October 19, 1591. Published in Vicente Maroto and Esteban Piñeiro, *Aspectos de la ciencia*, 127.

33. After three years of service, Arias de Loyola had still not presented anything to the Council of Indies, and the Council's bursar refused to pay his salary. Arias de Loyola alleged that although he had Velasco's papers, these were "many, very varied, and confusing" and that he had been trying to study them before compiling the history. The Council warned him that if he did not present something by the following year his salary would be denied. Arias failed to comply and was promptly removed from the post. After his dismissal, the duties of the chronicler briefly reverted to Ondériz. AGI, IG-742, N. 153, "Informe del Consejo de Indias dirigido a Felipe II sobre el cumplimiento de sus obligaciones por el licenciado Arias de Loyola," April 8, 1594.

34. "[Y] que como tal Cosmógrafo Mayor de las dichas Indias entendais en ordenar, disponer y ejecutar las cosas de la cosmografía y descripciones de las dichas Indias." AGI, IG-874, "Nombramiento de Ondériz como Cronista Mayor del Consejo de Indias y su asiento en los libros del citado Consejo de Indias," September 16, 1595.

In contrast to the chronicler, the cosmographer was under no obligation to present yearly proof of having carried out his duties but had to present to the Council "all the things [he] invent[ed] and the general or particular descriptions that [he] might make." These would be reviewed at the Council of Indies and could not be "practiced or communicated" without its approval.

Before they could garner their salaries, the chronicler and cosmographer were also required to teach at the Royal Mathematics Academy and had to solicit a quarterly letter from Juan de Herrera certifying that they had given their lectures. The supervisory provisions in the appointment orders suggest that the chronicler and cosmographer reported to Herrera in relation to their duties at the Academy but responded to the Council of Indies on all other matters. Nonetheless, Herrera's influence is quite evident in the new list of duties pertaining to the office of cosmographer major. Cosmography at the Council now emphasized cartography, instrumentation, and navigation practices. For Herrera this was the chance to reformulate cosmographical practice at the Council so that it focused on the mathematical aspects of the discipline. The purpose of this redirection was clear. With this new orientation, the Council of Indies would now be capable of asserting its fiduciary duty to oversee cosmographical activity and navigation at the Casa de la Contratación. The key was placing in the position of cosmographer major an individual capable of exerting such influence over the traditionally unruly Casa cosmographers and pilots. Herrera thought that Ondériz, the young mathematician he had groomed in Portugal, was the perfect candidate.

During his tenure, López de Velasco had never conducted an audit of the *padrón real* at the Casa de la Contratación, despite being instructed under the *Instructions* (article 69) to "make and correct all the navigation charts given to pilots of the *carrera de Indias*" as new discoveries and investigations came to light. This process, the article explained, should take place while the cosmographer-chronicler conducted an audit (*visita*) of the Casa and after meeting to gather the opinions of the Casa's cosmographers and pilots. Velasco ignored the matter until 1575, when he was reminded of his duties by the cosmographer of the armada, Alonso Álvarez de Toledo, who raised a complaint to the monarch when he was prevented from attending pilot examinations.[35]

In early 1575, Velasco requested a royal order admonishing the Casa for not maintaining up to date the *padrón real*.[36] The royal order instructed

35. AGI, IG-1968, L. 20, f. 94–94v, "Real Cédula a los oficiales de la Casa de la Contratación para que, mientras resida en Sevilla Alonso Álvarez de Toledo, cosmógrafo de la armada de la Carrera de Indias, pueda estar presente en los exámenes que para piloto y maestre, se llevan a cabo en la Casa de la Contratación," February 2, 1575, Madrid.

36. AGI, IG-1968, L. 20, f. 93v, "Por orden del Rey debe haber padrón por escrito

the Casa that in addition to the *padrón*—a map—there should be a book in which the longitudes and latitudes of everything described in the *padrón* were kept in writing. The order noted that this measure would address the problem with the *padrón*'s condition. Apparently the *padrón* was maintained as a map drawn on parchment, and this material tended to shrink, thus making readings of coordinates by compass from the chart inaccurate. The order instructed the catedrático Ruiz (Zamorano's predecessor) to prepare the book of latitudes and longitudes and send a copy to the Council of Indies so that it could be kept there for safekeeping—and as a potential source for Velasco's ongoing cosmographical work. It is not unreasonable to suppose that Velasco's request came about after he examined the state of the current official rutter to the Indies and found it unsatisfactory, particularly as a source of geographical coordinates.

Another order reminded officials at the Casa that pilots had to file reports and descriptions of their navigations—the kinds of firsthand sources Velasco cherished—as they had been instructed to do before Alonso de Santa Cruz in 1536.[37] This was the extent of Velasco's involvement with the Casa and remained so for the rest of his tenure, even after a *visita* to the Casa de la Contratación in 1578 by a council member, the Lcdo. Benito Pérez de Gamboa, found some serious problems with the condition of the *padrón real* and the classes and examination of pilots.[38] In that instance, Alonso Álvarez de Toledo once again issued the harshest critique of the cosmographical situation at the Casa.

In contrast, once appointed cosmographer major, Ondériz wasted no time undertaking reform of the navigation charts, instruments, and navigation practices at the Casa de la Contratación, a project known as the *enmienda del padrón*.[39] He began by first conducting a study of the navigational charts and instruments certified by the Casa de la Contratación for use in the *carrera de Indias*. By the end of 1593, Ondériz had presented the Council with a comprehensive report, the result of his investigation and of consultations with other cosmographers. He pointed out a number of problems with the construction and graduation of the navigation instruments in use. Some of

---

en libro, y a instancia de Juan López de Velasco, cosmógrafo y coronista mayor, se mando que le hubiese," February 27, 1575, Madrid. The order also instructs the Casa de la Contratación to inform if any problems (*inconvenientes*) could result from this order.

37. AGI, IG-1956, L. 1, f. 266–266v, "Descripción de viajes por maestres y pilotos," March 14, 1575, Madrid.

38. Barrera-Osorio, *Experiencing Nature*, 49–55.

39. Esteban and Vicente have studied Ondériz's work on the *enmienda*. Vicente Maroto and Esteban Piñeiro, *Aspectos de la ciencia*, 390–404.

these, particularly the Jacob's staff, he considered completely inadequate for the purpose of finding latitude. The navigation charts fared no better. Not only was the universal chart incorrect in a "general" sense—it showed the Moluccas on the Portuguese side of the line of demarcation—but it also had a number of "particular" errors that were in urgent need of correction. He noted that the *padrón real* dated from 1567 and had not been corrected since.

Ondériz, a faithful disciple of Herrera, considered that properly correcting the master chart required that "qualified persons" travel with the upcoming fleet expressly for the purpose of collecting the geographical, geodesic, and hydrographic information that was lacking. The Casa objected to this measure and instead proposed collecting the necessary information by asking all the pilots scheduled to depart with the fleet in 1594 to complete a questionnaire. The pilots were also asked to use new nautical instruments—made to Ondériz's specifications—and follow a set of instructions when determining the coordinates of the geographical features. The Casa's plan had a predictably disappointing outcome. Upon the fleet's return the following year, only three of twenty-two pilots had complied with the instructions, and most of these poorly.

Over the summer of 1595, Juan de Herrera proposed a solution well in keeping with the style of cosmographical practice he championed. He suggested that two small ships be sent to conduct a hydrographic survey of the American Atlantic coast. These should have an expert—meaning a mathematician cosmographer—on board to oversee the survey. As the expedition readied to depart, Ondériz, now also pilot major of the Casa de la Contratación, was to move to Seville along with Luis Jorge and, while in the city, collaborate with Rodrigo Zamorano and Domingo de Villaroel in preparing the new *padrón*. Despite what appears to be Philip II's endorsement of the venture, the expedition did not take place, for Ondériz became ill later that summer and died in January 1596. Work on the *enmienda* would resume in earnest, however, in June of 1596, when Andrés García de Céspedes was appointed cosmographer major of the Council of Indies.

In Andrés García de Céspedes (ca. 1545–1611) the Council found a uniquely qualified cosmographer.[40] Céspedes proudly hailed from the Valle de Tobalina in the region of the Montaña de Burgos, more specifically from the

40. Biographical notes and documents concerning García de Céspedes are found in Pérez Pastor, *Bibliografía madrileña*, 3:103–7, and Felipe Picatoste y Rodríguez, *Apuntes para una biblioteca científica española del siglo* XVI (Madrid: Impr. de M. Tello, 1891), 120–27. Additional biographical notes on Céspedes's life and career are in Vicente Maroto and Esteban Piñeiro, *Aspectos de la ciencia*, 145–53.

town of Villanueva del Grillo, today deserted. Son of *hidalgos*, he entered the clergy after university studies, earning the degree of *licenciado*. Most of what is known about his early career is gleaned from autobiographical comments contained in his works. Thus we know that he spent time at the castle in Burgos while the fortress contained an artillery school and cannon foundry.[41] In 1582–83 he went to Lisbon in the service of Archduke Albert when the king's nephew served as viceroy of Portugal (1583–93).[42] We do not know in what capacity Céspedes served the archduke, but it was a very productive period for him as a mathematical practitioner and astronomer. He mentions discussing astronomical navigation and cartography with Portuguese pilots and studying the works of Portuguese cosmographer Pedro Nuñez. For his astronomical observations, some of which he carried out with the king's confessor, Dr. Sobrino, he built a very large quadrant to observe planetary and solar positions. He became convinced that the solar declination tables currently in use—a critical component for calculating latitude while at sea—were incorrect and in need of reform. He also immersed himself in the study of a number of technical and scientific problems, the results of which he collected into several works composed in Castilian. As of 1594, he listed the following works:[43]

- A book on how to determine latitude[44]
- A book on theoretical and practical perspective[45]

41. "Letter to the reader," in Andres García de Céspedes, *Libro de instrumentos nuevos de geometría muy necesarios para medir distancias, y alturas, sin que intervengan numeros como se demuestra en la práctica* (Madrid: Juan de la Cuesta, 1606).

42. Céspedes states in his *Regimiento de tomar la altura de Polo* that he lived in Lisbon for twelve years. BN, MS 3036, f. 2.

43. BN, MS 3036, f. 2v–3. Céspedes would refer a number of times to his body of work. This list, written in 1594, is very important, not only to trace the evolution of his scientific career and thought but also to identify works that Céspedes authored. As mentioned in chapter 2, Céspedes presented to Philip III as his own two works by Alonso de Santa Cruz: a commentary on Sacrobosco titled the *Astronómico real* and the *Islario general*. Céspedes sought and was granted licenses for both, but they were never published. As would be expected, this list of works dating from 1594 does not refer to the Sacrobosco commentary or to the *Islario*.

44. BN, MS 3036, Andrés García de Céspedes, "Regimiento de tomar la altura de Polo en la mar y cosas tocantes a la navegación. Dirigido al Rey don Felipe Tercero [segundo] deste nombre."

45. This does not survive as an individual work, but Céspedes may have included parts in the manuscript BAH, 9/2711, "Libros de reloges de Sol que hizo Andrés García de Céspedes," and in his published *Libros de instrumentos nuevos de geometría*; both have short sections on perspective.

- A book on planetary theory according to Copernicus with Céspedes's observations[46]
- A book on "the theory and construction of the astrolabe with a discussion of its ingenious uses in astronomy and perspective"[47]
- A book on sundials[48]
- A book on mechanics and the theory of machines which discusses the construction of a number of the machines[49]
- A book on "how to move water"[50]

46. This work is lost. Over the years, Céspedes apparently continued working on this early version of his book on Copernican theory. In 1606 he described the work as containing three parts: the first being a planetary theory according to Copernicus; the second explaining why, based on Céspedes's personal observations, Copernicus's theory of the sun and moon, as well as the Alphonsine tables, was incorrect; and a third that discussed the stations of the planets with a treatise on parallax. Although the work was never published, Céspedes was granted a printing license in 1603 for a work described as "Planetary theory, where the doctrine according to Copernicus and Ptolemy are discussed, including many observations whereby the many errors of both are discovered" ("Teórica de Planetas, donde se declaraba la doctrina de Copérnico y de Ptolomeo, y se ponían, muchas observaciones que en este tiempo tenia de hechas, por las cuales se averiguaban muchos errores que se hallaban así en la una doctrina como en la otra"). The printing license appears in his *Regimiento de navegación*. In the preface to his 1606 *Libro de instrumentos nuevos de geometría*, he also mentions having written a commentary on Peuerbachs's planetary theory and another work discussing some equations or theories (*equatorios o teóricas*) to determine a planet's latitude and longitude without the need to use tables and including an instrument "to know eclipses" (*con que saber los eclipses*). Esteban and Vicente located the commentary on Peuerbach at the BAH, manuscript 9/5630, "Theóricas de los Planetas de Jorge Puerbachio con el comento de Andrés García de Céspedes." This document also includes a section indicating how to calculate to position of planets using the Alphonsine and Prutenic tables.

47. This work is lost. It was also given a printing license in 1603, then described as "Teórica, práctica y uso del Astrolabio."

48. BAH, 9/2711, Andrés García de Céspedes, "Libros de reloges de Sol que hizo Andrés García de Céspedes Cosmógrafo mayor del Rey nro. Señor y natural del Valle de Tobalina montaña de Burgos en el qual se enseña como se ecribiran Reloges en qualquiera superficie q. sea que el extremo de la sombra del estilo muestre varios círculos del primer mobil asi otras muchas curiosidades." There is another manuscript copy at BUS, MS 2639, Andrés García de Céspedes, "Tratado de relojes solares." Céspedes apparently never sought a printing license for this book.

49. This work is lost. In a later reference to it, he mentions that he shows thirty such machines. It was also granted a publication license in 1603. In the license, the book was described as "a book on mechanics with theory for engineers and all kind of people" (*un libro de Mecánicas, con Teórica, práctica para ingenieros, y todo género de gente*).

50. The book on *como mover las aguas* does not survive, but Céspedes later added a

None of these works, at least as he conceived them in 1594, was ever published. Céspedes would publish only two works, a book of surveying instruments and a two-part book written as the result of the *enmienda*. He dedicated the *Libro de instrumentos nuevos de geometría* (Madrid, 1606) to his old patron, now the ruler of the Low Countries, Archduke Albert. In the dedication he announces that he has selected this particular book from among other more important ones given that the archduke is "occupied in matters of war," and he hopes that two of the instruments he designed (an improved quadrant and Jacob's staff) could be of use to the archduke's soldiers in the field. Céspedes's practical little book explains the use and construction of a quadrant and a Jacob's staff, as well as that of a level he "saw in the house of Juan de Herrera." The book also contains a study on how to determine and control the flow of water in aqueducts, as well as an artillery treatise. This book places Céspedes squarely in the tradition of early modern mathematical practitioners, interested in appealing to a nonexpert but curious audience.

From his list of works, it is clear that Céspedes was most interested in astronomy, and it was as an astronomer that he first tried to get the attention of Philip II. In late 1593, Céspedes was tending the king's astronomical clocks at the royal palace in Madrid. A few years later, 1595, we find him filling in during the absence of one of the professors at the mathematics academy.[51] It was likely during this period that Céspedes wrote to Philip II proposing an astronomical observatory for El Escorial.[52] In the document, Céspedes

---

hydrology section titled "Tratado de conducir las aguas de un lugar a otro cosa bien importante para los que tratan de semejante oficio," in García de Céspedes, *Libro de instrumentos nuevos de geometría*, 25–43.

51. We know of two other works Céspedes wrote and might have used when he taught these courses. They were copied by Jehan Lhermite in his memoir of the years spent at the courts of Philip II and Philip III. *Le Passetemps de Jehan Lhermite* at Bibliothèque royale Albert Ier (Royal Library of Belgium) MS II 1028. The *Tratado de astronomía y astrología* is in f. 230–44 and another on sundials at f. 293–308v.

52. Fernández de Navarrete was the first historian to identify the document at the library of El Escorial and described its location as "Codice j. Lib. 16 en la Biblioteca alta." My several attempts to locate the document at El Escorial proved unsuccessful. While I believe it is only through Navarrete's paraphrase in the *Disertación sobre la historia de la náutica*, 77:356–57, that historians learned of Céspedes's proposal, a similar document is described in a catalog made in 1994 of the family archives of the Marquis of Legarda and cited in Julio Fernando Guillén y Tato, *Inventario de los papel pertenecientes al Excmo. Señor D. Martín Fernández de Navarrete existente en Abalos, en el Archivo del Marques de Legarda* (Madrid: Ediciones Cultura Hispánica, 1944), 35. Through the gracious assistance of the current Marquis of Legarda, Don Francisco

portrayed himself as someone who combined mechanical ability and mathematical knowledge and offered to build a number of large astronomical instruments to equip the proposed astronomical observatory. The instruments consisted of "two large celestial and terrestrial globes made of golden metal, the first one imitating the movements of the sun, moon and other planets, a large quadrant, eight hands in size, and an astronomical ring ten hands in length."[53] These would be used to "observe and investigate the true location of the sun and moon." In addition, Céspedes proposed building some large armillary spheres made of metal "to correct the locations of the fixed stars," another showing the theory of the sun, moon, and eight spheres, and a number of smaller spheres representing the theory of the planets.[54] Céspedes envisioned astronomers from all over Europe coming to El Escorial to conduct astronomical observations, just as "Hipparchus had traveled from Rhodes to Alexandria." The primary objective of the observatory, however, was to correct the Alphonsine tables, "for these no longer show the true locations of the planets nor of the fixed stars." This, he added to ease the mind of the perpetually cash-strapped Philip II, could be done for a fraction of what it had cost Alphonse the Wise. Along with the proposal, Céspedes submitted a number of mathematical treatises and some instruments as a proof of his capabilities as an instrument maker and mathematician.

Céspedes may have been aware of Tycho Brahe's observatory, Uraniborg, on the island of Hven, and we might speculate that he was trying to establish a similar observatory at the monastery/palace of El Escorial. It would be only a few years later, 1597, that Brahe left Hven in search of a new patron.[55] Yet it is unlikely that the Lutheran Danish astronomer would have considered the Spanish Habsburgs as patrons or that he would have

---

Fernández de Navarrete, I was able to review the document. It is an eighteenth-century copy made by one of Fernández de Navarrete's copyists and verified by him from the original at El Escorial. It is a memorandum summarizing Céspedes's proposal and was probably prepared in order to consult the matter with someone at court.

53. "[D]os grandes globos, celeste y terrestres, de metal dorado, imitando en el primero los movimientos del sol, luna y demás planetas; un gran cuadrante de ocho palmos y un radio astronómico de diez para observar y averiguar los verdaderos lugares del sol y la luna" (in ibid., 77:356).

54. "[U]nas armillas de seis palmos de diámetro para rectificar los lugares de las estrellas fijas; una esfera grande de metal con la teórica del sol, luna y octava esfera, y otras teorías de planetas en globos pequeños cubiertos con sus círculos" (in ibid., 77:356).

55. The tantalizing question was addressed by José Manuel Sánchez Ron in, "Felipe II, El Escorial y la ciencia Europea," in *La ciencia en el Monasterio del Escorial. Actas del simposium, 1 al 4 de noviembre de 1993* (Madrid: Ediciones Escurialenses, 1994), 67–69.

been welcomed at the Escorial. As it turns out, Brahe found his new patron in Prague, one who turned out to have close ties to Spain. His new patron, Rudolph II, was educated in the court of his uncle Philip II, as was his brother, Archduke Albert (Céspedes's long-time patron). It was likely at the Spanish court that Rudolph developed his lifelong passion for astrology, alchemy, and natural philosophy.

Upon Ondériz's death, the Council did not have far to look for an eminently well qualified cosmographer and, even more important, one whose preparation and interest lay in the mathematical and experimental aspects of the science rather than its narrative and descriptive sides. Céspedes's first duty, clearly outlined in his appointment order, was to continue and conclude the reform of the maps, instruments, and rutters used at the Casa de la Contratación. Céspedes's appointment listed duties not unlike those of Ondériz ("organize, decide upon, and execute cosmographical matters and descriptions of the Indies")[56] but without the additional duties of the chronicler that had reverted to Ondériz after Arias de Loyola had been dismissed from his post.

There were, however, some differences between Ondériz's and Céspedes's appointments. Céspedes, for example, was not required to teach at the Royal Mathematics Academy and thus was not under the direction of Juan de Herrera.[57] He was, in contrast, under obligation to submit to the Council of Indies at the end of every year "some work concerning the descriptions" before he could collect his salary.

Céspedes immediately turned his attention to seeing the reform of navigation charts and nautical instruments completed. He was also handed a lengthy set of instructions that listed the Council's expectations for

56. AGI, IG-874, "Provisión de Felipe II con el nombramiento y título de Cosmógrafo Mayor del Consejo de Indias a favor de Andrés García de Céspedes," May 15, 1596. Published in Vicente Maroto and Esteban Piñeiro, *Aspectos de la ciencia*, 181–82.

57. He would formally take up these responsibilities once again in 1607. In a move that betrays his preference for the theoretical aspects of the discipline, Céspedes changed the curriculum to a three-year program to resemble that used in the astrology chair at the University of Salamanca. The first two years were dedicated to gaining the necessary mathematical skills: theory of the sphere, planetary theory, Alphonsine tables, Euclid's geometry, and Ptolemy's *Almagest*. Céspedes's curriculum culminated in the third year in courses on cosmography, navigation, and the use of its instruments. Ibid., 150–51.

the specifics of the *enmienda*.[58] The reform would consist of reviewing and creating master versions of the principal instruments used in navigation: the astrolabe, the Jacob's staff (*ballestilla*), and two types of compass needles. Further, he was to compose a universal chart showing inland geographical details (*carta universal reformada con tierra adentro*) and six specific charts that would serve as the *padrón real*. These were to be accompanied by a navigation manual (*regimiento de navegación*) that included corrected tables of solar declination and new tables for using the Jacob's staff and the regiment of the North Star. He was also asked to issue an opinion on two specific matters: First, should the posts of pilot major and professor of navigation (*catedrático*) at the Casa de la Contratación remain in the possession of a single person? Both positions were being held by Rodrigo Zamorano. Second, Céspedes was to assess the lessons and examinations given to would-be pilots of the *carrera de Indias*. Céspedes's appointment was followed by a series of royal orders instructing a number of cosmographers to assist him in the reform.[59]

By early 1597, Céspedes was in Seville attempting to restart work on the reform of the *padrón*, but things were not going well. He wrote the Council of Indies, clearly exasperated at the officials at the Casa de la Contratación, accusing them of delays and impeding his work. In addition, the Casa's financial officer had not paid him the moneys he had been promised, which had forced him to sell some jewels to finance his stay. He also complained that the recent floods had caused great damage and illness in the city and that he had been unable to leave his house.[60] Despite these difficulties and as a testament to his tenacity, Céspedes would finish the *enmienda* by the end of the following year.

58. AGI, P-262, R. 2, "Instrucción de Felipe II sobre lo que debe hacer Andrés García de Céspedes para la enmienda de las cartas e instrumentos de la navegación," June 13, 1596. Also in IG-1957, L. 6, fol. 139v–141v. Published in Vicente Maroto and Esteban Piñeiro, *Aspectos de la ciencia*, 435–38.

59. Among the cosmographers asked to assist Céspedes were Rodrigo Zamorano and Domingo de Villarroel from the Casa de la Contratación, Sevilian cosmographer Simón de Tovar, and Luis Jorge de la Barbuda, who was instructed to take leave of his duties at the royal court in Madrid and travel with Céspedes to Seville. Of these, Jorge fell ill and never made the trip to Seville. By the time Céspedes got to Seville, Villarroel had defected to France, accused by Zamorano of being a spy, and Tovar had died. Only Zamorano assisted Céspedes. See orders dated June 13, 1596 at AGI, IG-1957, L. 6, f. 143.

60. AGI, IG-744, N. 119v, "Petición de Andrés García de Céspedes," January 20, 1597.

Once in Seville, he printed a brief questionnaire requesting the opinion of pilots concerning the accuracy and usefulness of the rutters and charts in use. The questionnaire also asked whether pilots found the size of the astrolabes currently in use adequate and whether the compass should be mounted on an adjustable frame so that the pilots could correct for the variation from true north.[61] Most of the pilots who replied—forty-two replies survive—insisted that there was nothing wrong with the current *padrón* or the navigation charts and showed little enthusiasm for changing the size of the astrolabe or the compass. A few offered some helpful observations as to the locations of dangerous shallows, but most insisted that changing the rutter to the Indies could be very dangerous. Their reaction was very much in keeping with a conservative and protective attitude assumed by the equivalent of a Sevilian pilots' guild, the Universidad de Mareantes.[62] At this point the cosmographer major realized that for the reform to bear fruit, he had adopt a conservative stance himself.

In late 1598 Céspedes considered his work completed, and the Council submitted the work for review to a panel of cosmographers in Madrid. The first panel, although laudatory of Céspedes's work, recommended that a panel of "learned men of which, luckily, there are many in these parts" study his work further.[63] As Vicente and Estéban have pointed out, this initial opinion elicited a flurry of activity on Céspedes's part, including a letter in which he accused some members of that panel, specifically Arias de Loyola (the former chronicler major of the Council of Indies), Luis Jorge, and Doctor Osma, of "malice and of responding out of [personal] interests" and also of being persons who had "never seen the sea nor have dealt with pilots."[64] He then asked three mathematicians from Seville, who "not only know theory but are acquainted with practice through daily communication with pilots," to independently evaluate his work.[65] They, as would

61. AGI, P-262, R. 2\1\27–120, "Memorial de lo que han de advertir los Pilotos de las Carreras de Indias acerca de la reformación del padrón de las Cartas de Marear, y los demás instrumentos de que usan, para saber las alturas y derrotas de sus viajes," 1598.

62. Alison D. Sandman, "Educating Pilots: Licensing Exams, Cosmography Classes, and the Unversidad de Mareantes in Sixteenth Century Spain," in *Ars nautica: Fernando Oliveira and His Era; Humanism and the Art of Navigation in Renaissance Europe* (1450–1650), ed. Inácio Guerreiro and Francisco Contente Domingues, Ninth International Reunion for the History of Nautical Science and Hydrography (Cascais, Portugal: Patrimonia, 1999).

63. AGI, P-262, R. 2\6\1, "Parecer de todos los cosmógrafos sobre reforma del padrón," 1598.

64. AGI, P-262, R. 2\1\3–6. "Memorial sobre junta de cosmográfos," January 8, 1599. See also Vicente Maroto and Esteban Piñeiro, *Aspectos de la ciencia*, 408–9.

65. They were Diego Pérez de Mesa, Antonio Moreno, and Gerónimo Martínez Pradillo.

be expected, found it to be a sound work. Perhaps more important, they explicitly backed Céspedes's conservative strategy: they stated that pilots would use the new charts and instruments since they did not deviate much from accepted practice. They fully endorsed Céspedes's use of *cartas arrumbadas* or charts drawn using a square projection rather than one using "curved meridians," and testified that the cosmographer had spent many hours going over old and new rutters and interviewing pilots.

Céspedes's effort to have his work on the *enmienda* recognized and implemented did not end with appeasing the cosmographical junta and obtaining the letters of endorsement. In an unprecedented move, he also sought to have the results of his efforts published. The Council of Indies praised him for a job well done and instructed all pilots chartered by the Casa de la Contratación to base future navigation charts on Céspedes's new *padrón*. This in itself was not new, as Céspedes had since 1597 also been serving as pilot major and thus had the authority to define the contents of the charts used by the Casa de la Contratación. The Council also recommended that the navigation manual discussing the new instruments, navigation methods, and astronomical tables be published.[66] This, again, was not a novelty. Note that as cosmographer of the Casa de la Contratación, Rodrigo Zamorano had published in 1581 his *Arte de Navegar*, although it was a much simpler sailor's manual. What is noteworthy about Céspedes's petition is the remarkable change that the Council's recommendation underwent when the matter was referred to Philip III. The king agreed with all of the Council's suggestions and issued a royal order to that effect on May 3, 1599.[67] The royal order now made provisions for printing all the material Céspedes had prepared: "It is ordered that the [nautical] charts should, from now on, be made according to this master chart and that it be printed, and that the manual that said Andrés García de Céspedes wrote explaining their use and construction also be printed, reducing to a separate [work] what Pilots need and that it be also printed, as well as the rest, so that Pilots can easily benefit from it."[68] The king had thus authorized that the once

66. "Y que también se debe mandar que las cartas se hagan conforme al padrón que él trae pues en él están enmendados los que se han hecho hasta ahora y mejor declarado todo por diligencia que se ha puesto para averiguar lo que se a podido y de ello se puede resultar mayor noticia y utilidad sin perjuicio de nadie." AGI, IG-745, f. 202, "Sobre emmienda del padrón e instrumentos hechos por Andrés García de Céspedes," January 31, 1599, Madrid.

67. AGI, P-262, R.2 \5\1, "Orden a los oficiales de la Casa de la Contratación de que se impriman los regimientos," May 3, 1599.

68. "[S]e debe mandar que las cartas se hagan de aquí adelante conforme al dicho

secret and carefully guarded *padrón real* circulate as printed material. Moreover, a book that discussed the geographical arguments used to construct the chart would accompany it.

With his book nearing publication, Céspedes moved to assert complete authority over the navigation principles and astronomical values used and taught at the Casa de la Contratación. Céspedes set his sights on Rodrigo Zamorano and his *Arte de navegar*. Although Zamorano resumed his duties as pilot major in 1598, the pilot's major's authority to determine what constituted the *padrón real* and suitable navigation instruments had by then passed to the cosmographer of the Council of Indies. In 1603, Céspedes recommended that Zamorano step down as pilot major, that he teach only navigation, and that he refrain from making maps and instruments—all apparently to no avail.[69] Céspedes also issued a joint statement with the principal lecturer at the Royal Mathematics Academy, Dr. Firrufino, recommending that someone more "curious and careful" than Zamorano be selected to teach the pilots. When Céspedes's book with its new declination tables and rutters was ready, he asked the Council of Indies to prohibit the use of Zamorano's book, because "the declination table and of the stars are wrong . . . which causes many ships to be lost."[70]

The two-part structure of the book, the *Regimiento de navegación e hydrografía* (Madrid, 1606), followed closely the parameters established in the monarch's decree. The first part consisted in a sophisticated navigation manual, while the geographical material and the explanation of the *padrón real* were included in the *Hydrografía*. Both parts of the book, for reasons that are as yet unclear but likely have to do with the comprehensive approach Céspedes chose, would not appear in print until 1606. It seems that the navigation charts were never printed, likely due to the problems such a step would have caused with the cadre of Sevilian cartographers

padrón, y que se imprima, y use del regimiento que para el uso, y gobierno de la cartas ha hecho el dicho Andrés García de Céspedes, reduciendo lo necesario para los Pilotos aparte, y que se imprima de por si, y que lo demás también se imprima, para que con mas facilidad saquen los Pilotos provecho de ello," May 3, 1599. Published as a preface in García de Céspedes, *Regimiento de navegación*.

69. Biblioteca Palacio Real (BPR), II/175, f. 311–12, "Apuntamientos necesarios para la buena navegación por Andrés García de Céspedes y el Dr. Firrufini," (1603). See also Pulido Rubio, *Piloto mayor*, 698.

70. "[H]abiéndose impreso nuevo regimiento que nadie use el de Zamorano que la tabla de declinación y las estrellas está defectuosa . . . por la que causa que se pierdan muchos navios." BPR, II/175, f. 311v.

who made a living copying the *padrón* and drawing the maps used in the *carrera de Indias.*

The first part of Céspedes book, the *Regimiento de navegación*, followed the structure that had become typical of the Spanish navigation manual, yet he far surpassed what the genre typically included in terms of supporting theoretical explanations. The first part of the book contains the "de rigueur" introduction on the theory of the sphere and a discussion of the need to reform the official navigational charts and itineraries at the Casa de la Contratación. From that point, Céspedes's book set a new and ambitious standard for the navigation manual. Rather than explaining as clearly and succinctly as possible the rules for astronomical navigation and the construction and use of navigation instruments—what his processors in the genre had aspired to do—Céspedes approached each topic by first examining methodically all the procedures, instruments, charts, and astronomical tables used in navigation. The results of this exercise were redesigned astrolabes, Jacob's staffs, and compass needles, as well as new procedures for determining latitude, either during the day or at night, using the regiment of the North Star. Noting that the inclination of the ecliptic used to compose the Alphonsine and Copernican tables was no longer valid, he computed new solar declination tables based on a series of observations he carried out personally and corroborated with those of other contemporary observers, including Pedro de Retes in Bogota.

Céspedes was deeply skeptical of the validity of solar, lunar, and stellar tables, whether they were computed by Ptolemaic or Copernican theories. He repeatedly reiterated his distrust by pointing out how on countless occasions his observations (made, as he notes, with very large and precise instruments and always corroborated by other observers) showed the canonical tables to be incorrect. For Céspedes, reform of the *padrón* and of navigation practices used at the Casa de la Contratación necessitated a fundamental reevaluation of the astronomical data upon which latitude, and potentially longitude, computations were based. It was to this end that he had proposed an astronomical observatory at El Escorial before he was appointed to the Council of Indies. Yet his duties at the Council limited his astronomical activities to those associated directly with navigation and geography: solar declination tables and observations to determine the true position of the North Star. Céspedes's new tables of solar declination, however, were a significant improvement over Zamorano's and Labaña's previous efforts.[71]

71. Navarro Brotóns compares the declination tables of Zamorano, Céspedes, and

Mathematics and Cosmographical Epistemology

The *enmienda* and Céspedes's *Regimiento* marked the definitive divorce of the descriptive and mathematical aspects of cosmographical practice at the Council of Indies. At the heart of this disassociation was a shift in cosmographical epistemology characterized by an increasing emphasis on using mathematical arguments to ascertain matters of cosmographical fact. This approach was fundamentally at odds with the methods Velasco employed that relied principally on establishing the credibility of eyewitness testimony and ensuring comprehensiveness. The evolution of the epistemology adopted for cosmographical practice at the Council of Indies is evident in documents dating from before the publication of the *Regimiento de navegación e hydrografía*. They help establish the trajectory of Céspedes's epistemological approach and reveal how malleable the epistemic foundation of cosmography proved to be once in the hands of a mathematical practitioner.

Upon accepting his appointment to reform the *padrón real*, Céspedes was faced with a delicate problem. The guidelines under which the reform of the *padrón* was to take place instructed the cosmographer of the Council of Indies to collect and consider the opinions of other cosmographers, as well as those of sailors and pilots, to determine the data that would yield the definitive rutters and maps to be used by the Casa de la Contratación. In addition, the purpose of the project was to create a series of products—charts and instruments—that would be used by these same pilots. These requirements were bound to clash with Céspedes's mathematical rationalism.

Two years before becoming cosmographer major, Céspedes wrote a treatise on how to determine latitude, the *Regimiento de tomar la altura del Polo*, in which he adamantly rejected the validity of pilots' accounts when it came to matters of determining position at sea, just as Jaime Juan had done.[72] He based his opinion on having observed and conversed with sailors over the twelve years he lived in Lisbon. This had convinced him that pilots differed in their methods and lacked a systematic approach (*no proceden con orden*) when conducting their observations. It is through a series of marginal annotations on this document made in Céspedes's hand after he became the cosmographer of the Council of Indies, that we are able to witness how his

Labaña in Víctor Navarro Brotóns, "Astrología y cosmografía entre 1561 y 1625," *Cronos* 3, no. 2 (2000).

72. References in folios 2 and 103 suggest that the document was originally written circa 1594. BN, MS 3036, *Regimiento de tomar la altura de Polo*.

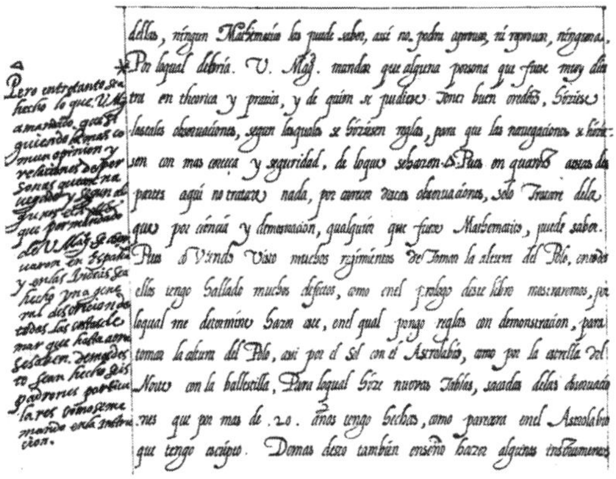

Figure 7.1. Excerpt with marginal annotations from Céspedes, *Regimiento de tomar la altura del polo*. MS 3036, f. 2v. Biblioteca Nacional de España.

new responsibilities led him to moderate his opinion concerning utilizing pilot narratives.

When Céspedes first wrote the manuscript, he was particularly critical of "others" who used pilot observations to "give rules" without having verified these accounts with the appropriate "experiences." In such an instance he had begged the king to employ a person "skilled in theory and practice" to prepare this information. On a later date, perhaps 1598 (the date on the manuscript's cover), and thus after he had carried out the *enmienda*, he revisited this section of the manuscript. He added a marginal note indicating that in the meantime—that is, before a skilled person could be sent to collect this information—a new general description and six rutters had been made "following the most common opinion and accounts of persons that have sailed and according to some eclipses that following Your Majesty's order were observed in Spain and in the Indies"[73] (see figure 7.1). In an even later note, likely written after the *Regimiento* was prepared for publication, Céspedes changed the conclusion of the introduction completely. He deleted the request to have a skilled person prepare this information and explained that when he had found it necessary in the treatise to rely on pilots' testimonies, he selected out of them only those

73. "[S]iguiendo la mas común opinión y relaciones de personas que han navegado y según algunos eclipses que por el mandato de V. Mag. se observaron en España y en las Indias se ha hecho una descripción de todas las costas de mar que hasta ahora se saben." BN, MS 3036, f. 2v.

facts that were held to be the common and reasonable opinion. In all other instances where a proof (*demostración*) was required, he stressed, the matter was treated mathematically.[74]

It was not until Céspedes had to defend the results of the *enmienda* before the panel of cosmographers summoned by the Council of Indies that we see evidence of how his epistemological approach had changed to accommodate pilots' opinions. Among the most interesting aspects of the controversy surrounding the cosmographical review were the arguments Céspedes deployed to defend his new maps. He resorted to three different sources of evidence: pilot accounts, eclipses, and *relaciones* from experts. Each served to buttress a particular aspect of the methods he had employed to conduct the *enmienda*. For example, he used the pilot accounts and eclipses to support his chart of the western navigation: "As far as correcting the navigation chart, not a single thing should be changed of the western navigation, since pilots find it to be so, based on one hundred years of experience sailing [the route], and eclipses observed here and there [both] coincide with the description shown in the chart, save for some shallows that are being discovered daily, as shown by the opinions of forty pilots I have submitted."[75] The fact that pilots had confidence in the chart and found it to be accurate based on their experience (*experiencia*) was now reason enough to leave the chart as it stood. The results of the eclipses simply confirmed what the pilots had established as fact. Céspedes, speaking as "someone who has sailed and has studied mathematics for forty years," warned that creating new charts based solely on

74. Annotations continue throughout the manuscript toning down his original criticism of pilots' observations. For example, after discussing an instrument that corrected for the deviation of the compass needle and that he considered essential to amend "with the necessary perfection" navigational charts, he added that the instrument was necessary because pilots tended to differ so much in their accounts of their navigations that it was difficult to give then credit. Céspedes rubbed out the rest of the sentence and added the caveat, "for now we shall make do with what they agree on the most and is closer to the truth." (*Porque son tan varias las opiniones de los Pilotos que no se puede tener buen crédito de ellas pero por ahora se ha de pasar con lo que es mas semejante y llegado a la verdad.*) BN, MS 3036, f. 103.

75. "[E]n lo que toca a la enmienda de la carta, no debe mudarse cosa alguna en la parte de la navegación occidental, por hallarla así los pilotos con experiencia de cien años que se navega, y los eclipses que allá y acá se han observado conforman con la descripción de la carta, salvo algunos bajos que cada día se van descubriendo como lo aprecian los pareceres de cuarenta pilotos que aquí presento." This and the discussion that follows refer to AGI, P-262, R. 2\1\3–6, Andrés García de Céspedes, "Memorial sobre junta de cosmógrafos," January 8, 1599.

the opinion of men "who had never seen the sea"—a biting reference to the lack of practical experience of some members of the panel—would endanger current navigations and result in losses of life and property. He argued that pilots were familiar with the rutters currently in use and used them to navigate to the Indies, something they would be reluctant to do if the charts were drastically changed. But by adding to his epistemological criteria the weight of the results of Velasco's lunar eclipse project, he was able to demonstrate mathematically the validity of key latitudes and longitudes and add a dimension of mathematical coherency to the pilot narratives.[76] His new charts, he maintained, were designed to satisfy both pilots and cosmographers.

In the same document he conceded that in the case of the rutters to northern Europe and the Mediterranean, the charts might have certain "parts not placed according to their latitude and longitude." He permitted this, he explained, because sailors in these waters navigated using distance and bearing rather than latitudes and changing the charts would make them unusable. He suggested amending only the Universal Map to reflect the true latitude and longitudes. Céspedes addressed the cosmographers' concerns about his map of the Indian Ocean and the location of the western line of demarcation by explaining (as Gesio and Santa Cruz had done before him) that the Portuguese were notorious for having reduced distances in their rutters of the South Atlantic and Indian oceans so as to place the Moluccas in their side of the line of demarcation. To correct for this error, he chose to place the demarcation line through Malacca based "principally" on the opinion reached in Badajoz in 1524 and supported by the work of Alonso de Santa Cruz, Juan López de Velasco, Pedro Ruiz de Villegas, Jacobo Castaldo, Sebastian Cabot, and Miguel López de Legazpi. Therefore, Céspedes concluded, based on "the *relaciones* that have been presently found and eclipse observations and with the approval of so many pilots," unless someone presented evidence to refute them, the new charts and rutters should be implemented at the Casa de la Contratación.

Céspedes expanded these arguments in the second part of the *Regimiento de navegación*, the *Hydrografía*, when he presented the evidence and the methods he used for preparing the new universal map and the six regional charts (figure 7.2). He prefaced his new world map with an explanation of the criteria he used for selecting latitude and longitude data and

76. For more on the use of the implied certainty of mathematics during the early modern era as a tool to resolve "epistemological concerns," see Dear, *Discipline and Experience*, 31–62.

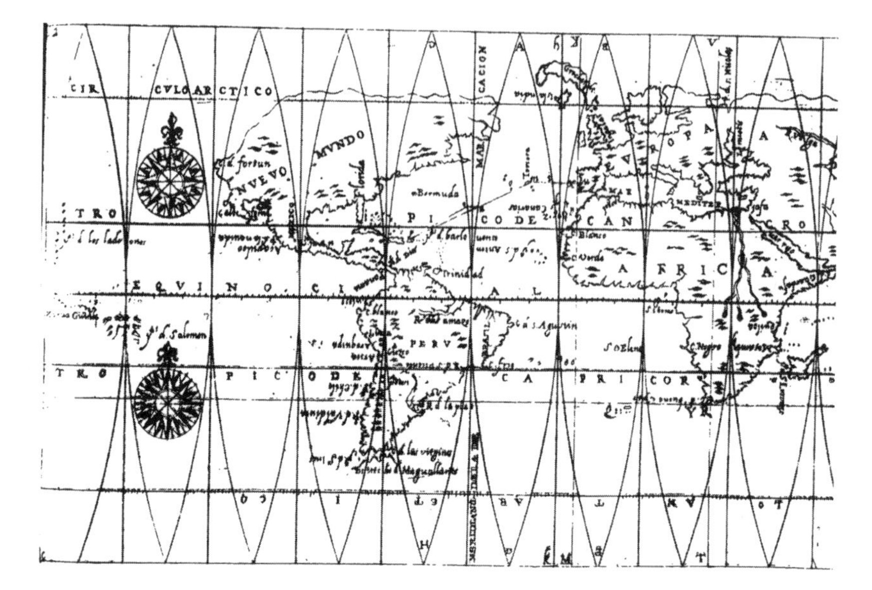

Figure 7.2. Illustration from the *Regimiento de navegación e hydrografía* (1606) based on Céspedes's universal map.

made it clear that although he relied on many mathematical operations and astronomical observations for his conclusions, he had not ignored the opinions of pilots and experienced navigators—particularly in cases where eclipse observations were not available. In instances where both were available, he had "accommodated" (*acomodado*) the pilot's rutter to the eclipses. For this astronomer, the stars never ceased to be the final arbiter.[77] Yet in instances where eclipse observations were not available, as in the case of the route to the East Indies, Céspedes preferred to rely on the testimony of Portuguese pilots, since they had the most experience sailing that route.[78] Otherwise, he relied entirely on Spanish accounts and the results from the lunar eclipse observations generated by Velasco's project.[79]

77. "[N]o apartándome de la común opinión de los navegantes. He tomado por fundamento algunas de ellas, para describir las partes donde no había observación de Eclipse, acomodando las tales derrotas, con la observación de los Eclipses." García de Céspedes, *Regimiento de navegación*, 117v–118.

78. For the eastern navigation, he used primarily the account of João de Castro and some testimonies given by pilots when Archduke Albert attempted to reform the Portuguese *padrón*. Ibid., 131–39.

79. "[M]uchas observaciones de Eclipses de la Luna, con las cuales se averigua la

Throughout the book he systematically examined all the available lunar eclipse data until he was satisfied that he could determine key longitudes with certainty. These longitude coordinates then served to establish the reference points from which rutters and coastal outlines originated. Using this approach, Céspedes justified a major revision to the map of South America which included locating the entrance to the Strait of Magellan and Puerto Viejo in Peru in the same meridian. He explained that on September 4, 1588, per Velasco's instructions a number of prominent persons (*personas de la justicia y Regimiento*) observed a lunar eclipse in Puerto Viejo, on the coast of Peru. The eclipse began at 20:54 and ended at midnight. He, "very diligently," observed the same eclipse in Lisbon and noted that it had begun at 1:49. Comparing the times suggested that there was a longitudinal distance between Lisbon and Puerto Viejo of 73° 45′. The same event was also observed in the Cuidad de los Reyes (Lima) and in Arequipa, and the results, which he does not include "to avoid prolixity," confirmed that the coast of the South American continent ran due north-south until the entrance to the Strait of Magellan at 52° S. This coincided, he noted, with the opinion of Pedro Sarmiento de Gamboa and his pilot. As a result, Céspedes placed the entrance to the Strait of Magellan and Puerto Viejo on the same meridian.[80]

For the coastal details in the specific maps, Céspedes relied exclusively on the accounts given by "the most expert pilots found here in Seville," which he corroborated with other pilots familiar with the routes. For example, in the case of Florida he used the coastal description prepared by "Captain Pedro Bernal Cermeño and pilot Juan de Coy, who, sent by the Governor of Cuba in the year 1595, surveyed [*costearon*] all the keys on the coast of Florida."[81]

As the marginalia examined above indicates, Céspedes's willingness to "accommodate" the pilots was the result of having been *ordered* to consider their opinions when reforming the *padrón*. He had not been disposed toward giving pilot accounts much credit before he took on his responsibilities at the Council of Indies. Even once he acquiesced in terms of conferring with pilots, Céspedes had a tool at his disposal that he wielded to trump any pilot account: the results of Velasco's eclipse project. These

---

longitud de las partes donde se hicieron: y las longitudes de las partes donde no hubiere observación de Eclipse, se procederá por la distancia de camino que los navegantes han hallado" (ibid., 140).

80. Ibid., 153v–154.

81. Ibid., 160–61.

observations, and not pilot accounts, were used throughout the *Hydrografía* to establish key geographical coordinates. He used the account of one nameless pilot for the rutter between Cebu (Philippines) and Navidad (Mexico), but Céspedes reassured the reader that the longitudes calculated from the pilot's account coincide with those determined by Velasco. As far as the specific charts were concerned, Céspedes was indeed willing to go with what he considered to be most common opinion of pilots. But for the pragmatic Céspedes this decision was based on a careful review of the quality of the geographical material before him and not on deference to the pilots' *experiencia*. Aware that most pilot descriptions of local hydrographic features had been collected using unreliable methods, he knew he lacked the type of information needed to prepare accurate regional maps. If pilots trusted the current maps and found them to be true representations (*representación verdadera*), he simply chose to leave well enough alone. Still, the cosmographer made sure that key geographical coordinates used in his universal map, which unfortunately is now lost, either were derived from eclipse observations or had been vetted by other cosmographers.

Céspedes's use of eclipse observations taken by others might suggest that he was working within an epistemic structure that relied on eyewitness accounts in order to compose a cosmographical description. This is true only in the broadest sense, however. Céspedes understood the value of establishing the credibility of his eclipse observers and the fact that they had followed a theoretically sound approach. But whereas Velasco's *relaciones* questionnaire and eclipse projects were designed to establish the credentials of the collected cosmographical information by ensuring that they remained as unmediated expressions of eyewitness experiences, Céspedes's approach was to act as interpreter and translate the observations, as well as the pilots' account, into the language of mathematical cartography.

The descriptive accounts intrinsic to Renaissance cosmography in general fell short of what mathematician-cosmographers preferred as sources of facts. For although Velasco had attempted to comply with the descriptive and mathematical aspects of Renaissance cosmography with his questionnaires and eclipse observations, he never produced a cohesive body of cosmographical knowledge that gave just testimony of his effort. Velasco's important contribution to cosmography was the development of a rationalized program for collecting knowledge about the New World. It would be up to Céspedes to implement a cosmography at the Council of Indies that insisted on an empirical and mathematical approach and privileged as sources of geographical information only those descriptive accounts that could be shown to be mathematically coherent. This was not, however, a mathematical coherency to be achieved for its own sake but rather one deployed selectively with the final purpose of producing a useful result. Céspedes's epistemic approach was molded by his desire to demonstrate the validity of his results by using mathematical proofs, but it was in equal measure shaped by the pragmatism demanded of his position as cosmographer major of the Council of Indies. Operating within the new economy of patronage that surrounded the court of Philip III, Céspedes took full advantage of the opportunity afforded by the publication of the *Regimiento de navegación e hydrografía* to cement his reputation as an astronomer and mathematical practitioner.

In the *Hydrografía*, the cosmographer of the Council of Indies had some harsh words for his predecessor. Much to Céspedes's dismay, neither the printed instructions nor Velasco's papers explained the method for carrying out the necessary—and complicated—mathematical computations to translate into longitudes the drawings made by observers showing the moon's shadow cast on the instrument of the Indies at the beginning and end of the eclipse. Neither did Céspedes find among Velasco's documents any evidence that Velasco had interpreted the observation results. Céspedes went so far as to hint that Velasco might not have known how to do the computations! Why take such pains to point out Velasco's mistakes and shortcomings, even while he found the results so valuable?

By criticizing Velasco, Céspedes was making two points about the methods used in the eclipse project. Not only were the mathematical calculations difficult and beyond the reach of most, but even the trusty astrolabe in the hands of a cosmographer could yield incorrect results. The motivation for his critique becomes clear when he proposes to solve these two problems with a new calculation instrument of his own invention[82] (figure 7.3). With the eclipse project no longer subject to the secrecy provisions, Céspedes took the opportunity to publicly promote his own inventions and increase his prestige as a mathematical practitioner.

While his position as the cosmographer major carried with it an implied authority over cosmographical matters—one Céspedes had been willing to

82. The instrument is intended as a computational aid to assist in solving astronomical problems involving ecliptic and horizontal coordinates. It consists of a revolving transparent disk with a universal stereographic projection, mounted on a disk also engraved with a universal stereographic projection. It uses the projection typical of the Islamic *saphea* astrolabe, reintroduced to Europe by Gemma Frisius as the "Catholic astrolabe." Céspedes does not include in either section an explanation of the theory behind the instrument but refers the reader to his (now lost) treatise on the astrolabe.

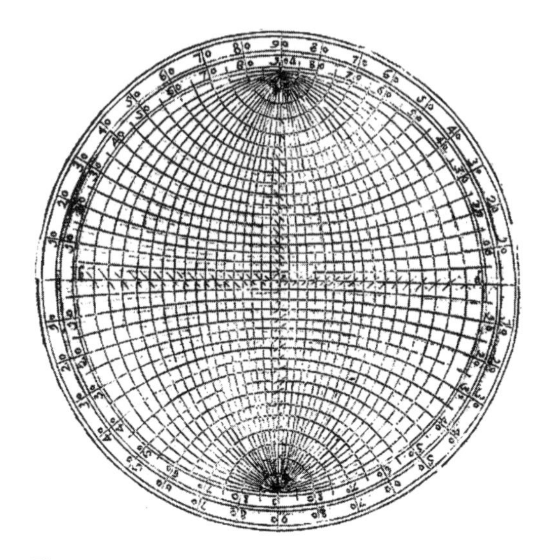

Figure 7.3. Computational instrument from the *Regimiento de navegación e hydrografía* (1606).

assert during the *enmienda*—it was an authority that could be easily undermined and that Céspedes took pains to protect. He was often called upon to review projects by inventors and explorers alike. For example, in 1597 Fernando de los Ríos Coronel, solicitor general of the *audiencia* of Manila, wrote a lengthy letter to the Council of Indies discussing several projects. In addition to his invention of a new astrolabe, he presented a proposal for an expedition to seek a northern passage connecting the northern Atlantic with the Pacific through the Strait of Anian.[83]

Céspedes was also asked by the Council of Indies to review the results of the expedition of Sebastian Vizcaíno, who was sent to explore the coast of-California from Mexico to beyond Cape Mendocino. Based on the observations collected during the voyage, Céspedes issued an opinion endorsing the use of the port of Monterey as a suitable spot for the Manila galleons to stop for resupply and repairs after the arduous Pacific crossing. When the project

83. Ríos Coronel based his proposal on an account found among Martín de Rada's documents that claimed the existence of such a passage. "Memorial de Hernando de los Ríos sobre navegación a las Filipinas, incluye noticias de Martín de Rada, 1597," in Fernández de Navarrete, *Colección Fernández de Navarrete*, 18:633–42. The original is AGI, Filipinas-35, N. 15, with supporting documentation in AGI, Filipinas-18B, R.7, N. 68, "Carta de Fernando de los Ríos Coronel dando cuenta del astrolabio que había inventado," June 27, 1597, Manila.

to explore more fully the area around Monterey got under way, Céspedes prepared "a set of questions regarding the situation and condition of the port which were to be answered by Vizcaíno or by whoever accompanied the expeditions and was competent to do so."[84] The relevant authorities were asked to instruct all who visit the new port over the next couple of years to send the royal cosmographer accounts and descriptions of the place.[85]

By far the most notorious of the projects Céspedes evaluated concerned those submitted by projectors claiming to have solved the problem of calculating longitude both on land and at sea. The topic was not limited to Spain; English and Continental projectors—including Galileo—made similar claims during the same period.[86] In Spain, however, the quest to determine longitude at sea entered a particularly vigorous phase marked by a renewed interest in using the "fixed" compass needle (*aguja fija*—a term alternatively used to denote a compass needle that always pointed to the true north or one that employed the effect of magnetic declination to determine longitude at sea). Juan Arias de Loyola, the Portuguese Luis de Fonseca Coutinho, and others presented proposals for instruments or methods based on this concept that the cosmographer major had to evaluate. A skeptical Céspedes quickly dismissed some and recommended that others be subjected to careful sea trials and experiments.[87] Céspedes, however, had little faith that anyone would solve the problem of longitude at sea during his lifetime. His many years of observing the heavens and building instruments had convinced him that longitude at sea could not be calculated "using either the movements of the heavens or the movements of elemental things."[88]

Chroniclers and Historians

While Céspedes focused on getting the *Regimiento* and *Hydrografía* published, the Council of Indies did not entirely abandon collecting descriptions of

84. The content of the questionnaire is unknown. Wagner, *Spanish Voyages*, 275.

85. BL, Add 13976, f. 469–72, "Carta del rey a Pedro de Acuña, Gobernador y Capitán General de las Islas Filipinas sobre descubrimientos en las costas de California para resguardo de los galeones de Manila," August 19, 1606.

86. Cormack, *Charting an Empire*, 37.

87. Their records are at AGI, P-262, R. 3 for Arias and R. 4 for Fonseca. Fernández de Navarrete summarized the documents in his *Disertación sobre la historia de la náutica*.

88. To his arguments against using the compass needle for the purpose of determining longitude, he added a reasoned explanation of the problems of using solar, lunar, and stellar position tables to calculate longitude, as well as the mechanical difficulties inherent in sand, water, and mechanical clocks. García de Céspedes, *Regimiento de navegación*, 106v–10.

the Indies. The Council found in its new president, the Count of Lemos, a champion of the questionnaire, and the practice was briefly revived. In 1604–5, the Council of Indies issued a new questionnaire, composed of 355 questions printed in a pamphlet of eight folios.[89] It is unclear to what extent Céspedes participated in this effort, for the questionnaire was promulgated to the Indies under the signature of the Count of Lemos.[90] Lemos may have personally directed the questionnaire's preparation or delegated it to one of the men then serving as chronicler of the Indies, Antonio de Herrera y Tordesillas or Gil González Dávila. Pedro de Valencia did not start serving in that capacity until 1607.[91]

The 1604 questionnaire yielded a rather meager number of responses, likely the result of local administrators' complaints about its length and difficulty. Thirty-one responses have been identified, fourteen pertaining to South America and seventeen to New Spain.[92] Historian Silvia Vilar finds in the new questionnaire's length and format an effort to get the respondent to give concise, preferably "numerical and quantitative" answers.[93] She also finds a greater use of American terms in the questions, which indicates sensitivity on the part of the author to express the queries in a manner comprehensible to the respondents. The new questions reflect an increased preoccupation with economic, communications, and demographic issues. They only halfheartedly request geographical information. The questions never address latitudes and longitudes, asking only for estimated distances from nearby towns and their names; similarly, they do not ask for maps.[94] Answers to the questionnaire could not be used to comprehensively describe the geographical features of a territory using mathematical car-

89. BN, MS 3035, "Interrogatorio para todas las Ciudades, Villas y Lugares de Españoles y Pueblos de Naturales de las Indias Occidentales, Islas y Tierra Firme: al cual se ha satisfacer, conforme a las preguntas siguientes, habiendolas averiguado en casa pueblo con puntualidad y cuidado." Published in Solano, *Cuestionarios*, 97–111.

90. AGI, IG- 428, L. 32, f. 89, "Real Cédula al Conde de Monterrey, virrey del Perú, enviándole para que conteste un interrogatorio sobre la descripción de aquellas tierras," January 25, 1605, Valladolid.

91. AGI, IG-874, "Título a Pedro de Valencia de Historiographo y Choronista general de las Indias Occidentales," May 11, 1607.

92. The number of replies is based on León Pinelo's inventory of 1636. Cline, " *Relaciones geográficas* of the Spanish Indies, 1577–1648," 230.

93. Vilar, "Trajectoire des curiosités Espagnoles sur les Indes," 258–67.

94. There is an ambiguous query that might ask for latitudes: "125. In what graduation is this town by the part of south or north parts" (*125. En q- graduación esta este pueblo por la parte del Sur, o del Norte*). BN, MS 3035. f. 2v.

tography. Likewise, there is no interest in inquiring about a region's history or that of its native inhabitants, and the questions that pertain to natural history deal exclusively with the current and potential extractive use of natural resources.

These 355 questions, rather, sought to collect quantifiable information upon which to base administrative decisions. The author of this questionnaire was not concerned with collecting material for composing a complete cosmographical description or even collecting the information deemed appropriate for one, as Velasco had continued to do with his questionnaires. Collecting this information and presumably composing the corresponding brief for the Council of Indies did not require the expertise of a humanist cosmographer, or for that matter that of a mathematical cosmographer.

The genesis of the questionnaire lies in a renewed effort on the part of the monarchy to increase revenue from the Indies.[95] Lemos, aware that any efforts to increase revenue had to target untapped sources of wealth overseas, hoped that this questionnaire would supply such information. He stated his plan in the dedication to a work actually authored by Pedro de Valencia but published under Lemos's name, the *Descripción de la Provincia de los Quixos* (Madrid, 1608).[96] Pedro de Valencia apparently prepared it as a new model for regional descriptions, based on the replies to the 1604 questionnaire and intended for the Council's use. In the book's dedication Lemos expressed the hope that these would provide the members of the Council of Indies with timely information about the territories they governed, information they had heretofore lacked (an echo of Ovando's complaints of forty years before).

From Lemos's remarks it is evident that the material collected by Velasco's 1577 and 1584 questionnaires was now considered dated. Lemos explained that a "very copious questionnaire" was sent to the Indies to inquire about "the finances, occupations, and commerce of the natives, navigation, and other details of this sort."[97] He noted that he ordered local administrators to update the replies every five years, and remarked that the replies from Quito and Panama had arrived and "give clear testimony of

95. Since 1600 the Council of Indies had been under pressure by the Duke of Lerma to look into ways of increasing receipts from the Indies. Schäfer, *Consejo de Indias*, 1:178.

96. A recent edition of Pedro de Valencia's work includes an excellent introductory study and transcribes all the *relaciones de Indias* attributed to the author at BN 3064 and BN 594. Pedro de Valencia et al., *Obras completas: Relaciones de Indias: Nueva Granada y Virreinato de Perú*, 2nd ed., vol. 5/1 (León: Universidad de León, 2001).

97. "[L]as haciendas, ocupaciones y comercio de los naturales, la navegación de sus Mares y otras cosas menudas de este género" (ibid., 110).

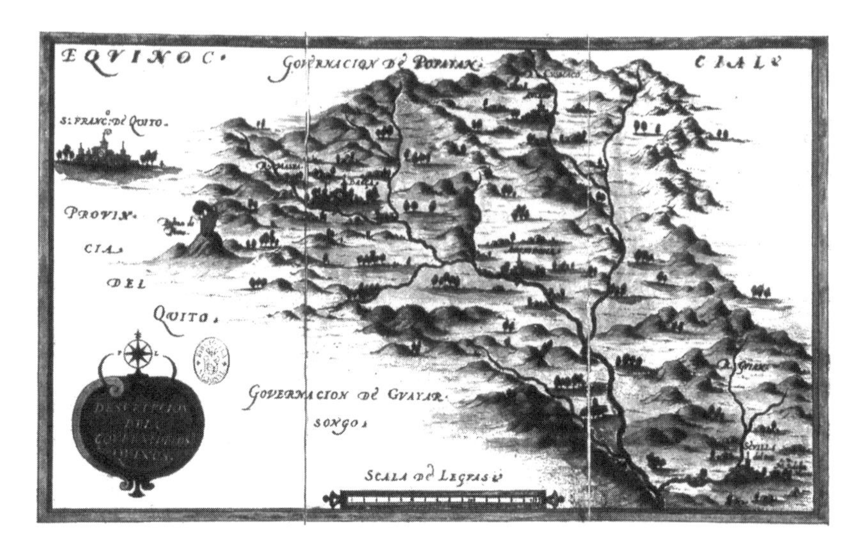

Figure 7.4. Map of the *Descripción de la gobernación de los Quixos*. BNE, MS 594, f. 5. Biblioteca Nacional de España.

the utility" of this mode of collecting information. Lemos explained that the book, and presumably other descriptions prepared using it as a model, included only those aspects "that the experience of my post has taught me to note and that are important for the good administration."[98]

The description of the Quixos is divided into four sections: *lo natural, lo moral, lo eclesiástico, lo militar*, followed by short descriptions of principal Spanish cities. It also included a chorographic map of the region (figure 7.4). It was the first of twenty-one descriptions written by Valencia that survive. All are similar in approach, style, and content to the Quixos description. In contrast to Velasco, Valencia used a number of responses to the 1604 questionnaire to prepare what I categorize as descriptive briefs for the Council of Indies. Valencia, however, used a different thematic organization from the one used in the 1604 questionnaire, placing at the beginning of the narration the natural and productive aspects of the territory for emphasis. Once again, the act of composing consisted in extracting information from the questionnaire, only this time the author's focus was not in presenting a comprehensive cosmographical survey of the territory but on framing the natural and extractive potential of the territory within each discrete

98. "De lo demas he recogido aquello solamente que la experiencia del oficio me ha enseñado a notar, que es de importancia para la buena expedición de los negocios" (ibid., 111).

administrative unit. Valencia does not pretend to be writing a cosmography but rather a well-organized and synthetic report. He considered the project completed in 1613.[99]

Like Céspedes's *Regimiento*, this 1604 questionnaire and Valencia's *Relaciones de Indias* mark the complete rupture within the Council of Indies with the humanistic cosmographical heritage that had yielded Ovando's reforms and still found expression in Velasco's questionnaire. Even in the hands of Pedro de Valencia, one of Spain's leading humanists, albeit a member of the last generation that would claim this moniker,[100] the geographical descriptions retained the factual atomatization that originated with Velasco's questionnaire. By the early seventeenth century, the comprehensive, all-inclusive approach of Renaissance cosmography (geography, ethnography, natural history, and history)—the approach Ovando thought would provide a method and taxonomy to encompass the reality of the New World—was no longer used at the Council to set the thematic agenda regarding what constituted knowledge about the Indies. Rather than trying to understand the whole, the Council now focused on having at hand the expertise to support its immediate administrative duties and responsibilities. These works, with their characteristic pragmatism and utilitarian aims, were the last vestiges of practices associated with Renaissance cosmography at the Council.

With the mathematical aspects of cosmography in the hands of a qualified mathematical practitioner, the descriptive aspects of cosmography—those aspects of the discipline that necessitated words in order to find expression—became the domain of the chronicler major of the Council of Indies. When Antonio de Herrera y Tordesillas assumed the post in 1596, he began work on what would become the Spanish monarchy's official history of the discovery and conquest of the New World. The first part of the *Historia general de los hechos de los castellanos en las Islas y Tierra Firme del Mar Océano*, known as the *Décadas*, appeared in 1601. In the manner of an introduction for the whole work, the first volume contained a geographical section titled *Descripción de las Indias Ocidentales* that consisted essentially in a recast version of Velasco's *Sumario*.[101] And in a move that reflected the change

99. Pedro de Valencia et al., *Obras completas*, vol. 5, *Relaciones de Indias*, no. 2, *México*, 2nd ed. (León: Universidad de León, 2001), 52.

100. Bustamante García, "Círculos intelectuales," 45.

101. Herrera added some historical material and apparently consulted some *relaciones geográficas* to supplement Velasco's work, as in the case of Diego García de Palacio's *Descripción de Guazacapán, Los Izalcos, Cuzcatln y Chiquimula (Guatemala)* (Acuña, ed., *Relaciones geográficas*, 1:252). Cuesta Domingo identified most of Herrera y Tordesilla's sources

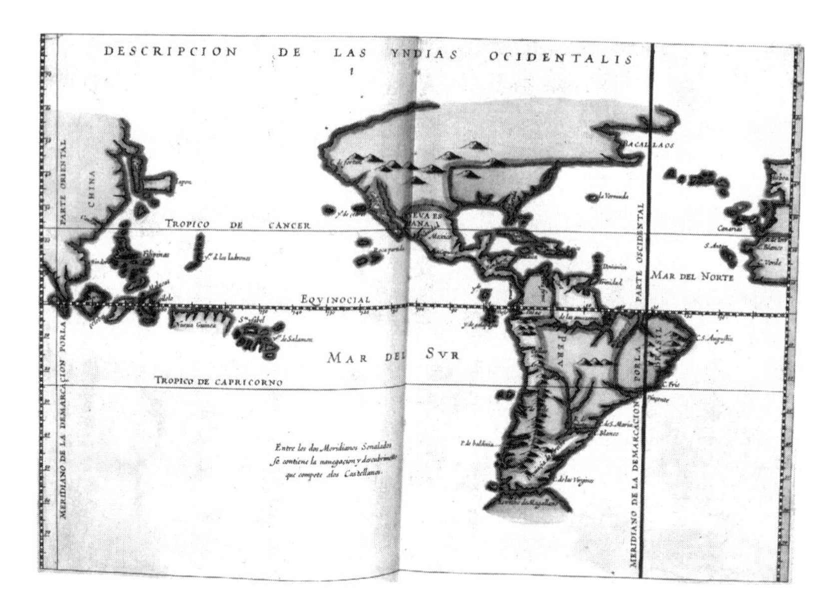

Figure 7.5. *Descripción de las Indias Occidentales*, after the one published in Hererra's *Décadas*. Library of Congress.

in secrecy policy concerning the geography of the New World, Herrera included maps based on those Velasco had so conscientiously kept within the confines of the Council chambers[102] (figure 7.5). Herrera y Tordesillas's selection of these maps for the geographical introduction to his book—from what must have been a rich cartographic collection at his disposal—was wise. Velasco's maps were intended to function as subordinate to the text, and Herrera's maps fulfilled the same function in his book. Geography in Herrera y Tordesillas's text was secondary, merely a backdrop on which to paint the heroic deeds of the Spanish in the New World.

---

in the introduction to Antonio de Herrera y Tordesillas, *Historia general de los hechos de los castellanos en las Islas y Terra Firme del Mar Océano, o, Décadas de Antonio de Herrera y Tordesillas (1601–1615)*, ed. Mariano Cuesta Domingo, 4 vols. (Madrid: Universidad Complutense de Madrid, 1991), 1:57–59.

102. The maps vary somewhat from the geographical coordinates in Velasco's *Geografía* and the *Sumario*. Although the general appearance of the American continent is similar, the Asian part has been redrawn, perhaps to satisfy Gesio's concerns. One toponymic coincidence, however, suggests that Herrera used Velasco's maps as guide. The River Ganges still appears on the coast of China, a glaring mistake that Gesio found outrageous. For a discussion of the relationship between these two sets of maps, see Cuesta, "Cartografía grabada en la obra de Antonio de Herrera."

Céspedes recognized the maps' function as a complement to a descriptive text, rather than as a definitive cartography of the Indies, when he endorsed the publication of the *Décadas*. He issued an opinion certifying that Herrera y Tordesillas's geography and cosmography were prepared "according to what is commonly practiced and is accepted among the persons who sail, and according to what happened during the first discovery and what has been found later. . . . [P]rinting it will result in much utility and honor the Castilian nation."[103] Recall that when Céspedes issued this opinion he was involved in preparing what he and the Council considered to be the documents that defined definitively the geography of the New World: the *padrón real* and the cartographic corpus that accompanied it. Herrera's use of Velasco's old maps to illustrate a historical narrative did not impinge on the cosmographer's parallel—but now conceptually distinct—project.

The desire to write a universal cosmography, that chimerical dream of Renaissance humanists, had dissolved when confronted with the challenge of incorporating the New World into its conceptual framework. Its dissolution, however, did not put an end to practices that its adepts had been refining for over a century. The resulting divorce between the descriptive and mathematical aspects of cosmography opened the door to specialization, particularly among practitioners of the mathematical aspects of the discipline. Historians have identified similar changes in the discipline of cosmography among French and German cosmographers, while by the 1620s there was a similar drive toward specialization among English geographers.[104]

Andrés García de Céspedes brought to his post the firm conviction that the value of cosmography rested on the mathematical aspect of the discipline. The perspicacious cosmographer, however, understood and accommodated to the institutional settings of the Council of Indies and the Casa de la Contratación. He knew the limitations of both the instruments then available (astrolabes, astronomical tables, and maps) and the users of these tools (pilots and bureaucrats). By developing an epistemic approach

103. "[T]odo está conforme a lo que comúnmente se platica y está más recibido entre todos los que navegan y conforme a lo que pasó en el primer descubrimiento y a lo que después acá se ha hallado y que adonde quiera parecerá bien, y que se puede muy bien imprimir y que de la impresión resultará mucha utilidad y honra a la nación Castellana, y lo firmé de mi nombre a 3 de enero de 1599" (Herrera y Tordesillas, *Décadas*, 1:122).

104. Lestringant, *Mapping the Renaissance World*, 129, and Cormack, *Charting an Empire*, 37–42.

rooted in mathematical rationalism that bowed to the experience of sailors and the exigencies of bureaucrats, he bridged the gap between the eyewitness-based episteme of Renaissance cosmography and the mathematical and experimental approach of the seventeenth century.

With Philip III, the interest in making geographical information public introduced a new role for geographical information about the New World that went beyond its strictly utilitarian use by the Council of Indies and the Casa de la Contratación. Making this information public did not mean that geographical knowledge about the New World had ceased to be important to the Spanish monarchy. Cosmographical knowledge was now repackaged strategically for public consumption. Rather than being kept secret for fear of usurpation or for defensive reasons, as had been the approach taken during most of the reign of Philip II, geographical knowledge had won a new currency. Its value now consisted in how well it served to buttress the Spanish monarchy's territorial claims. To achieve this, it had to be publicly deployed. It became one of the tools through which the Spanish monarchy announced to the world the extent of its empire, and thus cosmography ceased to be the secret science of the Spanish empire.

CONCLUSION

When Ptolemy's humanist imitators laid down the foundation of Renaissance cosmography during the fifteenth century, they adopted methodologies and modes of representation that best served their intended objective: a bookish attempt at a synthetic description of the world. In their attempt to answer Ptolemy's call to describe all the known world, they compiled elegant—but static—images of the world that inscribed new territories and peoples within the worldview of early modern Europe. The discovery of the New World and a century of continuous geographical discoveries tested the capability of the relatively new discipline to serve as a conceptual approach and a set of practices from which to explicate hitherto unknown lands and oceans. For while the discoveries acted as a catalyst for cosmographical practice and elevated the profiles of its practitioners, it also strained the epistemic approach, methodologies, and textual products associated with Renaissance cosmography.

In Spain, the discovery obligated royal cosmographers to abandon the humanistic practices that had given Renaissance cosmography its original character in favor of new practices that accommodated the fluidity of cosmographical information emanating from the New World. For these cosmographers, practice took place at the very frontier between what was known and what was not known. Each new *relación* and pilot chart, each new letter from the Indies and new account of a discovery, presented the cosmographer with a difficult epistemic problem that required the development of new methods for firmly establishing matters of cosmographical fact. To this end they designed new ways of collecting and organizing knowledge and developed ways to test the validity of the information, all the while ensuring that the new facts served a utilitarian purpose.

In the Spain of the 1570s, the practices associated with Renaissance cosmography entered a period of reconceptualization that required the development of new epistemological approaches, modes of representation, and methodologies that fundamentally altered the discipline. Renaissance cosmography, as the case of Alonso de Santa Cruz demonstrates, indeed provided a comprehensive—and perhaps too ambitious—conceptual framework from which to approach the New World. Yet aspects of the discipline, particularly those associated with humanism, proved inadequate when pressed into the service of the Spanish empire. In the eyes of Juan de Ovando, despite its apparent limitations, cosmography still provided the comprehensive—we might even say holistic—image of the world he thought essential to gain an understanding of the land and people of the New World. Ovando's legal approach to the discipline was a well-intentioned effort to ensure that the members of the Council of Indies, who were largely unaware of the realities of the new lands, had at their disposal the information necessary to govern the New World responsibly. Notwithstanding the law's attempts to respect the practices associated with cosmography, it narrowed the scope of cosmographical production only to work deemed "useful" by members of the Council of Indies. This utilitarian mandate, institutionalized and codified, would transform cosmographical practice at the Council of Indies so that it shed its humanistic heritage in favor of quantifiable and discrete modes of collecting, compiling, and presenting the New World.

Juan López de Velasco, reluctant at first to abandon the mode of representation associated with Renaissance cosmography—the all-encompassing magnum opus—would eventually acquiesce. His questionnaires and eclipse project in effect renovated cosmographical methodology to account for the fluidity inherent in cosmographical information during the century of discovery. Velasco's work, however, remained rooted in an epistemology that privileged descriptive firsthand accounts above other sources. Instead of treating these as sources of cosmographical knowledge that required interpretation, he treated them as legal depositions and applied to them a legally inspired criterion that placed the value of the document on a par with the sworn testimony of a witness before a judge. Within this epistemic framework, any manipulation of an eyewitness account meant corrupting the evidence given by a witness and therefore had inherently negative connotations.

In adopting this arm's-length approach, Velasco removed himself from mediating between the facts as stated by the eyewitness and the facts as they appeared compiled in a cosmography, effectively abandoning his cardinal task as cosmographer. I interpret his actions as the necessary culmination of an epistemic approach that placed inordinate value on the testimony of eyewitnesses. Velasco was fulfilling what was essentially a descriptive program whose effectiveness would be judged by how close the description resembled reality—a reality that could be challenged by the arrival of news with the next fleet from the Indies. Within this program, leaving the eyewitness account integral—unmediated—was as close to honoring the legal definition of what constituted truth as possible.

As the cosmographer-chronicler major of the Council of Indies, Velasco became the unwitting exemplar of official cosmography, and his descriptive approach to the discipline became a target for other cosmographers who preferred a wholly different set of practices. For what by the mid-sixteenth century could be described as a discipline that embraced practitioners with different cosmographical styles had by the end of the century become associated mostly with mathematical practitioners. I attribute this change to a revalorization of cosmography's epistemological foundation, that is to say, a reevaluation of the methods employed and the criteria used to ascertain matters of cosmographical fact. Mathematical cosmographers, such as Andrés García de Céspedes, proposed moving beyond a strict reliance on eyewitness accounts; they were convinced that submitting narrative accounts to mathematical analysis would produce new—and more certain—knowledge. Descriptive *relaciones*—from pilots, government officials, or learned observers—generally fell short of what these mathematically oriented cosmographers preferred as sources of facts. It would be up to Céspedes to implement a cosmography at the Council of Indies that insisted on an empirical and mathematical approach and privileged as sources of geographical information only those descriptive accounts that could be shown to be mathematically coherent. The descriptive part of geography became the domain of the historian and chronicler.

Ovando's *Ordinances* had also formalized within the Council of Indies a secrecy policy that determined the public dissemination, or rather lack thereof, of cosmographical knowledge about Spain's overseas empire during the closing decades of the sixteenth century. As the first cosmographer-chronicler subject to the provision of the *Ordinances*, Velasco, perhaps overzealously but definitely with his accustomed efficiency, applied the censorship this secrecy policy implied to any work that discussed the history, geography, natural history, or native peoples of the New World. In his hands, cosmography became the secret science of the Spanish empire. We must not lose sight, however, of the fact that he coordinated a vast and unprecedented information-gathering network, a difficult task given the logistical difficulties imposed by cultural and geographical distance. The storehouse of knowledge about the New World Velasco kept secure in the secret archive of the Council of Indies informed the works of Antonio de Herrera and Pedro de Valencia and a cadre of other seventeenth-century chroniclers of the Council of Indies. It is only when we view Velasco's actions from a presentist perspective that lays on science the burden of openness and collaboration that his actions appear conspiratorial.

As the military and economic advantage gained by keeping this material secret began to wane during the 1590s, this storehouse of knowledge found a new strategic mission. During the reign of Philip III, the geography of the Indies—one aspect of cosmography that did not threaten the internal politics of the colonies—was repackaged and deployed by the cosmographer and the chronicler of the Council of Indies for the purpose of constructing a public image of imperial power and control.

This study of the Spanish royal cosmographers challenges the conclusions of recent studies of patronage and science. Historical work in this area is inclined to associate the emergence of the court philosopher during this time with a gradual weakening of Aristotelian natural philosophy because it allowed scientific practitioners to operate in a court context and outside the restrictive scholasticism of the universities.[1] Yet for Spanish cosmographers working as bureaucrats in an institutional environment, their source of patronage, whether demanding that practice take place secretly or requiring that knowledge circulate in the public sphere, did not produce any challenges to the philosophical underpinnings of Aristotelianism. The institutionalization of cosmographical practice, whether at the *casa*, *consejo*, or *corte*, instead produced a focused effort to develop aspects of the discipline that promised solutions to the pressing problem of an expanding empire.

The works of Santa Cruz, Velasco, and Céspedes are almost entirely devoid of natural philosophical speculation, let alone any challenging of the accepted Aristotelian natural philosophical explanation of natural phenomena. This absence posits an interesting question: was cosmographical practice in Spain so bounded by the cosmographers' role as servants of the empire that little room was left for philosophical speculation? The

1. Westman, "Astronomer's Role in the Sixteenth Century," 122–27. For a case study of a mathematical practitioner who refashioned himself into a natural philosopher, see Biagioli, *Galileo, Courtier*.

question addresses a central theme of the traditional narrative of the history of science, which considers that alongside Baconianism and an interest in mathematizing natural phenomena, a surge in speculative natural philosophy during the early seventeenth century contributed to the intellectual climate that led to the Scientific Revolution. This school of thought suggests that Cartesian and other mechanistic explanations of nature opened the doors to mathematical formulations of natural phenomena, such as Boyle's laws on gases and Newton's gravitational laws.

The material I have presented suggests that it was the exigencies of empire and its associated utilitarian demands that steered Spanish royal cosmographers away from speculative natural philosophy, by rewarding work that represented nature as an inventory of the real, visible, and tangible world. Speculative natural philosophy of the kind that might have led to challenges to Aristotelian natural philosophy promised Spanish royal cosmographers little knowledge that could ultimately be put to practical use and therefore had little value within a system of patronage that rewarded utilitarian results.[2] In this context, it is not surprising to find that the only works of Céspedes that were published under the auspices of the Crown were those with intrinsic utilitarian value. His astronomical works on planetary motions and alternate cosmological systems remained unpublished, not because these might have challenged Aristotelian natural philosophy (which I suspect they did not, given Céspedes's instrumentalist approach) but because they did not have any immediate practical application.

In complying with the duties dictated by their official posts, Spanish royal cosmographers developed new practices that relied on observation, and in some instances experimentation, and that resembled practices

2. The utilitarian aspect of the scientific enterprise in early modern Spain has been the subject of scrutiny over the years. In the late nineteenth century, Marcelino Menéndez y Pelayo, in the context of a critique of science education, blamed Spain's scientific decadence during the seventeenth century on an "obstinate" emphasis on utilitarianism "that reduced astronomy to navigation and mathematics to fortification and artillery" and thus removed Spain from the "chain of theoretical discoveries." See his introduction to Acisclo Fernández Vallín, *Cultura científica en España en el siglo* XVI (Sevilla: Padilla Libros, 1893), facsimile ed., ed. Marcelino Menéndez y Pelayo (Sevilla: Padilla Libros, 1989), xlv. Recently, Navarro Brotóns gives credit to the role of practice in the Spanish scientific enterprise, concluding "that the science of the cosmographers, under the crown's patronage and because it was seen as a means, was limited in its very essence to be a science restricted to technology." Víctor Navarro Brotóns, "Cartografía y cosmografía en la época del descubrimiento," in *Mundialización de la ciencia y cultura nacional: Actas del Congreso Internacional "Ciencia, descubrimiento y mundo colonial,"* ed. A. Lafuente, A. Elena, and M. L. Ortega (Madrid: Doce Calles, 1991), 72.

advocated early in the seventeenth century by Sir Francis Bacon as fundamental for constructing a new philosophy. But in effect a wide gulf separated our cosmographers from Bacon. For Bacon, the frontier of knowledge lay in constructing the tenets of a new philosophy and a new investigative methodology that would explain the natural world. As we have seen throughout this study, Spanish cosmographers indeed developed new investigative methodologies and systems of organizing knowledge in response to their country's encounter with the New World, yet the goals of Spanish royal cosmographers were fundamentally different from Bacon's. They believed—with a conviction that did not need articulation—that Providence had placed in Spain's care a New World that needed to be accounted for and described so that its peoples could be converted to Christianity and governed effectively. This transcendental mission required developing technological tools that facilitated reaching the New World: cartography and navigation. It also required knowing about the natural resources of both land and sea, and finally it meant learning about the history and the people of this new land. Fulfilling this monumental mission did not inherently require the development of a new natural philosophy. The cosmographer's mission was to encompass rather than to explain; Aristotelian natural philosophy sufficed for both.

The work of Spanish royal cosmographers was directed at identifying and organizing useful knowledge for the benefit of the Spanish empire, not at uncovering hidden secrets of nature. They did not seek a causal analysis of the observed phenomena and thus did not develop an epistemic approach that elucidated causality. (Which is, curiously enough, itself anti-Aristotelian, since Aristotle claimed that real knowledge was causal knowledge.)[3] Spanish royal cosmographers operated in a cognitive environment that featured an unquestioned commitment to Aristotelian natural philosophy but that was in effect undermining the very epistemic foundation of that philosophy, the search for causality. By developing refined empirical practices and sophisticated epistemic approaches to elucidate matters of fact, the cosmographers contributed to the viability of an alternative path into the inquiry of nature that skirted the search for causality.[4] Spanish royal cosmographers naturally favored the path that led to

3. For an interesting study of how distinct national styles affected the search for causality in France and England during the Scientific Revolution, see John Henry, "National Styles in Science," in *Geography and Revolution*, ed. David N. Livingstone and Charles W. J. Whithers (Chicago: University of Chicago Press, 2005), 43–74.

4. Skirting speculative thinking was not uncommon, however, among sixteenth-century European mathematical practitioners. For the case of English practitioners,

patronage and the professional satisfaction that came from contributing to the imperial project.

In contrast, natural philosophical speculation of the sort that yielded mechanistic explanations of nature during the seventeenth century came about precisely as the result of a search for causality. In that case, it was the theoretical foundation of Aristotelian natural philosophy that was undermined, not its epistemic approach. This path, once it integrated with experimentalism and refined empirical practices, was one of the roads that led to modern science. Mapping these different approaches suggests to me a starting point for conceptualizing the work of Spanish practitioners of science relative to that of their European counterparts during the seventeenth century and one that might yield new historical explanations of the effects of the discovery of the New World on the development of modern science.

A flood of new knowledge about the world's geography, natural history, and peoples, unprecedented in recorded history, followed the discovery of the New World. Living in Spain during the century of discovery, Spanish royal cosmographers were in an enviable position relative to this new knowledge. For whereas other areas in Europe learned about the New World mostly from often novelized secondhand sources, Spanish cosmographers had direct access to firsthand accounts about the new lands and their peoples. They were closer, figuratively speaking, to the New World than were other European cosmographers, with an access the others could only envy. (For example, after the mid-sixteenth century, geographical information about new discoveries became so difficult for German geographers to obtain that they turned to constructing better maps of Germany instead.[5]) This access allowed Spanish cosmographers to apply to their discipline the rigor Ptolemy envisioned for geography and gave then the confidence to design an epistemic program that established firmly within the realm of reality what constituted a cosmographical fact. With such a wealth of material at their disposal, it is not surprising that the speculative aspects of scientific practice failed to entice them. Writing from his Eurocentric perspective, Pedro Cieza de León reminded his king circa 1556 that if not the Spanish, then "[w]ho will describe the different and great things of [these Indies]: the

---

see Bennett, "'Mechanics' Philosophy," 11. (I thank one of my anonymous reviewers for pointing this out.)

5. Gallois, *Géographes allemands de la Renaissance*, 238.

tall mountains and deep valleys where the discovery and conquest began, the many rivers, so large and so deep, the variety of provinces, all of such different qualities, the differences among its cities and the many cultures, rites, and strange ceremonies of its people, so many birds and animals, trees and fish, all so different and unknown?"[6]

6. "¿[Q]uién podrá decir las cosas grandes y diferentes que en él son, las sierras altísimas y valles profundos por donde se fue descubriendo y conquistando, los ríos tantos y tan grandes, de tan crecida hondura; tanta variedad de provincias como en él hay, con tantas diferentes calidades; las diferencias de pueblos y gentes con diversas costumbres, ritos y ceremonias extrañas; tantas aves y animales, árboles y peces tan diferentes y ignotos?" Pedro Cieza de León, *La crónica del Perú*, ed. Manuel Ballesteros (Madrid: Dastin, 2000), 56.

Bibliography

Acosta, José de. *Historia natural y moral de las Indias, en que se tratan de las cosas notables del cielo, y elementos, metales, plantas y animales dellas: y los ritos, y ceremonias, leyes y gobierno, y guerras de los Indios.* Sevilla: Juan de León, 1590. Facsimile ed., edited by Barbara Beddall. Valencia: Hispaniae Scientia, 1977.

———. *Historia natural y moral de las Indias, en que se tratan de las cosas notables del cielo, y elementos, metales, plantas y animales dellas: y los ritos, y ceremonias, leyes y gobierno, y guerras de los Indios.* Edited by Edmundo O'Gorman. 2nd ed. Mexico: Fondo de Cultura Económica, 1962.

———. *Obras del P. José de Acosta.* Edited by Francisco Mateos. Biblioteca de autores españoles. Madrid: Atlas, 1954.

Acosta Rodríguez, Antonio, Adolfo Luis González Rodríguez, and Enriqueta Vila Vilar, eds. *La Casa de la Contratación y la navegación entre España y las Indias.* Sevilla: Universidad de Sevilla, CSIC, Fundación El Monte, 2003.

Acuña, René, ed. *Relaciones geográficas del siglo XVI.* 10 vols. Mexico: Universidad Nacional Autónoma de México, 1982.

Albares Albares, Roberto. "El humanismo científico de Pedro Ciruelo." In *La Universidad Complutense Cisneriana*, 177–205. Madrid: Editiorial Complutense, 1996.

Albuquerque, Luis de. "Acerca de Alonso de Santa Cruz y de su 'Libro de las longitudes.'" In *América y la España del siglo XVI: Homenaje a Gonzalo Fernández de Oviedo, cronista de Indias, en el V centenario de su nacimento*, edited by F. de Solano and F. del Pino, 189–204. Madrid: CSIC, Instituto Gonzalo Fernández de Oviedo, 1982.

———. *Astronomical Navigation.* Lisbon: Comissão Nacional para as Comemorações dos Descobrimentos Portugueses, 1988.

Alejo Montes, Francisco Javier. *La Universidad de Salamanca bajo Felipe II, 1575–1598.* Burgos: Editiorial Aldecoa, 1998.

Alvar Ezquerra, Alfredo, María Elena García Guerra, and María de los Angeles Vicioso Rodríguez, eds. *Relaciones topográficas de Felipe I: Madrid.* 4 vols. Madrid: CSIC, 1993.

Álvarez Peláez, Raquel. *La conquista de la naturaleza Americana.* Madrid: CSIC, 1993.

———. "Etnografía e historia natural en los cuestionarios oficiales del siglo XVI." *Asclepio* 41, no. 2 (1989): 103–26.

———. "La historia natural en los siglos XVI y XVII." Paper presented at the Jornadas sobre España y las expediciones científicas en América y Filipinas, Ateneo Científico, Literario y Artístico de Madrid, 1991.

———. "Las relaciones de Indias." In *Felipe II, la ciencia y la técnica*, edited by E. Martínez Ruiz, 291–315. Madrid: Actas, 1999.

Andrés, Gregorio de. "Juan Bautista Gesio, cosmógrafo de Felipe II y portador de documentos geográficos desde Lisboa para la Biblioteca de El Escorial en 1573." *Publicaciones de la Real Sociedad Geográfica*, ser. B, 478 (1967): 1–12.

———. "Viaje del humanista Alvar Gómez de Castro a Plasencia en busca de códices de obras de S. Isidoro para Felipe II (1572)." In *Homenaje a don Agustín Millares Carlo*, 608–21. Las Palmas: Caja Insular de Ahorros de Gran Canaria, 1975.

Andrewes, William J. H., ed. *The Quest for Longitude: The Proceedings of the Longitude Symposium, Harvard University, Cambridge, Massachusetts, November 4–6, 1993*. Cambridge, Mass.: Harvard University Press, 1993.

Anonymous. *Vida del Lazarillo de Tormes castigado*. Edited by Gonzalo Santoja. Madrid: Sociedad Estatal España Nuevo Milenio, 2000.

Apian, Peter. *Cosmographicus liber Petri Apiani mathematici studiose collectus*. Landshut, Germany: P. Apiani, 1524. Microform.

Arellano Moreno, Antonio, ed. *Relaciones geográficas de Venezuela: Recopilación, estudio preliminar y notas de Antonio Arellano Moreno*. Caracas: Academia Nacional de la Historia, 1964.

Argensola, Bartolomé Leonardo. *Conquista de las Islas Maucas*. Madrid: Alonso Martín, 1609.

Arribas Lázaro, Angeles. "Unas cartas de Alonso de Santa Cruz." *Asclepio* 26–27 (1974–75): 257–66.

Ash, Eric H. *Power, Knowledge, and Expertise in Elizabethan England*. Baltimore: Johns Hopkins University Press, 2004.

Ashworth, William B. "Natural History and the Emblematic World View." In *Reappraisals of the Scientific Revolution*, edited by David Lindberg and Robert Westman, 301–32. Cambridge: Cambridge University Press, 1990.

Barona, Josep Lluís, and Xavier Gómez i Font, eds. *La correspondencia de Carolus Clusius con los científicos españoles*. Valencia: Seminari d'Estudis sobre la Ciència, 1998.

Barrera-Osorio, Antonio. "Empire and Knowledge: Reporting from the New World." *Colonial Latin American Review* 15, no. 1 (2006): 39–54.

———. *Experiencing Nature: The Spanish American Empire and the Early Scientific Revolution*. Austin: University of Texas Press, 2006.

Barros, João de. *Décadas*. Edited by António Baião. 4 vols. Colecção de clássicos Sá da Costa. Lisbon: Livraria Sá da Costa, 1945–46.

Baudot, Georges. *Utopia and History in Mexico: The First Chroniclers of Mexican Civilization (1520–1569)*. Translated by Bernard R. Ortíz de Montellano and Thelma Ortíz de Montellano. Niwot: University Press of Colorado, 1995.

Beltrán y Rózpide, Ricardo. "América en el tiempo de Felipe II según el cosmógrafo cronista Juan López de Velasco." *Publicación de la Real Sociedad Geográfica*, 1927, 1–48.

Bennett, J. A. "The Challenge of Practical Mathematics." In *Science, Culture, and Popular Belief in Renaissance Europe*, edited by Maurice Slawinski, Paolo L. Rossi, and Stephen Pumfrey, 176–90. Manchester: Manchester University Press, 1991.

———. *The Divided Circle: A History of Instruments for Astronomy, Navigation, and Surveying*. Oxford: Phaidon, 1987.

———. "The 'Mechanics' Philosophy and the Mechanical Philosophy." *History of Science* 24 (1986): 1–28.

Bergevin, Jean. *Déterminisme et géographie: Hérodote, Strabon, Albert le Grand et Sebastian Münster.* Travaux du Département de géographie de l'Université Laval 8. Sainte-Foy, Québec: Presses de l'Université Laval, 1992.

Berthe, Jean-Pierre. "Juan López de Velasco (ca. 1530–1598), cronista y cosmógrafo mayor del Consejo de Indias: Su personalidad y su obra geográfica." *Relaciones* 19, no. 75 (1998): 141–72.

Biagioli, Mario. *Galileo, Courtier: The Practice of Science in the Culture of Absolutism.* Chicago: Chicago University Press, 1993.

Blair, Ann. *The Theater of Nature: Jean Bodin and Renaissance Science*. Princeton, N.J.: Princeton University Press, 1997.

Bouza, Fernando. *Corre manuscrito: Una historia cultural del Siglo de Oro*. Madrid: Marcial Pons, 2001.

Bowen, Margarita. *Empiricism and Geographical Thought: From Francis Bacon to Alexander von Humboldt*. Cambridge: Cambridge University Press, 1981.

Broecke, M. P. R. van den. *Ortelius Atlas Maps: An Illustrated Guide*. Goy-Houten, Netherlands: HES Publishers, 1996.

Brotton, Jerry. *Trading Territories: Mapping the Early Modern World*. Ithaca, N.Y.: Cornell University Press, 1998.

Browne, Walden. *Sahagún and the Transition to Modernity*. Norman: University of Oklahoma Press, 2000.

Buchwald, Jed Z., ed. *Scientific Practice: Theories and Stories of Doing Physics*. Chicago: University of Chicago Press, 1995.

Buisseret, David, ed. *Monarchs, Ministers, and Maps: The Emergence of Cartography as a Tool of Government in Early Modern Europe*. Chicago: University of Chicago Press, 1992.

Burkholder, Mark A., ed. *Administrators of Empire*. Aldershot, UK: Ashgate/Variorum, 1998.

Bustamante García, Jesús. "Los círculos intelectuales y las empresas culturales de Felipe II: Tiempos, lugares y ritmos del humanismo en la España del siglo XVI." In *Élites intelectuales y modelos colectivos: Mundo ibérico (siglos XVI–XIX)*, edited by Mónica Quijada and Jesús Bustamante García, 33–58. Madrid: CSIC, 2003.

———. "El conocimiento como necesidad de estado: las encuestas oficiales sobre Nueva España durante el reinado de Carlos V." *Revista de Indias* 60, no. 218 (2000): 33–55.

———. "De la naturaleza y los naturales americanos en el siglo XVI: Algunas cuestiones críticas sobre la obra de Francisco Hernández." *Revista de Indias* 52 (1992): 297–328.

———. "Francisco Hernández, Plinio del Nuevo Mundo: Tradición clásica, teoría nominal y sistema terminológico indígena en una obra renacentista." In *Entre dos mundos: Fronteras culturales y agentes mediadores*, edited by B. Ares Queija and S. Gruzinski, 243–68. Sevilla: Escuela de Estudios Hispano-Americanos, 1997.

———. *La obra etnográfica y lingüística de Fray Bernardino de Sahagún*. Madrid: Editorial de la Universidad Complutense de Madrid, 1989.

Bustos Tovar, E. "La introducción de las teorías de Copérnico en la Universidad de Salamanca." *Real Academia de Ciencias Exactas, Físicas y Naturales* 67–68 (1973): 236–52.

Butzer, Karl W. "From Columbus to Acosta: Science, Geography, and the New World." *Annals of the Association of American Geographers* 82, no. 3 (1992): 543–65.

Cañizares Esguerra, Jorge. *Nature, Empire, and Nation: Explorations of the History of Science in the Iberian World*. Stanford, Calif.: Stanford University Press, 2006.

———. "New World, New Stars: Patriotic Astrology and the Invention of Indian and Creole Bodies in Colonial Spanish America, 1600–1650." *American Historical Review* 104, no. 1 (1999): 33–68.

———. "Renaissance Iberian Science: Ignored How Much Longer?" *Perspectives on Science* 12, no. 1 (2004): 86–125.

Carabias Torres, Ana María. "Los conocimientos de cosmografía en Castilla en la época del Tratado de Tordesillas." In *El Tratado de Tordesillas y su época*, edited by Luis Antonio Ribot García, 959–41. Madrid: Junta de Castilla y León, 1995.

———. "La medida del espacio en el Renacimiento: La aportación de la Universidad de Salamanca." *Cuadernos de Historia de España* 76 (2000): 185–202.

Casado Soto, José Luis. *Discursos de Bernardino de Escalante al Rey y sus ministros (1585–1605): Presentación, estudio y transcripción por José Luis Casado Soto*. Santander: Servicio de Publicaciones de la Universidad de Cantabria, 1995.

Ceballos-Escalera Gila, Alfonso. "Una navegación de Acapulco a Manila en 1611: El Cosmógrafo Mayor Juan Bautista Labaña, el inventor Luis de Fonseca Coutinho, y el problema de la desviación de la aguja." *Revista de la Historia Naval* 17, no. 65 (1999): 7–42.

Cerezo Martínez, Ricardo. *La cartografía náutica española en los siglos XIV, XV y XVI*. Madrid: CSIC, 1994.

———. "El meridiano y el ante meridiano de Tordesillas en la geografía, la náutica y la cartografía." *Revista de Indias* 54, no. 202 (1994): 509–42.

Cervantes de Salazar, Francisco. *Crónica de la Nueva España*. Edited by Agustín Millarer Carlo. Biblioteca de autores españoles 244–45. Madrid: Atlas, 1971.

Cervera Vera, Luis. "Instrumentos náuticos inventados por Juan de Herrera para determinar la longitud de un lugar." *Llull* 20, no. 38 (1997): 143–60.

———. *Inventario de los bienes de Juan de Herrera*. Valencia: Albatros, 1977.

Chaves, Jerónimo de. *Chronographía o Repertorio de los tiempos, el más copioso y preciso que hasta ahora ha salido à luz*. Sevilla: En casa de Fernando Diaz en la calle de la Sierpe, 1581.

Cieza de León, Pedro de. *La crónica del Perú*. Edited by Manuel Ballesteros. Madrid: Dastin, 2000.

———. *Obras Completas*. Edited by Carmelo Sáenz de Santa María. 3 vols. Madrid: CSIC, 1985.

Cline, Howard F. "*Relaciones geográficas*: Revised and Augmented Census of *Relaciones geográficas* of New Spain, 1579–1585." In *Handbook of Middle American Indians*. Washington, D.C.: Library of Congress, 1966.

———. "The *relaciones geográficas* of the Spanish Indies, 1577–1586." *Hispanic American Historical Review* 44, no. 3 (1964): 341–74.

———. "The *relaciones geográficas* of the Spanish Indies, 1577–1648." In *Handbook of Middle American Indians: Guide to Historical Sources*, 183–242. Austin: University of Texas Press, 1964.

Consejo de Indias. *Ordenanzas reales del Consejo de Indias: Gobernación y estado temporal*. Madrid: Casa de Francisco Sanchez, 1585.

Cormack, Lesley B. *Charting an Empire*. Chicago: University of Chicago Press, 1997.

Cortés, Martín. *Breve compendio de la sphera y de la arte de navegar: Introducción por Mariano Cuesta Domingo*. Sevilla: Casa de Antón Álvarez, 1551. Reprint, Madrid: Editiorial Naval, 1990.

Cortesão, Armando. *Cartografia portuguesa antiga*. Lisbon: Comissão Executuva das Comemorações do Quinto Centernário da Morte do Infante D. Henrique, 1960.

Cortesão, Jaime. "The Pre-Columbian Discovery of America." *Geographical Journal* 89, no. 1 (1937): 29–42.

Cosgrove, Denis E. "The Geometry of the Landscape: Practical and Speculative Arts in Sixteenth-Century Venetian Land Territories." In *The Iconography of Landscape: Essays on the Symbolic Representation, Design, and Use of Past Environments*, edited by Denis E. Cosgrove and Stephen Daniels, 254–77. Cambridge: Cambridge University Press, 1988.

Cotter, Charles H. *A History of Nautical Astronomy*. London: Hollis and Carter, 1968.

Covarrubias Horozco, Sebastián de. *Tesoro de la lengua castellana o española: Edición integral e ilustrada de Ignacio Arellano y Rafael Zafra*. Madrid: Iberoamericana; [Frankfurt am Main]: Vervuert.

Crombie, Alistair C. "Science and the Arts in the Renaissance: The Search for Truth and Certainty, Old and New." In *Science and the Arts in the Renaissance*, edited by John W. Shirley and F. David Hoeniger, 15–26. London: Folger Books, 1985.

Crosby, Alfred W. *The Measure of Reality: Quantification and Western Society, 1250–1600*. Cambridge: Cambridge University Press, 1997.

Cuesta Domingo, Mariano. "La cartografía grabada en la obra de Antonio de Herrera." In *Descubrimiento y cartografía en la época de Felipe II*, edited by Mariano Cuesta Domingo, 71–114. Valladolid: Universidad de Valladolid, 1999.

———. " 'Tierra nueva e cielo nuevo,' navegación, geografía y mundo nuevo." *Boletín de la Real Sociedad Geográfica* 128 (1992): 15–37.

Dainville, François de. *La géographie des humanistes: Les Jésuites et l'éducation de la société française*. Paris: Beauchesne et Ses Fils, 1940.

Daston, Lorraine. "Baconian Facts, Academic Civility, and the Prehistory of Objectivity." *Annals of Scholarship* 8 (1991): 337–64.

———. "Strange Facts, Plain Facts, and the Texture of Scientific Experience in the Enlightenment." In *Proof and Persuasion: Essays on Authority, Objectivity, and Evidence*, edited by Suzanne Marchand and Elizabeth Lunbeck, 42–59. Turnhout, Belgium: Brepols, 1996.

Dear, Peter R. *Discipline and Experience: The Mathematical Way in the Scientific Revolution*. Chicago: University of Chicago Press, 1995.

Del Pino Díaz, Fermín. 1999. "La historia natural y moral de las Indias como género: Orden y génesis literario de la obra de Acosta." In *Primer Congreso Internacional de Peruanistas en el Extranjero, April 29–May 1, 1999*, http://www.fas.harvard.edu/~icop/fermindelpino.html. (accessed April 5, 2007).

d'Olwer, Luis Nicolau. *Fray Bernardino de Sahagún, 1499–1590*. Translated by Mauricio J. Mixco. Salt Lake City: University of Utah Press, 1987.

Domínguez Ortiz, Antonio. *Desde Carlos V a la paz de los Pirineos, 1517–1660*. Barcelona: Ediciones Grijalbo, 1974.

Edgerton, Samuel Y. "From Mental Matrix to Mappamundi to Christian Empire: The Heritage of Ptolemaic Cartography in the Renaissance." In *Art and Cartography*, edited by David Woodward, 10–50. Chicago: University of Chicago Press, 1987.

Edwards, Clinton R. "Mapping by Questionnaire and Early Spanish Attempts to Determine New World Geographical Positions." *Imago Mundi* 23 (1969): 17–28.

Elliott, John H. *Empires of the Atlantic World: Britain and Spain in America, 1492–1830*. New Haven, Conn.: Yale University Press, 2006.

———. "A Europe of Composite Monarchies." *Past and Present* 137 (1992): 48–71.

———. *Imperial Spain, 1469–1716*. London: Penguin Books, 1990.

———. *The Old World and the New, 1492–1650*. Cambridge: Cambridge University Press, 1970.

———. *Spain and Its World, 1500–1700*. New Haven, Conn.: Yale University Press, 1989.

Escalante de Mendoza, Juan de. *Itinerario de navegación de los mares y tierras occidentales, 1575*. Madrid: Museo Naval, 1985.

Esteban Piñeiro, Mariano. "Los cosmógrafos al servicio de Felipe II." *Mare Liberum* 10 (1995): 525–39.

———. "Los cosmógrafos del rey." In *Madrid, ciencia y corte*, edited by Antonio Lafuente and Javier Moscoso, 121–33. Madrid: CSIC, 1999.

———. "Elio Antonio de Nebrija y la búsqueda de patrones universales de medida." In *El Tratado de Tordesillas y su época*, edited by Luis Antonio Ribot García, 569–82. Madrid: Junta de Castilla y León, 1995.

———. "La primera versión castellana de *De Revolutionibus Orbium Caelestium*: Juan Cedillo Diaz (1620–1625)." *Asclepio* 43, no. 1 (1991): 131–62.

Esteban Piñeiro, Mariano, María I. Vicente Maroto, and Félix Gómez Crespo. "La recuperación del gran tratado científico de Alonso de Santa Cruz: El astronómico real." *Asclepio* 44 (1992): 3–32.

Etayo-Piñol, Marie Ange. "Medina y Cortés o el aprendizaje de las técnicas de navegación en Europa en el siglo XV." *Revista de Historia Naval* 16, no. 64 (1998): 41–47.

Euclid. *La perspectiva, y specularia de Euclides: Traduzidas en vulgar Castellano, y dirigidas a la S.C.R.M. del Rey don Phelippe nuestro Señor*. Translated by Pedro Ambrosio Onderíz. Madrid: Imp. casa de la viuda de Alonso Gómez, 1585.

Falero, Francisco. *El tratado de la esphera y del arte de marear (1535)*. Facsimile ed. Edited by Ministerio de Defensa and Ministerio de Agricultura. Borriana, Spain: Ediciones Histórico Artísticas, 1989.

Feingold, Mordechai, and Víctor Navarro Brotóns, eds. *Universities and Science in the Early Modern Period*. Dordrecht, Netherlands: Springer, 2006.

Fernández Álvarez, Manuel. *Copérnico y su huella en la Salamanca del Barroco*. Salamanca: Universidad de Salamanca, 1974.

Fernández-Armesto, Felipe. *Columbus*. New York: Oxford University Press, 1991.

Fernández de Navarrete, Martín. *Biblioteca marítima española*. 2 vols. Madrid: Viuda de Calero, 1851. Reprint, New York: Burt Franklin, 1968.

———. *Colección de documentos y manuscritos compilados por Martín Fernández de Navarrete*. Edited by Museo Naval. 32 vols. Nendeln, Liechtenstein: Kraus-Thomson, 1971.

———. *Colección de opúsculos del excmo. sr. d. Martín Fernández de Navarrete*. 2 vols. Madrid: Impr. de la Viuda de Calero, 1848.

———. *Disertación sobre la historia de la náutica y de las ciencias matemáticas que han contribuido a sus progresos entre los españoles*. Edited by Carlos Seco Serrano. Biblioteca de autores españoles 77. Madrid: Ediciones Atlas, 1954–55.

Fernández de Navarrete, Martín, et al., eds. *Colección de documentos inéditos para la historia de España*. 113 vols. Vaduz, Liechtenstein: Kraus Reprint, 1964–75.

Fernández Vallín, Acisclo. *Cultura científica en España en el siglo XVI*. Foreword by Marcelino Menéndez y Pelayo. Madrid: Sucesores de Rivadeneyra, 1893. Facsimile ed. Sevilla: Padilla Libros, 1989.

Feros, Antonio. *El Duque de Lerma: Realeza y privanza en la España de Felipe III*. Madrid: Marcial Pons, 2002.

———. "El viejo monarca y los nuevos favoritos: Los discursos sobre la privanza en el reinado de Felipe II." *Studia Historica: Historia Moderna* 17 (1997): 11–36.

Finé, Oronce. *Protomathesis: Opus varium*. . . Paris: Gerardi Morrhij et Ioannis Pet, 1532.

———. *Quadratura circuli*. . . *De invenienda longitudinis locorum differentia*. . . Paris: S. Colinaeum, 1544.

Flórez Miguel, Cirilo. "Cosmógrafos salamantinos de Renacimiento y cambio de paradigma." In *Ciencia, vida y espacio en Iberoamérica*, edited by José Luis Peset, 379–87. Madrid: CSIC, 1989.

Flórez Miguel, Cirilo, Pablo García Castillo, and Roberto Albares Albares. *El humanismo científico*. Salamanca: Caja de Ahorros y Monte de Piedad, 1988.

Friede, Juan. "La censura española del siglo XVI y los libros de historia de América." *Revista de Historia de América* 47 (1959): 45–94.

Frisius, Gemma. *De principiis astronomiae et cosmographiae*. Antwerp: Ioannis Steelfii, 1553. A facsimile reproduction, with an introduction by C. A. Davids. Reprint, Delmar, N.Y.: Scholars' Facsimiles and Reprints for the John Carter Brown Library, 1992.

Frisius, Gemma, et al. *Cosmographia, siue Descriptio uniuersi orbis*. Antwerp: Apud Ioan. Bellerum, 1584. Microform.

Galilei, Galileo. "Discourse on the Tides (1616)." In *The Galileo Affair: A Documentary History*, edited by Maurice Finocchiaro, 119–33. Berkeley: University of California Press, 1989.

Gallois, Lucien Louis. *Les géographes allemands de la Renaissance*. Paris: E. Leroux, 1890.

García Camarero, Ernesto, ed. *La polémica de la ciencia española*. Madrid: Editorial Alianza, 1970.

García Cárcel, Ricardo. *La leyenda negra: Historia y opinión*. Madrid: Alianza Editorial, 1998.

García de Céspedes, Andres. *Libro de instrumentos nuevos de geometría muy necesarios para medir distancias, y alturas, sin que intervengan numeros como se demuestra en la práctica*. Madrid: Juan de la Cuesta, 1606.

———. *Regimiento de navegación e hydrografía*. Madrid: Casa de Juan de la Cuesta, 1606.

García Pimentel, Luis. *Descripción del arzobispado de México hecha en 1570*. Mexico: J. J. Terrazas, 1897.

———, ed. *Relación de los obispados de Tlaxcala, Michoacán, Oaxaca y otros lugares en el siglo XVI: Manuscrito de la colección del señor don Joaquín García Icazbalceta*. Edited by Luis García Pimentel. 2 vols. Documentos históricos de Méjico. Mexico, 1904.

García Tapia, Nicolás, and M. I. Vicente Maroto. "Juan de Herrera, un científico en la corte española." In *Instrumentos científicos del siglo XV: La corte española y la Escuela de Lovaina*, 41–54. Madrid: Fundación Carlos de Amberes, 1997.

Garza, Mercedes de la, ed. *Relaciones histórico-geográficas de la gobernación de Yucatán*. Mexico: Universidad Nacional Autónoma de México, 1983.

Gerbi, Antonello. *Nature in the New World: From Christopher Columbus to Gonzalo Fernández de Oviedo*. Translated by Jeremy Moyle. Pittsburgh: University of Pittsburgh Press, 1985.

Gerhard, Peter. A *Guide to the Historical Geography of New Spain*. Cambridge Latin American Studies 14. Cambridge: Cambridge University Press, 1972.

Gil, Juan. "Pedro Mártir de Anglería, intérprete de la cosmografía colombina." *Anuario de Estudios Americanos* 39 (1982): 487–502.

Gingerich, Owen. "Sacrobosco as a Textbook." *Journal for the History of Astronomy* 19 (1988): 269–73.

Golinski, Jan. "The Theory of Practice and the Practice of Theory: Sociological Approaches in the History of Science." *Isis* 81 (1990): 492–505.

Gómez Gómez, Margarita, and Isabel González Ferrín. "El archivo secreto del Consejo de Indias y sus fondos bibliográficos." *Historia, Instituciones, Documentos* 19 (1992): 187–214.

González de Mendoza, Juan. *Historia de las cosas mas notables, ritos y costumbres del gran reino de la China*. Introduction by P. Felix García. Madrid: M. Aguilar, 1944.

González Echevarría, Roberto. "The Second Discovery of America." *Yale Review* 86, no. 1 (1998): 136–53.

González González, Francisco Javier. "Martín Cortés de Albácar, Cádiz y el *Breve compendio de la sphera y de la arte de navegar* (1551)." *Gades* 22 (1997): 311–26.

Goodman, David C. *Power and Penury: Government, Technology, and Science in Philip II's Spain*. Cambridge: Cambridge University Press, 1988.

Gracián, Antonio. "Diurnal de Antonio Gracián, secretario de Felipe II." In *Documentos para la historia del monasterio de San Lorenzo el Real de El Escorial*, edited by Gregorio de Andrés. El Escorial: Imprenta del Real Monasterio, 1962.

Grafton, Anthony. *Defenders of the Text: The Traditions of Scholarship in an Age of Science, 1450–1800*. Cambridge, Mass.: Harvard University Press, 1991.

Grafton, Anthony, April Shelford, and Nancy Siraisi, eds. *New World, Ancient Texts: The Power of Tradition and the Shock of Discovery*. Cambridge: Belknap Press of Harvard University Press, 1992.

Grant, Edward. *The Foundations of Modern Science in the Middle Ages*. Cambridge: Cambridge University Press, 1996.

Grendler, Paul F. *The Universities of the Italian Renaissance*. Baltimore: Johns Hopkins University Press, 2002.

Guillén y Tato, Julio Fernando. *Inventario de los papel pertenecientes al Excmo. Señor D. Martín Fernández de Navarrete existente en Abalos, en el Archivo del Marques de Legarda*. Madrid: Ediciones Cultura Hispánica, 1944.

Harley, J. B. *The New Nature of Maps*. Edited by Paul Laxton. Baltimore: Johns Hopkins University Press, 2001.

Harley, J. B., and D. Woodward, eds. *The History of Cartography*, vol. 1. 4 vols. Chicago: University of Chicago Press, 1987.

Harris, Steven J. "Confession-Building, Long-Distance Networks, and the Organization of Jesuit Science." *Early Science and Medicine* 1, no. 3 (1996): 287–318.

Harvey, David. "Between Space and Time: Reflections on the Geographical Imagination." *Annals of the Association of American Geographers* 80, no. 3 (1990): 418–32.

Henry, John. "National Styles in Science." In *Geography and Revolution*, edited by David N. Livingstone and Charles W. J. Whithers, 43–74. Chicago: University of Chicago Press, 2005.

———. *The Scientific Revolution and the Origins of Modern Science*. New York: St. Martin's, 1997.

Heredia Herrera, Antonia, ed. *Catálogo de las consultas del Consejo de Indias, 1529–1591*. 2 vols. Madrid: Dirección General de Archivos y Bibliotecas, 1972.

Hernández, Francisco. *The Mexican Treasury: The Writings of Dr. Francisco Hernández*. Edited by Simon Varey. Stanford, Calif.: Stanford University Press, 2000.

Hernando, Agustín. *La imagen de un país: Juan Bautista Labaña y su mapa de Aragón (1610–1620)*. Zaragoza: Institución Fernando el Católico, 1996.

Herrera, Juan de. *Institución de la Academia Real Mathemática*. Edited by José Simón Díaz y Luis Cervera Vera. Madrid: Instituto de Estudios Madrileños, 1995.

Herrera y Tordesillas, Antonio de. *Historia general de los hechos de los castellanos en las Islas y Terra Firme del Mar Océano, o, Décadas de Antonio de Herrera y Tordesillas (1601–1615)*. Edited by Mariano Cuesta Domingo. 4 vols. Madrid: Universidad Complutense de Madrid, 1991.

Hessels, Jan Hendrik. *Abrahami Ortelii et virorum eruditorum ad eundem at ad Iacobum Colium Ortelianum epistolae*. 4 vols. Ecclesiae Londino-Batavae Archivum. Cambridge: Academiae sumptibus Ecclesiae Londino-Batavae, 1887–97.

Hewson, J. B. *A History of the Practice of Navigation*. Glasgow: Brown, 1951.

Hillgarth, J. N. *The Mirror of Spain, 1500–1700: The Formation of a Myth*. Ann Arbor: University of Michigan Press, 2000.

Holmes, Frederic L., Jürgen Renn, and Hans-Jörg Rheinberger, eds. *Reworking the Bench: Research Notebooks in the History of Science*. Dordrecht, Netherlands: Kluwer Academic, 2003.

Hurtado Torres, Antonio. "La 'Esphera' de Sacrobosco en la España de los siglos XVI y XVII: Difusión bibliográfica." *Cuadernos Bibliográficos* 44 (1982): 49–58.

Jalón, Mauricio. "Empresas científicas: Sobre las políticas de la ciencia en el siglo XVII." In *Madrid, ciencia y corte*, edited by Antonio Lafuente and Javier Moscoso, 155–75. Madrid: CSIC, 1999.

Jiménez de la Espada, Marcos. *Relaciones geográficas de Indias: Perú*. Madrid: Real Academia Española, 1881–1897. Reprint, Biblioteca de Autores Españoles 183–85. Madrid: Ediciones Atlas, 1965.

Juderías, Julián. *La leyenda negra: Estudios acerca del concepto de España en el extranjero*. Castilla y León: Junta de Castilla y León, Consejería de Educación y Cultura, Caja Salamanaca y Soria, 1997.

Kagan, Richard L. "Arcana Imperii: Mapas, sabiduría y poder en la corte de Felipe IV." In *El atlas de Rey Planeta: La descripción de España y de las costas y puertos de sus reinos de Pedro Texeira*, 49–70. Madrid: Editoral Nerea, 2002.

———. "Clio and the Crown: Writing History in Habsburg Spain." In *Spain, Europe, and the Atlantic World: Essays in Honor of John H. Elliott*, edited by Richard L. Kagan and Geoffrey Parker, 73–99. Cambridge: Cambridge University Press, 1995.

———. *Lawsuits and Litigants in Castile, 1500–1700*. Chapel Hill: University of North Carolina Press, 1981.

———. "Philip II and the Geographers." In *Spanish Cities of the Golden Age: The Views of Anton van den Wyngaerde*, edited by Richard L. Kagan, 40–53. Berkeley: University of California Press, 1989.

———. *El rey recatado: Felipe II, la historia y los cronistas del rey*. Colección Síntesis. Valladolid: Universidad de Valladolid, 2004.

Karrow, R. W., Jr. "Intellectual Foundations of the Cartographic Revolution." PhD diss., Loyola University Chicago, 1999.

Kelley, Donald R., ed. *History and the Disciplines: The Reclassification of Knowledge in Early Modern Europe*. Rochester, N.Y.: University of Rochester Press, 1997.

Kremer, Richard L., and Jerzy Dobrzycki. "Alfonsine Meridians: Tradition versus Experience in Astronomical Practice, c. 1500." *Journal for the History of Astronomy* 29 (1998): 187–99.

Lamb, Ursula. *Cosmographers and Pilots of the Spanish Maritime Empire*. Aldershot, UK: Variorum/Ashgate, 1995.

Las Casas, Bartolomé de. *A Short Account of the Destruction of the Indies*. Introduction by Anthony Pagden. London: Penguin Books, 1992.

León Pinelo, Antonio de. *El epítome de Pinelo, primera bibliografía del Nuevo Mundo*. 1629. A facsimile ed., edited by Agustín Millares Carlo. Washington, D.C.: Unión Panamericana, 1958.

Leovitius, Cyprianus. *Eclipsium omnium ab anno Domini 1554 usque in annum Domini 1606: Accurata descriptio & pictura, ad meridianum Angustanmum ita supputata, ut quibus aliis facillimè accommodari possit, una cum explicatione effectuum tam generalium quàm particularium pro cuiusque genesi*. Augsburg: Philippus Ulhardus, 1556.

Lestringant, Frank. *Mapping the Renaissance World*. Translated by David Fausset. Cambridge: Polity, 1994.

Lindberg, David C., and Robert S. Westman, eds. *Reappraisals of the Scientific Revolution*. Cambridge: Cambridge University Press, 1990.

Livingstone, David N. *The Geographical Tradition: Episodes in the History of a Contested Enterprise*. Oxford: Blackwell, 1992.

Llaguno y Amirola, Eugénio, and Juan A. Ceán-Bermúdez. *Noticias de los arquitectos y arquitectura de España desde la restauración*, vol. 2. 4 vols. Madrid: Imprenta Real, 1829. Reprint, Madrid: Ediciones Turner, 1977.

Lloyd, G. E. R. "Saving the Appearances." *Classical Quarterly* 28, no. 1 (1978): 202–22.

Long, Pamela O. *Openness, Secrecy, and Authorship: Technical Arts and the Culture of Knowledge from Antiquity to the Renaissance*. Baltimore: Johns Hopkins University Press, 2001.

López de Gómara, Francisco. *La conquista de México*. Crónicas de América. Madrid: Dastin, 2000.

López de Velasco, Juan. *Geografía y descripción universal de las Indias.* Edited by Justo Zaragoza. Madrid: Real Academia de Historia, 1894.

———. *Geografía y descripción universal de las Indias.* Introduction by Doña María del Carmen González Muñoz. Biblioteca de autores españoles 248. Madrid: Ediciones Atlas, 1971.

———. *Orthographía y pronunciación castellana.* Burgos: Felipe de Junta, 1582.

López Piñero, José María. *El arte de navegar en la España del Renacimiento.* 2nd ed. Barcelona: Editorial Labor, 1986.

———. *Ciencia y técnica en la sociedad española de los siglos XVI y XVII.* Barcelona: Labor, 1979.

López Piñero, José María, and José Pardo Tomás. *La influencia de Francisco Hernández (1515–1587) en la constitución de la botánica y la materia médica modernas.* Valencia: Instituto de Estudios Documentales e Históricos sobre la Ciencia, Universitat de València, CSIC, 1996.

Lovett, A. W. "Philip II and Mateo Vázquez de Leca: The Government of Spain (1572–1592)." Geneva: Librairie Droz, 1977.

Maltby, William S. *The Black Legend in England: The Development of Anti-Spanish Sentiment, 1558–1660.* Durham, N.C.: Duke University Press, 1971.

Mancho Duque, María Jesús, ed. *Pórtico a la ciencia y a la técnica del Renacimiento.* Salamanca: Junta de Castilla y León, 2001.

Manzano Manzano, Juan. *Historia de las recopilaciones de Indias.* 2 vols. Madrid: Ediciones Cultura Hispánica, 1950.

———. "La visita de Ovando al Real Consejo de las Indias y el Código Ovandino." In *El Consejo de las Indias en el siglo XVI,* 111–23. Valladolid: Universidad de Valladolid, Secretariado de Publicaciones, 1970.

Marchetti, Giovanni. "Hacia la edición crítica de la Historia de Sahagún." *Cuadernos Hispanoamericanos* 396 (1983): 1–36.

Martin, Julian. *Francis Bacon, the State, and the Reform of Natural Philosophy.* Cambridge: Cambridge University Press, 1992.

Martín Merás, Luisa. *Cartografía marítima hispana.* Colección ciencia y mar. Barcelona: Lunwerg Editores, 1993.

Mayhew, Robert J. "Geography, Print Culture, and the Renaissance: 'The Road Less Travelled By.'" *History of European Ideas* 27 (2001): 349–69.

Medina, José Toribio. *Biblioteca hispanoamericana, 1493–1810.* A facsimile ed. Santiago de Chile: Fondo Histórico y Bibliográfico José Toribio Medina, 1958.

———. *Historia de la imprenta en los antiguos dominios españoles de América y Oceanía.* 2 vols. Santiago de Chile: Fondo Histórico y Bibliográfico José Toribio Medina, 1958.

———. *El veneciano Sebastián Caboto al servicio de España y especialmente de su proyectado viaje á las Molucas por el Estrecho de Magallanes y al reconocimiento de la costa del continente hasta la gobernación de Pedrarias Dávila.* 2 vols. Santiago de Chile: Imprenta y Encuadernación Universitaria, 1908.

Medina, Pedro de. *Libro de las grandezas y cosas memorables de España. Introducción y prólogo de Ángel González Palencia.* Madrid: CSIC, 1944.

Mela, Pomponius. *Pomponius Mela's Description of the World.* Translated by F. E. Romer. Ann Arbor: University of Michigan Press, 1998.

Mignolo, Walter D. "Cartas, crónicas y relaciones del descubrimiento y la conquista." In *Historia de la literatura hispanoamericana,* edited by Luis Iñigo Madrigal, 57–116. Madrid: Cátedra, 1982.

———. *The Darker Side of the Renaissance: Literacy, Territoriality, and Colonization.* Ann Arbor: University of Michigan Press, 1995.

Miguélez, Manuel. *Catálogo de los códices españoles de la Biblioteca del Escorial.* 2 vols. Madrid: Imprenta Heletica, 1917.

———. "Sobre el verdadero autor del 'Dialogo de las lenguas.' " *La Ciudad de Dios* 117 (1919): 441–57.

Millán de Benavides, Carmen. *Epítome de la conquista del Nuevo Reino de Granada: La cosmografía Española de siglo XVI y el conocimiento por cuestionario.* Bogotá: Pontificia Universidad Javeriana, Instituto de Estudios Sociales y Culturales Pensar, 2001.

Millares Carlo, Agustín, ed. *Cartas recibidas de España para Francisco Cervantes de Salazar (1569–1575).* Mexico: Antigua Librería Robredo, 1946.

Moran, Bruce T., ed. *Patronage and Institutions: Science, Technology, and Medicine at the European Court, 1500–1750.* Rochester, N.Y.: Boydell, 1991.

Mundy, Barbara M. *The Mapping of New Spain: Indigenous Cartography and the Maps of the "relaciones geográficas."* Chicago: University of Chicago Press, 1996.

Muñoz, Jerónimo. *Jerónimo Muñoz: Introducción a la astronomía y la geografía.* Edited by Víctor Navarro Brotóns. Colleció Oberta. Valencia: Consell Valencià de Cultura, 2004.

———. *Libro del nuevo cometa.* Edited by Víctor Navarro Brotóns. Valencia: Valencia Cultural, 1981.

Muro Orejón, Antonio. "Las ordenanzas de 1571 del Real y Supremo Consejo de las Indias: Reproducción facsimilar." *Anuario de Estudios Americanos* 14 (1957): 363–423.

Naudé, Françoise. *Reconnaissance du Nouveau Monde et cosmographie a la Renaissance.* Kassel, Germany: Edition Reichenberger, 1992.

Nauert, Charles G., Jr. "Humanists, Scientists, and Pliny: Changing Approaches to a Classical Author." *American Historical Review* 84, no. 1 (1979): 72–85.

Navarro Brotóns, Víctor. "Astrología y cosmografía entre 1561 y 1625." *Cronos* 3, no. 2 (2000): 349–80.

———. *Bibliographía physico-mathemática hispánica (1475–1900).* Valencia: CSIC, 1999.

———. "Cartografía y cosmografía en la época del descubrimiento." In *Mundialización de la ciencia y cultura nacional: Actas del Congreso Internacional "Ciencia, descubrimiento y mundo colonial,"* edited by A. Lafuente, A. Elena, and M. L. Ortega, 67–73. Madrid: Doce Calles, 1991.

Navarro Brotóns, Víctor, and William Eamon. "Spain and the Scientific Revolution: Historiographical Questions and Conjectures." In *Más allá de la leyenda negra: España y la Revolución Científica,* edited by Víctor Navarro Brotóns and William Eamon, 27–38. Valencia: Soler, 2007.

Navarro Brotóns, Víctor, and Enrique Rodríguez Galdeano. *Matemáticas, cosmología y humanismo en la España del siglo XVI: Los comentarios al segundo libro de la historia natural de Plinio de Jerónimo Muñoz*. Valencia: Instituto de Estudios Documentales e Históricos sobre la Ciencia, 1998.

Naylor, Ron. "Galileo's Tidal Theory." *Isis* 98 (2007): 1–22.

Nebrija, Elio Antonio de. *Elio Antonio de Nebrija, cosmógrafo: In cosmographiae libros introductorium*. Translated by Virginia Bonmatí Sánchez. Edited by Hermandad de los Santos de Lebrija. Cádiz: Agrija Ediciones, 2000.

Newman, William R., and Lawrence M. Principe. *Alchemy Tried in the Fire: Starkey, Boyle, and the Fate of Helmontian Chymistry*. Chicago: University of Chicago Press, 2002.

Nuñez, Pedro. "De erratis Orontii Finoei liber unus." In *Petri Nonii Salaciensis Opera*. Basel: Sebastianum Henricpetri, 1592.

Ortega Rubio, Juan. *Relaciones topográficas de los pueblos de España*. Madrid: Sociedad Española de Artes Gráficas, 1918.

Ortroy, Fernand Gratien van. *Bio-bibliographie de Gemma Frisius, fondateur de l'école belge de géographie, de son fils Corneille et de ses neveux les Arsenius*. Brussels: M. Lamertin, 1920.

Osler, Margaret J., ed. *Rethinking the Scientific Revolution*. Cambridge: Cambridge University Press, 2000.

Pacheco, Joaquín Francisco, Francisco de Cárdenas y Espejo, and Luis Torres de Mendoza, eds. *Colección de documentos inéditos relativos al descubrimiento, conquista y organización de las antiguas posesiones españolas de América y Oceanía, sacados de los archivos del reino, y muy especialmente del de Indias*. 42 vols. Vaduz, Liechtenstein: Kraus Reprint, 1964.

Padrón, Ricardo. "Mapping Plus Ultra: Cartography, Space, and Hispanic Modernity." *Representation* 79 (2002): 28–60.

Pardo de Guevara, Eduardo José, María del Pilar Rodríguez Suárez, and Dolores Barral. *Don Pedro Fernández de Castro, VII Conde de Lemos (1576–1622)*. 2 vols. Santiago de Compostela: Xunta de Galicia, 1997.

Pardo Tomás, José. *Ciencia y censura: La Inquisición española y los libros científicos en los siglos XVI y XVII*. Madrid: CSIC, 1991.

———. *Oviedo, Monardes, Hernández: El tesoro natural de América, colonialismo y ciencia en el siglo XVI*. Madrid: Nivola, 2002.

Pardo Tomás, José, and María Luz López Terrada. *Las primeras noticias sobre plantas americanas en las relaciones de viajes y crónicas de Indias, 1493–1553*. Valencia: Instituto de Estudios Documentales e Históricos sobre la Ciencia, Universitat de València, C.S.I.C, 1993.

Parks, George B. *Richard Hakluyt and the English Voyages*. New York: American Geographical Society, 1930.

Paso y Troncoso, Francisco del. *Epistolario de Nueva España, 1505–1818*. 16 vols. Mexico: Antigua Librería Robredo, 1939–42.

———, ed. *Relaciones geográficas de la Diócesis de Michoacán, 1579–1580.* 2 vols. Guadalajara: n.p., 1958.

Pedersen, Olaf. "In Quest of Sacrobosco." *Journal for the History of Astronomy* 16 (1985): 175–221.

Peña Camara, Jóse de la. "La copulata de leyes de Indias y las ordenanzas Ovandinas." *Revista de Indias* 6 (1941): 121–46.

Pérez de Oliva, Fernán. *Cosmografía nueva.* Bilingual edition prepared by Cirilo Flórez Miguel. Acta Salamanticensia. Salamanca: Ediciones Universidad de Salamanca, 1985.

Pérez Pastor, Cristóbal. *Bibliografía madrileña.* 3 vols. Madrid: Tipografía de los Huérfanos, 1891–1907.

Pérez-Mallaina Bueno, Pablo Emilio. "Los libros de náutica españoles del siglo XVI y su influencia en el descubrimiento y conquista de los océanos." In *Ciencia, vida y espacio en Iberoamérica,* edited by José Luis Peset, 457–84. Madrid: CSIC, 1989.

———. *Spain's Men of the Sea: Daily Life on the Indies Fleets in the Sixteenth Century.* Translated by Carla Rahn Phillips. Baltimore: Johns Hopkins University Press, 1998.

Pérez-Rioja, José Antonio. "Un insigne visontino del siglo XVI: Juan López de Velasco." *Celtiberia* 8, no. 15 (1958): 7–38.

Picatoste y Rodríguez, Felipe. *Apuntes para una biblioteca científica española del siglo XVI.* Madrid: Impr. de M. Tello, 1891.

Pickering, Andrew, ed. *Science as Practice and Culture.* Chicago: University of Chicago Press, 1992.

Pinon, Laurent. "Conrad Gessner and the Historical Depth of Renaissance Natural History." In *Historia: Empiricism and Erudition in Early Modern Europe,* edited by Gianna Pomata and Nancy G. Siriasi, 241–67. Cambridge, Mass.: MIT Press, 2005.

Pogo, A. "Gemma Frisius, His Method of Determining Differences of Longitude by Transporting Timepieces (1530), and His Treatise on Triangulation (1533)." *Isis* 22 (1935): i–xix, 469–85.

Pohl, Frederick J. *Amerigo Vespucci, Pilot Major.* New York: Columbia University Press, 1945.

Pomata, Gianna, and Nancy G. Siriasi, eds. *Historia: Empiricism and Erudition in Early Modern Europe.* Cambridge, Mass.: MIT Press, 2005.

Ponce Leiva, Pilar. "Los cuestionarios oficiales: ¿Un sistema de control de espacio?" In *Cuestionarios para la formación de las relaciones geográficas de Indias, siglos XVI–XIX,* edited by Francisco de Solano, xxix–xxxv. Madrid: CSIC, 1988.

———, ed. *Relaciones histórico-geográficas de la Audiencia de Quito, s. XVI–XIX,* vol. 1. 2 vols. Madrid: CSIC, 1991.

Poole, Stafford. *Juan de Ovando: Governing the Spanish Empire in the Reign of Phillip II.* Norman: University of Oklahoma Press, 2004.

Pozuelo Yvancos, José María. *López de Velasco en la teoría gramatical del siglo XVI.* Murcia: Universidad de Murcia, 1981.

Ptolemy. *Ptolemy's Geography: An Annotated Translation of the Theoretical Chapters.* Translated by J. Lennart Berggren and Alexander Jones. Princeton, N.J.: Princeton University Press, 2000.

Puente y Olea, Manuel de la. *Los trabajos geográficos de la Casa de Contratación.* Sevilla: Librería Salesianas, 1900.

Pulido Rubio, José. *El piloto mayor de la Casa de la Contratación de Sevilla: Pilotos mayores, catedráticos de cosmografía y cosmógrafos.* Sevilla: Publicaciones de la Escuela de Estudios Hispano-Americanos, 1950.

Randles, W. G. L. "Portuguese and Spanish Attempts to Measure Longitude in the Sixteenth Century." *Vistas in Astronomy* 28 (1985): 235–41.

———. "Science et cartographie: L'image de monde physique à fin du XVe siècle." In *El Tratado de Tordesillas y su época*, edited by Luis Antonio Ribot García, 935–41. Madrid: Junta de Castilla y León, 1995.

Rawson, Elizabeth. *Intellectual Life in the Late Roman Republic.* Baltimore: Johns Hopkins University Press, 1985.

Redondo, Augustín. "Censura, literatura y transgresión en época de Felipe II: El 'Lazarillo castigado' de 1573." *Edad de Oro* 18 (1999): 135–49.

———. "Exaltación de España y preocupaciones pedagógicas alrededor de 1580: Las reformas preconizadas por Juan López de Velasco, cronista y cosmógrafo de Felipe II." In *Felipe II, Europa y la monarquía católica*, edited by José Martínez Millán, 425–36. Madrid: Parteluz, 1998.

Reyes Gómez, Fermín de los. *El libro en España y América: Legislación y censura (siglos XV–XVIII).* Madrid: Arco/Libros, 2000.

Ribot García, Luis Antonio, ed. *El Tratado de Tordesillas y su época.* 2 vols. Madrid: Junta de Castilla y León, 1995.

Rodríguez-Sala, María Luisa. *El eclipse de Luna: Mision científica de Felipe II en Nueva España.* Huelva, Spain: Universidad de Huelva, 1998.

Sánchez Bella, Ismael. "El *Título de las descripciones* del código de Ovando." In *Dos estudios sobre el código de Ovando*, 91–217. Pamplona: Universidad de Navarra, 1987.

Sánchez Cantón, Francisco J. *La librería de Juan de Herrera.* Madrid: CSIC, 1941.

Sánchez Ron, José Manuel. "Felipe II, El Escorial y la ciencia Europea." In *La ciencia en el Monasterio del Escorial: Actas del simposium, 1 al 4 de noviembre de 1993*, 39–72. Madrid: Ediciones Escurialenses, 1994.

Sandman, Alison D. "Cosmographers vs. Pilots." PhD diss., University of Wisconsin, 2001.

———. "Educating Pilots: Licensing Exams, Cosmography Classes, and the Unversidad de Mareantes in Sixteenth Century Spain." In *Ars nautica: Fernando Oliveira and His Era; Humanism and the Art of Navigation in Renaissance Europe (1450–1650)*, edited by Inácio Guerreiro and Francisco Contente Domingues. Cascais, Portugal: Patrimonia, 1999.

Santa Cruz, Alonso. *Alonso de Santa Cruz y su obra cosmográfica*. Edited by Mariano Cuesta Domingo. 2 vols. Madrid: CSIC, 1983.

———. *Crónica de los Reyes Católicos*. Edited by Juan de Mata Carriazo. Sevilla: Publicaciones de la Escuela de Estudios Hispano-Americanos de Sevilla, 1951.

Sanz, Carlos. *Relaciones geográficas de España y de Indias: Impresas y publicadas en el siglo* XVI. Madrid: Bibliotheca Americana Vetustissima, 1962.

Sarmiento de Gamboa, Pedro. *Derrotero al Estrecho de Magallanes* (1580). Edited by Juan Batista. Madrid: Historia, 1987.

———. *The History of the Incas*. Introduction by Brian S. Bauer and Jean-Jacques Decoster. Austin: University of Texas Press, 2007.

Schäfer, Ernesto [Ernst]. *El Consejo Real y Supremo de las Indias: Su historia, organización y labor administrativa hasta la terminación de la Casa de Austria*. 2 vols. Madrid: Junta de Castilla y León, Marcial Pons, 2003.

———. "El cosmógrafo Jaime Juan." *Investigación y Progreso* 10 (1936): 10–15.

Schatzki, Theodore R., Karin Knorr Cetina, and Eike von Savigny, eds. *The Practice Turn in Contemporary Theory*. London: Routledge, 2001.

Sellés, Manuel. *Instrumentos de navegación: Del Mediterráneo al Pacífico*. Barcelona: Lunwerg Editores, 1994.

Seville, Isidore of. *De ecclesiasticis officiis*. Edited by Christopher M. Lawson. Turnhout, Belgium: Brepols, 1989.

Shapin, Steven. *A Social History of Truth: Civility and Science in Seventeenth-Century England*. Chicago: University of Chicago Press, 1994.

Shapiro, Barbara J. *A Culture of Fact: England 1550–1720*. Ithaca, N.Y.: Cornell University Press, 2000.

Sherman, William H. *John Dee: The Politics of Reading and Writing in the English Renaissance*. Amherst: University of Massachusetts Press, 1995.

Shirley, John W. "Science and Navigation." In *Science and the Arts in the Renaissance*, edited by John W. Shirley and F. David Hoeniger, 74–93. London: Folger Books, 1985.

Sieber, Harry. "The Magnificent Fountain: Literary Patronage in the Court of Philip III." *Cervantes: Bulletin of the Cervantes Society of America* 18, no. 2 (1998): 85–116.

Solano, Francisco de, ed. *Cuestionarios para la formación de las relaciones geográficas de Indias, siglos* XVI–XIX. Madrid: CSIC, 1988.

Strabo. *The Geography of Strabo*. Translated by Horace Leonard Jones and John Robert Sitlington Sterrett. London: Heinemann; New York: Putman's, 1917–33.

Taylor, E. G. R. *The Haven-Finding Art: A History of Navigation from Odysseus to Captain Cook*. London: Hollis and Carter, 1956.

Thomas, Werner, and Luc Duerloo, eds. *Albert and Isabella, 1598–1621*. Louvain, Belgium: Brepols, 1998.

Thorndike, Lynn. *The Sphere of Sacrobosco and Its Commentators*. Chicago: University of Chicago Press, 1949.

Torre Revello, José. *El libro, la imprenta y el periodismo en América durante la dominación española*. Buenos Aires: Casa Jacobo Peuser, 1940.

Torres Naharro, Bartolomé de. *Propalladia and Other Works of Bartolomé de Torres Naharro*. Edited by Joseph E. Gillet. 4 vols. Menasha, Wisc.: George Banta, 1943.

Trueba, Eduardo, and José Llavador. "Geografía conflictiva en la expanción marítima luso-española, siglo XVI." *Revista de Historia Naval* 15, no. 58 (1997): 19–38.

Turnbull, David. *Masons, Tricksters, and Cartographers*. Amsterdam: Harwood Academic, 2000.

Valencia, Pedro de, Jesús Paniagua Pérez, Francisco Javier Fuente Fernández, and Jesús Fuente Fernández. *Relaciones de Indias: Nueva Granada y Virreinato de Perú*. Vol. 5/1 of *Obras completas*. 2nd ed. León: Universidad de León, 2001.

Valencia, Pedro de, Jesús Paniagua Pérez, Rafael González Cañal, and Gaspar Morocho Gayo. *Relaciones de Indias: México*. Vol. 5/2 of *Obras completas*. 2nd ed. León: Universidad de León, 2001.

Vanden Broecke, Steven. "The Use of Visual Media in Renaissance Cosmography: The *Cosmography* of Peter Apian and Gemma Frisius." *Paedagogica Historica* 36, no. 1 (2000): 131–50.

Varey, Simon, Rafael Chabrán, and Dora B. Weiner, eds. *Searching for the Secrets of Nature: The Life and Works of Dr. Francisco Hernández*. Stanford, Calif.: Stanford University Press, 2000.

Vicente Maroto, M. I. "Alonso de Santa Cruz e el oficio de Cosmógrafo Mayor de Consejo de Indias." *Mare Liberum* 10 (1995): 509–23.

———. "El arte de navegar." In *Felipe II, la ciencia y la técnica*, edited by E. Martínez Ruiz. Madrid: Actas, 1999.

Vicente Maroto, M. I., and Mariano Esteban Piñeiro. *Aspectos de la ciencia aplicada en la España del Siglo de Oro*. 2nd ed. Valladolid: Junta de Castilla y León, Sever-Cuesta, 2006.

Vigón, Jorge. *Historia de la artillería española*, vol. 1. Madrid: Instituto Jerónimo Zurita, 1947.

Vilar, Sylvia. "La trajectoire des curiosités Espagnoles sur les Indes." *Mèlanges du Casa de Velázquez* 6 (1970): 247–308.

Vogel, Klaus A. "Cosmography." In *The Cambridge History of Science*, edited by Lorraine Daston and Katherine Park, 469–96. Cambridge: Cambridge University Press, 2006.

Wagner, Henry R. *Sir Francis Drake's Voyage around the World*. San Francisco: John Howell, 1926.

———. *Spanish Voyages to the Northwest Coast of America in the Sixteenth Century*. Amsterdam: N. Israel, 1966.

Westfall, Richard S. "Science and Patronage: Galileo and the Telescope." *Isis* 76, no. 1 (1985): 11–30.

Westman, Robert S. "The Astronomer's Role in the Sixteenth Century: A Preliminary Study." *History of Science* 18 (1980): 103–47.

Whithers, Charles W. J. "Geography, Science, and the Scientific Revolution." In *Geography and Revolution*, edited by David N. Livingstone and Charles W. J. Whithers, 75–105. Chicago: University of Chicago Press, 2005.

Wilkinson-Zerner, Catherine. *Juan de Herrera: Architect to Philip II of Spain*. New Haven, Conn.: Yale University Press, 1993.

Woodward, D. "Maps and the Rationalization of Geographic Space." In *Circa 1492: Art in the Age of Exploration*, edited by Jay A. Levenson. New Haven, Conn.: Yale University Press, 1991.

Wright, Edward. *Certaine errors in nauigation detected and corrected by Edw. Wright; with many additions that were not in the former edition as appeareth in the next pages*. London: Printed by Felix Kingsto[n], 1610.

Zamorano, Rodrigo. *Chronología y repertorio de la razón de los tiempos*. Sevilla: Imprenta de Francisco de Lyra, 1621.

———. *Compendio del arte de navegar*. Sevilla: Alonso de Barrera, 1581. Reprint, Colección primeras ediciones. Madrid: Instituto Bibliográfico Hispánico, 1973.

Zaragoza, Justo. *Historia del descubrimiento de las regiones austriales: Hecho por el general Pedro Fernández de Quirós, el Pacífico hispano y la búsqueda de la "Terra Australis."* Madrid: Impr. de Manuel G. Hernández, 1876–1882. Reprint, Madrid: Dove, 2000.

Index